Tissue Engineering in Musculoskeletal Clinical Practice

Published by the

American Academy of
Orthopaedic Surgeons

Tissue Engineering in Musculoskeletal Clinical Practice

Edited by
Linda J. Sandell, PhD
Washington School of Medicine
Department of Orthopaedics
St. Louis, Missouri

Alan J. Grodzinsky, ScD
Professor and Director
Massachusetts Institute of Technology
Center for Biomedical Engineering
Cambridge, Massachusetts

Workshop
Santa Fe, New Mexico
January 2003

Supported by the
American Academy of Orthopaedic Surgeons
and the
National Institute of Arthritis and Musculoskeletal and Skin
Diseases

Cosponsored by the
Orthopaedic Research and Education Foundation
and the
Orthopaedic Research Society

Supported in part by grants from
Zimmer, Inc.
and the
Wright Medical Technology, Inc.

Published by the
American Academy of Orthopaedic Surgeons
6300 North River Road
Rosemont, IL 60018

Tissue Engineering in Musculoskeletal Clinical Practice

American Academy of Orthopaedic Surgeons

First Edition
Copyright © 2004 by the
American Academy of Orthopaedic Surgeons

ISBN 0-89203-329-0

American Academy of Orthopaedic Surgeons

Cover illustration courtesy of Timothy M. Ritty, PhD

Contributors and Participants

Giovanni Abatangelo, MD
Institute of Histology and Embriology
Faculty of Medicine
University of Padova
Padova, Italy

H. Davis Adkisson, PhD
Senior Scientist
ISTO Technologies, Inc.
St. Louis, Missouri

Louis Almekinders, MD
Professor
North Carolina Orthopaedic Clinic
Duke University Medical Center
Durham, North Carolina

Kyriacos A. Athanasiou, PhD, PE
Professor
Rice University, Department of
    Bioengineering
Houston, Texas

Albert J. Banes, PhD
Professor
University of North Carolina School of
    Medicine
Orthopaedic Research Labs
Chapel Hill, North Carolina

Frank Barry, PhD
Director, Arthritis Research
Osiris Therapeutics, Inc.
Baltimore, Maryland

Oliver Betz, PhD
Center for Molecular Orthopaedics
Harvard Medical School
Boston, Massachusetts

Giordano Bianchi, MD
Dipartmineto Dioncologia, Biologia e
    Genetica
Universita Degli Studi Di Genova
Genova, Italy

Scott D. Boden, MD
Professor of Orthopaedic Surgery
Director, The Emory Spine Center
Emory University
Decatur, Georgia

John Bogdanske, BS
School of Veterinary Medicine
University of Wisconsin
Madison, Wisconsin

Lawrence J. Bonassar, PhD
Research Assistant Professor of
    Anesthesiology and Cell Biology
University of Massachusetts Medical
    School
Center for Tissue Engineering
Worcester, Massachusetts

Anna Borrione, BSc
Institito Clinico S. Ambrogio
Reparto di Ortopedia
Milano, Italy

Barbara Boyan, PhD
Professor and Deputy Director for Research
Georgia Institute of Technology
Department of Biomedical Engineering
Atlanta, Georgia

Domenico Brocchetta, MD
Institito Clinico S. Ambrogio
Reparto di Ortopedia
Milano, Italy

Scott P. Bruder, MD, PhD
Vice President DePuy Biologics
DePuy Spine Inc.
Raynham, Massachusetts

Joseph A. Buckwalter, MD
Professor and Chair of Orthopaedic Surgery
University of Iowa Hospital
Department of Orthopaedics
Iowa City, Iowa

Don Bynum, MD
Professor
Department of Orthopaedics
University of North Carolina at Chapel Hill
Chapel Hill, North Carolina

Ranieri Cancedda, MD
Professore
Lab. DiMedicina Rigenerativa
Instituto Nazionale Perla Ricerca sul Cancro
Universita Degli Studi Di Genova
Genova, Italy

Arnold Caplan, PhD
Professor of Biology
Case Western Reserve University
Cleveland, Ohio

Regina Cheung, BS
Department of Electrical Engineering and
    Computer Science
Massachusetts Institute of Technology
Cambridge, Massachusetts

James Cook, DVM, PhD
Comparative Orthopaedics Laboratory
University of Missouri
Columbia, Missouri

Christina Cosman, MS
Research Assistant
Department of Bioengineering/Mechanical
    Engineering
MIT
Cambridge, Massachusetts

Cristina Csimma, RPh, MHP
Senior Director
Experimental Medicine
Wyeth Research
Cambridge, Massachusetts

Bradford L. Currier, MD
Associate Professor
Mayo Medical School
Vice Chair Practice, Orthopaedic
    Department
Mayo Clinic
Rochester, Minnesota

Claudio deLuca, BSc
University of Bristol
Bristol, United Kingdom

Robert G. Dennis, PhD
Assistant Research Scientist
University of Michigan
Ann Arbor, Michigan

David Dines, MD
Chairman and Program Director
Clinical Professor of Orthopaedic Surgery
Albert Einstein College of Medicine
New Hyde Park, New York

Gary C. du Moulin, PhD
Vice President, Quality Systems
Genzyme Biosurgery
Cambridge, Massachusetts

Thomas A. Einhorn, MD
Professor and Chairman
Boston University Orthopaedic Surgical
    Associates
Boston, Massachusetts

Cynthia A. Entstrasser, BS
Director, Quality Assurance
Genzyme Biosurgery
Cambridge, Massachusetts

Christopher H. Evans, PhD
Robert Lovett Professor of Orthopaedic
    Surgery
Center for Molecular Orthopaedics
Harvard Medical School
Boston, Massachusetts

Joseph Feder, PhD
President
ISTO Technologies, Inc.
St. Louis, Missouri

Cyril Frank, MD
Professor and Division Head
Health Sciences Centre
Orthopaedic Surgery
Calgary, Alberta, Canada

Gary Friedlaender, MD
Professor and Chairman
Yale University School of Medicine
New Haven, Connecticut

Joanne Garvin, BS, MS
Department of Research
Bectin Dickinson
RTP, North Carolina

Steven C. Ghivizzani, PhD
Associate Professor
Orthopaedic Surgery
University of Florida
Gainesville, Florida

Jeffrey M. Gimble, MD, PhD
Professor
Stem Cell Department
Pennington Biomedical Research Center
Baton Rouge, Louisiana

Victor Goldberg, MD
Professor and Chair
Case Western Reserve University
Cleveland, Ohio

Steven A. Goldstein, PhD
Professor of Orthopaedic Surgery and
    Bioengineering
University of Michigan
Orthopaedic Research Labs
Ann Arbor, Michigan

Daniel Grande, PhD
Director of Orthopaedic Research
North Shore University Hospital
Department of Orthopaedics
Manhassett, New York

Alan J. Grodzinsky, ScD
Professor and Director
Massachusetts Institute of Technology
Center for Biomedical Engineering
Cambridge, Massachusetts

Marc Grynpas, PhD
Professor
Department of Laboratory Medicine and
    Pathobiology
Mount Sinai Hospital/University of Toronto
Toronto, Ontario, Canada

Farshid Guilak, PhD
Director of Research
Duke University Medical Center
Orthopaedic Research Labs
Durham, North Carolina

David A. Hart, PhD
Calgary Foundation-Grace Glaum Professor
Professor, Department of Surgery, Medicine
    and Microbiology and ID
Faculty of Medicine
University of Calgary
Calgary, Alberta, Canada

Hans J. Hauselmann, MD
Center for Rheumatology and Bone
    Diseases
Zurich, Switzerland

Anthony Hollander, PhD
University of Bristol
Bristol, United Kingdom

Keith Hruska, PhD
Children's Hospital
St. Louis, Missouri

Jerry C. Y. Hu, BS
Department of Bioengineering
Rice University
Houston, Texas

Johnny Huard, PhD
Associate Professor
University of Pittsburgh
Division of Orthopaedic Surgery, Molecular
    Genetics and Biochemistry and
    Bioengineering
Pittsburgh, Pennsylvania

Ernst B. Hunziker, MD
Professor, Director of the Mueller Institute
University of Bern
ITI Research Jashiluck
Bern, Switzerland

Mark Hurtig, DVM, MVSc
Associate Professor
Clinical Studies
Ontario Veterinary College
Ontario, Canada

Esmaiel Jabbari, PhD
Assistant Professor of Biomedical
    Engineering
Mayo Clinic School of Medicine
Mayo Clinic
Rochester, Minnesota

Brian Johnstone, PhD
Associate Professor of Orthopaedics
Case Western Reserve University
University Hospitals of Cleveland
Cleveland, Ohio

Rita Kandel, MD
Professor
Mt. Sinai Hospital
Pathology and Laboratory Medicine
Toronto, Ontario, Canada

Grace L. Kielpinski, BS
Genzyme Biosurgery
Cambridge, Massachusetts

John D. Kisiday, PhD
Biological Engineering Division
Massachusetts Institute of Technology
Cambridge, Massachusetts

Neil Kizer, PhD
ISTO Technologies, Inc
St. Louis, Missouri

Elisaveta Kon, MD
Researcher
Sport Traumatology and Orthopaedic
    Department
Istituti Ortopedici Rizzoli
Bologna, Italy

Paul H. Krebsbach, DDS, PhD
Associate Professor
Department of Oral
    Medicine/Pathology/Oncology
University of Michigan School of Dentistry
Ann Arbor, Michigan

David W. Levine, MD, MPH
Cambridge, Massachusetts

Jueren Lou, MD
Residential Assistant Professor of
    Orthopaedic Surgery
Washington University
Department of Orthopaedic Surgery
St. Louis, Missouri

Lichun Lu, PhD
Assistant Professor of Bioengineering
Department of Orthopaedic Surgery
Mayo Clinic
Rochester, Minnesota

Yan Lu, MD
Mayo Clinic
Rochester, Minnesota

Frank P. Luyten, MD, PhD
Professor and Chairman
Katholieke Universiteit Leuven
Belgium

Melissa Maloney, BE, MS
Engineer
Department of Research
Flexcell International Corporation
Hillsborough, North Carolina

William J. Maloney, MD
Chief of Surgery, Professor of Orthopaedic
    Surgery
Washington School of Medicine
Department of Orthopaedic Surgery
St. Louis, Missouri

Maurilio Marcacci, MD
Instituti Ortopedici Rizzoli
Laboratorio di Biomeccanica
Bologna, Italy

Mark Markel, DVM, PhD
School of Veterinary Medicine
University of Wisconsin
Madison, Wisconsin

James Mason, PhD
Director of Gene Therapy
Vector Laboratory
North Shore Research Institute
Manhassatt, New York

Maddalena Mastrogiacomo, PhD
Istituto Nazionale Perla Ricerca sul Cancro
Genova, Italy

Koichi Masuda, MD
Departments of Orthopaedic Surgery and
    Biochemistry
Rush Medical College
Chicago, Illinois

Bill McKay, ME
Medtronic Sofamor Danek
Memphis, Tennessee

Antonios G. Mikos, PhD
Professor
Bioengineering Department
Rice University
Houston, Texas

Thomas Minas, MD, MS
Clinical Instructor in Orthopaedic Surgery
Brigham and Women's Hospital
Brigham Orthopaedic Associates
Chestnut Hill, Massachusetts

Anita Muraglia, PhD
Istituto Nazionale Perla Ricerca sul Cancro
Genova, Italy

Brian Nussenbaum, MD
Assistant Professor Otolaryngology
Washington University
St. Louis, Missouri

Glyn Palmer, PhD
Center for Molecular Orthopaedics
Harvard Medical School
Boston, Massachusetts

James S. Panagis, MD, MPH
Orthopaedics Program Director
National Institute of Arthritis and
    Musculoskeletal & Skin Diseases
National Institute of Health
Bethesda, Maryland

Arnulf Pascher, MD
Center for Molecular Orthopaedics
BWH, Harvard Medical School
Boston, Massachusetts

Alessandra Pavesio, BSc
Director, Research & Development
Fidia Advanced Biopolymers
University of Padova
Padova, Italy

Hairong Peng, MD, PhD
Children's Hospital of Pittsburgh
Department of Orthopaedic Surgery
University of Pittsburgh
Pittsburgh, Pennsylvania

Robert M. Pilliar, BASc, PhD
Professor
Institute of Biomaterials and Biomedical
  Engineering
University of Toronto
Toronto, Ontario, Canada

Jie Qi, PhD
Research Scientist
Flexcell International Corporation
Hillsborough, North Carolina

Rodolfo Quarto, MD
Dipto Di Oncologia, Biologia e Genetica
Universita Degli Studi Di Genova
Genova, Italy

Anthony Ratcliffe, PhD
President and CEO
Synthasome, Inc
San Diego, California

Pasquale Razzano, MS
Research Tech/Laboratory Manager
Orthopaedics Research
North Shore Research Institute
Manhassett, New York

A. Hari Reddi, PhD
Professor and Director
University of California, Davis
Center for Tissue Regeneration and Repair
Sacramento, California

Randy N. Rosier, MD, PhD
Chairman, Department of Orthopaedics
Director, Center for Musculoskeletal
  Research
Department of Orthopaedics
University of Rochester
Rochester, New York

Sanaz Saatchi, MS
Department of Mechanical Engineering
Massachusetts Institute of Technology
Cambridge, Massachusetts

Robert L. Sah, MD, ScD
Professor
University of California, San Diego
La Jolla, California

Linda J. Sandell, PhD
Professor and Director of Research
Washington University School of Medicine
Department of Orthopaedic Surgery
St. Louis, Missouri

Howard Seeherman, PhD, VMD
Director of Orthopaedic Clinical
  Development
Wyeth Research
Cambridge, Massachusetts

Cheryle Seguin, MSc
Department of Laboratory Medicine and
  Pathobiology
Mount Sinai Hospital
Toronto, Ontario, Canada

Michael Sittinger, PhD
Laboratory for Tissue Engineering
Charité Universitätsmedizine Berlin
Berlin, Germany

Myron Spector, PhD
Professor and Director
Brigham and Women's Hospital
Harvard Medical School
Orthopaedic Research Laboratory
Boston, Massachusetts

Andre Steinert, MD
Research Fellow
Center for Molecular Orthopaedics
Harvard Medical School
Boston, Massachusetts

Bernard Stulberg, MD
Cleveland, Ohio

Eugene J. Thonar, PhD
Professor
Rush-St. Luke's Medical Center
Chicago, Illinois

Dave Toman, PhD
Atherton, California

Francesca Torasso, BSc
Policlinico di Monza
Reparto di Ortopedia
Monza, Italy

Stephen Trippel, MD
Associate Professor of Orthopaedic Surgery
Indiana University Medical Center
Orthopaedic Surgery
Indianapolis, Indiana

Rocky S. Tuan, PhD
Branch Chief, Cartilage Biology and
    Orthopaedics
Bethesda, Maryland

David Tuckman, MD
Fellow for Sports Medicine
North Shore Long Island Jewish Health
    System
New Hyde Park, New York

Alexandre Valentin-Opran, MD
Senior Director, Clinical Research
Clinical Research and Development
Wyeth
Cambridge, Massachusetts

Gordana Vunjak-Novakovic, PhD
Principal Research Scientist
Harvard-MIT
Division of Health Sciences & Technology
Cambridge, Massachusetts

Stephen Waldman, BASc, MSc, PhD
Assistant Professor
Departments of Chemical and Mechanical
    Engineering
Queen's University
Kingston, Ontario, Canada

Neal Watnik, MD
Attending Orthopaedic Surgeon
Orthopaedic Surgeons of Long Island, PC
Long Island Jewish Medical Center
New Hyde Park, New York

John Wozney, PhD
Assistant Vice President
Wyeth Research
Cambridge, Massachusetts

Michael J. Yaszemski, MD, PhD
Consultant, Department of Orthopaedic
    Surgery
Mayo Clinic
Orthopaedic Trauma Service
Rochester, Minnesota

Jung U. Yoo, MD
Associate Professor
Department of Orthopaedics
Case Western Reserve University
University Hospitals of Cleveland
Cleveland, Ohio

Stefano Zanasi, MD

Shuguang Zhang, PhD
Associate Director
Center for Biomedical Engineering
MIT
Cambridge, Massachusetts

Preface

In the last 10 years, there has been intense interest in tissue engineering. Both the public sector and the medical profession are aware of the critical clinical need for organs and tissues. This need is particularly evident in the rebuilding of the skeletal elements of the face and body where the enormous prevalence and societal costs of musculoskeletal disorders underscore the value of clinical application of new methodologies. Research scientists from the disciplines of molecular biology, bioengineering, and materials science have put great effort into designing potential products using cells, growth factors and DNA, alone, or in structural scaffolds, to recreate functional tissues. Clinicians, primarily in the surgical practices of orthopaedics, plastics, and dental surgery are deeply involved in transferring the knowledge gained from this basic research to clinical applications to help the patient.

The awareness of the potential for tissue engineering as a biological solution to the critical need for human tissues has been heightened by publication in the general press of articles relating to potential tissue engineered products. As this field is relatively new in the musculoskeletal area, there are questions from clinicians, patients, and researchers about the state of the art in orthopaedic practice. For example, the patient asks "when will I be able to replace my cartilage with cells or new cartilage?" The clinician wants to know "how close are these products to reality?" And the scientist questions "how can I get my idea to the market?" The goal of the workshop held January 16-19, 2003 in Santa Fe, New Mexico, was to address these questions and make the answers available to a wide population through publication of a text in print and on the website of the American Academy of Orthopaedic Surgeons. The leading authorities in the biological, bioengineering, clinical, and business aspects of tissue engineering were invited to participate. Focus of the workshop was placed on new musculoskeletal products and the potential for their availability to clinicians in the United States and internationally.

The organizers of the conference are grateful to the American Academy of Orthopaedic Surgeons for providing the leadership and financial resources for this workshop and publication, and also thank our partners, the National Institute of Arthritis and Musculoskeletal and Skin Diseases, the Orthopaedic Research and Education Foundation, and many private companies for their intellectual and financial support. We are grateful also to the publications staff of the American Academy of Orthopaedic Surgeons, in particular, our editor Gayle Murray and our desktop production staff David Stanley and Mike Bujewski.

Linda J. Sandell, PhD
Alan J. Grodzinsky, ScD

Table of Contents

Preface   xiii

Section 1   **"To the Patient": The Clinical and**
                                                            Marketing Challenges

1   Can Tissue Engineering Help Orthopaedic Patients? Clinical Needs
     and Criteria for Success   3
     Joseph A. Buckwalter, MD

2   The Clinical Market for Tissue Engineered Products
     in Orthopaedics   17
     Anthony Ratcliffe, PhD

3   Bringing Orthopaedic Tissue Engineering From Bench to Bedside:
     Clinical Research Challenges   25
     David W. Levine, MD, MPH

4   The Clinician's Prospective on Tissue Engineering   33
     William J. Maloney, MD

5   Cell-based Therapies: Toward an Integrated United States-European
     Vision?   41
     Frank P. Luyten, MD, PhD

6   An Industry Perspective on the Development of Osteogenic
     Agents Using Recombinant Human Bone Morphogenetic
     Protein-2 (rhBMP-2) as an Example   51
     *Howard Seeherman, PhD, VMD; Cristina Csimma, RPh, MHP; Alexandre Valentin-
     Opran, MD; John Wozney, PhD*

7   Commerial Approval of rhBMP-2 in Spinal Fusions:
     Bringing the Product to the Market   61
     Bill McKay, ME

8   Hyaluronan-based Scaffolds in the Treatment of Cartilage
     Defects of the Knee: Clinical Results   73
     *Alessandra Pavesio, BSc; Giovanni Abatangelo, MD; Anna Borrione, BSc;
     Domenico Brocchetta, MD; Claudio DeLuca, BSc; Anthony P. Hollander, PhD;
     Elisaveta Kon, MD; Francesca Torasso, BSc; Stefano Zanasi, MD;
     Maurilio Marcacci, MD*

9   The Role of the National Institutes of Health in Extramural
     Clinical Studies and Trials   85
     James S. Panagis, MD, MPH

Section 2   **Tissue Engineering of Bone**

10   Tissue Engineering of Fracture Repair   107
     Thomas A. Einhorn, MD

11 Engineered Cells in Scaffolds Heal Bone 115
Ranieri Cancedda, MD; Rodolfo Quarto, MD; Giordano Bianchi, MD;
Maddalena Mastrogiacomo, PhD, Anita Muraglia, PhD

12 Bone Engineering With Mesenchymal Stem Cells
and Gene Therapy 123
Jueren Lou, MD

13 Induction of Bone Formation by Stem Cells 131
Johnny Huard, PhD; Hairong Peng, MD, PhD

14 Clinical Trials in Bone Tissue Engineering:
Spine Applications Update 141
Scott D. Boden, MD

15 Practical Matters in the Application of Tissue Engineered
Products for Skeletal Regeneration in the Head and Neck
Region 151
Brian Nussenbaum, MD; Paul H. Krebsbach, DDS, PhD

Section 3   **Tissue Engineering of Cartilage**

16 Experimental Principles and Future Perspectives of
Skeletal Tissue Engineering 165
Rocky S. Tuan, PhD

17 Engineering Cartilage Structures 175
Michael Sittinger, PhD

18 The Role of the Microenvironment in Cartilage
Tissue Engineering 183
Ernst B. Hunziker, MD

19 Mesenchymal Stem Cells in Joint Therapy 195
Frank Barry, PhD

20 Autologous Chondrocyte Implantation of the Knee 201
Thomas Minas, MD, MS

21 Manufacture of Cartilage Tissue In Vitro 211
Eugene J. Thonar, PhD; Robert L. Sah, MD, ScD; Koichi Masuda, MD

22 The Promise of Chondral Repair Using Neocartilage 219
Joseph Feder, PhD; H. Davis Adkisson, PhD; Neil Kizer, PhD;
Keith A. Hruska, PhD; Regina Cheung, BS; Alan J. Grodzinsky, ScD,
Yan Lu, MD; John Bogdanske, BS; Mark Markel, DVM, PhD

23 Functional Assessment of Articular Cartilage 227
Jerry C. Hu, BS; Kyriacos A. Athanasiou, PhD, PE

Section 4   **Tissue Engineering of Ligament, Tendon, Meniscus, Intervertebral Disk, and Muscle**

24   Clinical Application of Tissue Engineered Tendon
and Ligaments   241
Cyril Frank, MD; David A. Hart, PhD

25   Bioartificial Tendons: An Ex Vivo Three-dimensional
Model to Test Tenocyte Responses to Drugs, Cytokines,
and Mechanical Load   257
*Albert J. Banes, PhD; Jie Qi, PhD; Melissa Maloney, BE, MS;
Joanne Garvin, BS, MS; Louis Almekinders, MD; Don Bynum, MD*

26   Meniscus Repair Through Tissue Engineering   267
Brian Johnstone, PhD; Jung U. Yoo, MD

27   Gene Enhanced-Tissue Engineered Repair of the Meniscus   273
*Daniel Grande, PhD; David Tuckman, MD; James Cook, DVM, PhD;
Neal Watnik, MD; James Mason, PhD; Pasquale Razzano, MS;
David Dines, MD*

28   Nucleus Pulposus Tissue Can Be Formed In Vitro   283
*Rita Kandel, MD; Cheryle Seguin, MSc; Mark Grynpas, PhD; Stephen Waldman,
BASc, MSc, PhD; Mark Hurtig, DVM; Robert Pilliar, BASc, PhD*

29   Tissue Engineering in Muscle: Current Challenges
and Directions   295
Robert G. Dennis, PhD

Section 5   **Methodologies Used in the Construction
of Engineered Tissues**

30   Quality Control in Producing Cells for Tissue Engineering   313
Gary C. du Moulin, PhD; Cynthia A. Entstrasser, BS; Grace L. Kielpinski, BS

31   Functional Tissue Engineering of Cartilage: Scaffolds
and Bioreactors   321
Gordana Vunjak-Novakovic, PhD

32   Injectable Polymers and Hydrogels for Orthopaedic and
Dental Applications   331
*Esmaiel Jabbari, PhD; Lichun Lu, PhD; Bradford L. Currier, MD; Antonios G.
Mikos, PhD; Michael J. Yaszemski, MD, PhD*

33   Collagen Uses in Current and Developing Orthopaedic Therapies   341
Dave Toman, PhD

34   Self-Assembling Peptide Hydrogel Scaffolds for Cartilage
Tissue Engineering   349
*John D. Kisiday, PhD; Shuguang Zhang, PhD; Christina Cosman, MS; Sanaz
Saatchi, MS; Alan J. Grodzinsky, ScD*

35   Design and Optimization of Scaffolds for Cartilage Tissue
Engineering   355
Lawrence J. Bonassar, PhD

36 Adipose Derived Adult Stem Cells for Musculoskeletal Tissue
Engineering   363
Farshid Guilak, PhD; Jeffrey M. Gimble, MD, PhD

37 A Bench to Bedside Case Study – The Development Pathway
of an Intraoperative, Autologous Progenitor Cell Preparation Kit   377
Scott P. Bruder, MD, PhD

38 Genetically Enhanced Tissue Engineering Without Cell Culture
or Manufactured Scaffolds   389
Christopher H. Evans, PhD; Glyn Palmer, PhD; Arnulf Pascher, MD;
Andre Steinert, MD; Oliver Betz, PhD; Steven C. Ghivizzani, PhD

39 In Situ Tissue Engineering: Localized Delivery of DNA   395
Steven A. Goldstein, PhD

Section One

"To The Patient": The Clinical and Marketing Challenges

Giovanni Abatangelo, MD
Anna Borrione, BSc
Domenico Brocchetta, MD
Joseph A. Buckwalter, MD
Cristina Csimma, RPh, MHP
Claudio DeLuca, BSc
Anthony P. Hollander, PhD
Elisaveta Kon, MD
David W. Levine, MD, MPH
Frank P. Luyten, MD, PhD
William J. Maloney, MD

Maurilio Marcacci, MD
Bill McKay, ME
James S. Panagis, MD, MPH
Alessandra Pavesio, BSc
Anthony Ratcliffe, PhD
Howard Seeherman PhD, VMD
Francesca Torasso, BSc
Alexandre Valentin-Opran, MD
John Wozney, PhD
Stefano Zanasi, MD

Chapter 1

Can Tissue Engineering Help Orthopaedic Patients? Clinical Needs and Criteria for Success

Joseph A. Buckwalter, MD

Abstract

Despite scientific and technical progress in producing engineered tissues, advances in clinical use of tissue engineered products to replace tissues have been limited. Current tissue bioengineering approaches include either stimulating in vivo tissue repair and regeneration or in vitro formation of tissues that can be implanted in patients. The formation of biologic implants in vitro offers the potential of replacing structurally important segments of bones, joints, tendons, ligaments, and intervertebral disks. The potential of either in vivo or in vitro tissue engineering to replace lost, damaged, diseased, or degenerated tissue with new living tissue is exciting and could open a new era in orthopaedics. However, the realization of this potential, especially the replacement of segments of the musculoskeletal system with biologic implants, will require ongoing evaluation of clinical needs, indications, and measures of clinical success.

Introduction

Few clinical disciplines are growing and changing as rapidly as orthopaedics, a specialty that employs medical, surgical, and physical methods to relieve pain, correct deformity, and improve mobility for patients with diseases, deformities, and injuries of the spine and limbs.[1] During the past 50 years, advances in the ability of orthopaedists to restore or improve musculoskeletal function and relieve pain has led to rapid expansion and diversification of the specialty as demonstrated by the formation, growth, and clinical impact of orthopaedic subspecialties. Examples of these subspecialties include sports medicine, hip and knee reconstruction, spine surgery, hand surgery, foot and ankle surgery, musculoskeletal oncology, and musculoskeletal trauma.

Despite many successes, orthopaedists lack effective treatments of a variety of musculoskeletal diseases and injuries. In particular, they have limited ability to replace lost or diseased bone, cartilage, dense fibrous tissue, or skeletal muscle with living functional tissue. Recently, a variety of methods have been shown to promote tissue repair and regeneration in vivo or to produce new musculoskeletal tissues in vitro that might be implanted into patients.[2-17] These reports have generated interest on the part of orthopaedists, scientists, and patients who now anticipate that lost and damaged tissues can be replaced with new tissues that restore or at least improve musculoskeletal structure and function.

The frequency and economic impact of musculoskeletal diseases and injuries[18] coupled with widespread enthusiasm for tissue engineering[4,6,8,13,14,17,19] has led to conferences, publications, research programs, and commercial enterprises dedicated to promoting clinical use of engineered tissues. More than one venture capitalist has invested in a company dedicated to producing tissues for orthopaedic patients. Yet, despite scientific and technical progress in producing engineered tissues, advances in clinical use of these tissues have been limited. Some orthopaedists, scientists, and health care company investors and directors have questioned how soon, or if, tissue engineering will help significant numbers of orthopaedic patients. This healthy skepticism has led to recognition that the clinical success of tissue engineering in orthopaedics presents challenges beyond discovering methods of forming new tissues.

Bioengineering and Orthopaedics

The most important improvements in the treatment of patients with musculoskeletal disorders resulted from basic biologic research, and not bioengineering.[20-23] Examples include antibiotics to cure tuberculosis, bacterial osteomyelitis, and arthritis, vaccines that prevent polio, diets that prevent and/or treat rickets, and antiseptic and aseptic techniques that prevent surgical infections. The development of methods of imaging the musculoskeletal system and safely anesthetizing patients facilitated the rapid growth of modern orthopaedic surgery.[20] However, applied musculoskeletal bioengineering, use of materials, forces and energy to restore the structure and function of the musculoskeletal system, have played a central role in the development of the specialty.[20,21] An appreciation of the evolution of applied bioengineering in orthopaedic practice helps clarify the potential of engineered tissues for treatment of patients with musculoskeletal disorders.

Brute Force Bioengineering

For more than 25 centuries, physicians have been devising strategies for applying force to mobilize stiff joints and to correct deformities due to joint injuries, arthritis, malunions of bones, and congenital or developmental abnormalities.[20] They have used ratchets, winches, levers, wrenches, ropes, wedges, screws, and other devices to increase the force they could apply to the musculoskeletal system.[20,21] These practices can be justifiably referred to as brute force bioengineering. For most of the last 2,500 years physicians had little or no understanding of the effects of the application of these forces on musculoskeletal structures such as bones, joints, tendons, nerves and blood vessels, or the tissues that form these structures. Clinical success or failure was assessed by gross correction of deformity or mobilization of stiff joints. The damage done to bones, joints, nerves, and blood vessels and the long-term outcomes of brute force bioengineering were either not assessed or were not recorded.

Structural Bioengineering

Less than a century ago, orthopaedics entered the era of structural bioengineering. Strategies included the fabrication of implants to replace damaged or lost structures and application of forces to specific structures to correct deformity or establish stability.[20,21] Progress has been made based on advances in the understanding of the structural components of the musculoskeletal system, biomechanics, and pathophysiology of diseases and injuries. Combined with progress in anesthesia, antisepsis, asepsis, biomaterials, and imaging of bones, joints, and soft tissues, these advances led to the development of instruments, implants, and procedures that orthopaedists use to apply forces to specific structures and replace damaged structures with internal prostheses.[20] These advances have made it possible to effectively treat displaced and unstable fractures, scoliosis, clubfoot deformity, joint dysplasias, and joint contractures and to replace joints and segments of bone with internal prostheses. Yet, structural bioengineering cannot replace missing or severely diseased tissues with living tissues that have the capacity to respond to biologic and mechanical stimuli.

Tissue Bioengineering

Over the last century orthopaedists have developed a variety of methods to promote or guide natural repair and regeneration responses.[24-26] These efforts represent the initial form of tissue bioengineering. Examples of these methods include autologous and allogeneic bone, cartilage, tendon, and fascia grafts and penetration of subchondral bone to stimulate formation of new articular surfaces. Orthopaedists have also found that controlled application of mechanical forces or joint and muscle tendon unit motion promotes bone, tendon, ligament, and joint healing.[27-30] Examples of these methods include active and passive motion, distraction osteogenesis, and joint distraction to stimulate formation of new articular surfaces.[26-34] More recently, scientists and surgeons have investigated the use of cytokines, cell transplants, synthetic or processed matrices, genetic therapies, and mechanical, ultrasonic and electromagnetic signals to promote the formation of new tissues.[3-5,7,8,11,13-15,29,35-48]

Current tissue bioengineering approaches fall into two overlapping categories: (1) stimulating in vivo tissue repair and regeneration, and (2) in vitro formation of tissues that can be implanted in patients. Application of electromagnetic fields, ultrasound, and mechanical stimuli, as well as injection of growth factors, matrices or matrix molecules, DNA, and genetically altered cells have been investigated as methods of promoting formation of new tissues in vivo.[4,6,8,9,13,14,19,29,35-46] In general, these approaches are intended to promote repair or regeneration of tissue in relatively small areas of lost or diseased tissue. Fabrication of tissues in vitro involves formation of a matrix and, in many applications, a matrix that contains cells or biologically active molecules.[4,13,14] Formation of biologic implants in vitro offers the potential of replacing structurally important segments of bones, joints, tendons, ligaments, and intervertebral disks.

The potential of either in vivo or in vitro tissue engineering to replace lost, damaged, diseased or degenerated tissue with new living tissue is exciting and could begin a new era in orthopaedics. However, the realization of this potential, especially the replacement of segments of the musculoskeletal system with biologic implants, will require ongoing evaluation of clinical needs, indications, and measures of clinical success.

Orthopaedic Clinical Needs

There is a clinical need for new interventions to treat musculoskeletal disorders that are not successfully treated with currently available therapies. Some of these disorders could be treated with engineered tissues. The principal disorders that cause clinically significant tissue damage or loss are age-related degenerative diseases, trauma, bone tumors, and congenital and developmental abnormalities. The number of patients having these disorders is large, at least many tens of millions in the United States alone.[18] Because orthopaedists and other physicians currently safely and effectively treat the majority of these people, using the total number of patients having these disorders overestimates the clinical need for tissue engineering treatments. For this reason, it is important to attempt to define the current populations of individuals having age-related degenerative diseases, traumatic tissue defects, tumors, and congenital and developmental defects who are most likely to benefit from tissue engineering approaches. Good information regarding this select patient population is not available, so estimates must be based on the number of individuals who present to physicians,[18] but excluding those who can currently receive safe, effective treatment including symptomatic treatment. Given the inaccuracy of such a calculation and the multiple factors that will affect the clinical use of tissue engineering approaches, it is best to underestimate the number of patients who could or would currently be treated using engineered tissues in the United States. Factors that will affect use of tissue engineered products include patient and physician acceptance, cost, availability, and benefits of using alternative therapies.

Age-related Degenerative Diseases

The need for new musculoskeletal tissues will be greatest in patients having age-related and other degenerative disorders. Age-related diseases include osteoarthritis and intervertebral disk degeneration. Degenerative disorders of the dense connective tissues include rotator cuff tears and meniscal degeneration.[49-57] Of the approximately half a million patients treated with knee, hip, shoulder, wrist, and ankle joint arthroplasties in the United States each year,[18] the majority have such advanced degenerative disease that it is unlikely any engineered tissue being developed could provide more effective treatment than current prosthetic joint arthroplasties. Indeed, improvements in joint arthroplasties have increased the indications and longevity and decreased the complications of these procedures.[58-60] Furthermore, drugs that may slow or prevent the progression of osteoarthritis are being tested.[61-64] Thus, ongoing advances in other therapies may decrease the number of patients with osteoarthritis that could benefit from treatments based on tissue engineering. Nonetheless, an important fraction of the people who have replacements of the knee and possibly other joints could be treated with implantation of an engineered articular surface or a method of promoting repair or regeneration of an articular surface. In addition, a significant number of patients with symptomatic osteoarthritis of the hand, wrist, foot, shoulder, knee, and hip cannot be effectively treated by any current surgical approach. Considering patients who are currently receiving surgical and nonsurgical treatments of hand, wrist, ankle, foot, shoulder, knee, and hip osteoarthritis,[65,66] it is likely that those who could benefit from tissue engineered chondral or osteochondral tissue is in excess of 200,000 per year.

Intervertebral disks undergo dramatic age-related changes that decrease disk height, cause fissuring of the disk, and, ultimately, decrease the ability to absorb loads.[51,53,67] These changes lead to a decline in spinal mobility and, in some individuals, an increased mechanical loading of the facet joints causing degeneration of these joints and an increased probability of spinal stenosis. In other patients, degeneration of the disks and facet joints leads to spondylolisthesis. Currently, the most common surgical treatment of symptomatic degenerative disease of the spine is stabilization by spinal fusion. Spinal fusion may relieve pain but cannot restore normal function. For these reasons, several patients might benefit from implantation or in vivo generation of new disk tissue early in the course of disk degeneration. Given the large number of patients with symptomatic back stiffness and pain associated with early intervertebral disk degeneration, it is likely that those who could benefit from an effective tissue engineered disk tissue is more than 100,000 people per year. Patients having degenerative tears of the dense fibrous tissues, including the knee menisci and the rotator cuff, could also benefit from restoration of the mechanical and biologic properties of these tissues. The number of these individuals is probably more than 25,000 per year.

Traumatic Tissue Defects

Most of the millions of bone fractures that occur annually[18] can be effectively treated with available methods. However, severe trauma that causes segmental bone loss is extremely difficult to treat. Osteopenia due to osteoporosis increases the risk of complex fractures that are also extremely difficult to treat with available methods. Despite treatment, a small number of fractures fail to heal and are known as nonunions. It is likely that the number of patients with either segmental bone loss, loss of bone substance due to osteoporosis, or nonunion of fractures that could benefit from a tissue engineered material exceeds 100,000 annually. Traumatic focal articular surface defects occur both with intra-articular fractures and chondral fractures. Based on the number of intra-articular fractures and chondral fractures treated by arthroscopy, microfracture, and other methods, an estimated 10,000 to

50,000 patients per year might benefit from a tissue engineered product. The incidence of meniscus, tendon, ligament, and joint capsule loss or rupture is not well documented, but the number of patients who might benefit from engineered products for these conditions is more than 25,000 a year. Less than 5,000 patients per year need reconstruction of muscle, nerve, and/or blood vessels.

Bone Tumors

Reconstruction of skeletal tissues following resection, curettage, or other treatment of benign and malignant bone lesions presents another clinical need for engineered tissue. In particular, tissues that can fill bone defects or reconstruct segments of bone are needed for these indications. Considering benign bone cysts, benign tumors, and other similar defects such as fibrous dysplasia, it is likely that the number of patients who could benefit from an effective tissue engineered bone material is more than 10,000 per year. Metastatic tumors are common and lead to substantial loss of bone tissue, often requiring complex surgical reconstruction. Many of these problems can be solved with currently available implants, but there are an estimated 5,000 patients per year who could benefit from an engineered tissue based on their expected longevity and extent of skeletal destruction. Primary bone malignancies are rare, but many of these patients can anticipate a normal lifespan. Biologic reconstruction of skeletal tissues of these patients would be advantageous compared with available implants. These individuals number more than 1,000 per year.

Congenital and Developmental Disorders

Congenital and developmental defects of the musculoskeletal system including tibial pseudarthrosis, femoral, tibial, and upper extremity deficiencies, joint dysplasias, and spinal deformities are relatively rare but cause significant disablity.[18] It is difficult to estimate how many patients could benefit from engineered tissues, but it is likely to be at least several thousand each year.

Criteria for Success

A tissue engineering approach that helps a small number of patients with severe complex musculoskeletal problems may be considered a success even if the cost may be high relative to the benefit. However, approaches that benefit small numbers of selected patients at high cost will not fundamentally change orthopaedic practice. To be considered a clinical success, a tissue engineering approach should benefit a population of at least 10,000 patients per year. A successful approach must also meet high standards of safety, efficacy, and clinical utility as determined by its mechanical and biologic properties. And finally, cost effectiveness and profitability should be considered as determinants of success. Treatment of 500,000 patients a year would establish tissue engineering as an integral part of orthopaedic practice.

The production of engineered tissues is not sufficient to achieve clinical success. Tissues that are fabricated in vivo and then implanted must meet stringent criteria for safety, efficacy, mechanical characteristics, and biologic properties. Most tissue engineering approaches that attempt to stimulate tissue repair or regeneration in vivo have less risk of complications and do not need to meet the same standards for the biologic and mechanical properties of an implant. However, these approaches must be safe, efficacious, cost effective, and profitable.

Safety and Efficacy
Initial work with some engineered tissues has shown that the procedures and engineered tissues can cause complications such as inflammatory and immunologic reactions, failure to bond to normal tissue, rapid resorption or loosening, damage to surrounding tissues, and mechanical failure. These events need to be studied more carefully. Clinical use of engineered tissues must have few complications, ideally less than 2%. Tissue engineering approaches should have low complication rates because most musculoskeletal disorders are not life or limb threatening. In most instances these disorders can be treated with methods that cause few adverse events. For example, if an engineering tissue designed to restore articular surfaces caused synovitis and arthrofibrosis, not only would this make the patient worse, it could make the result of a later total joint arthroplasty less satisfactory. The same principles apply to use of engineered tissue to restore bone, ligaments, and intervertebral disk.

In addition to having few complications, clinically useful engineered tissues need to produce good or excellent function. Ideally, this restored function would last 5 years or more in more than 95% of patients. Current treatment of fractures, severe osteoarthritis of the hips and knees, and most ligament and joint capsule injuries produce good or excellent function for more than 95% of patients at 5 years or more. It would be difficult to justify lower levels of efficacy except in patients with conditions for which no effective treatments are available.

Clinical Utility: Biologic and Mechanical Properties
To be useful to surgeons and patients, engineered tissues that are formed in vitro need to meet high standards for their biologic and mechanical properties. They must have structural integrity, durability, and stability to allow handling and implantation in patients. Ideally, engineered tissues should be in a form that allows them to be implanted or injected by minimally invasive techniques. They must be in a form that can be easily and securely stabilized to the host tissue and they should rapidly integrate with normal tissue, ideally within 8 weeks or less. Normal bone, ligament, and osteochondral repair tissue forms a stable relationship with the surrounding tissue within 6 to 8 weeks.[25,26] Lack of immediate stability and rapid integration of an engineered tissue increase the risk of loosening and mechanical failure. Slow integration of engineered tissue also imposes the requirement that patients protect the tissue from loading, a requirement that is often difficult to meet. Like normal repair tissue, engineered tissue should remodel to improve its

biologic and mechanical function. Engineered tissues should continue to respond to their biologic and mechanical environment and maintain or improve their properties.

Cost Effectiveness and Profitability

In vitro engineered tissues and in vivo promotion of tissue repair and regeneration should also be cost effective and profitable. Meeting the health care needs and expectations of the population without creating an excessive economic burden on society is a significant challenge. The costs of new treatments receive intense scrutiny. Patients, physicians, and payers for treatments are not likely to accept a new treatment that costs significantly more than alternative treatments, unless the new treatment provides dramatically greater and proven benefits. Physicians are also concerned with the time required for a treatment, its complexity, its risk of complications, and the extent to which physicians are reimbursed for each procedure. Physicians want the best results for their patients. Most physicians will adopt new treatments that are more expensive if the benefits to patients are clear and substantial. Physicians will not adopt or continue treatments if the benefits to patients are not reflected in increased patient satisfaction or if complications occur frequently. Hospitals can only support treatments of which they are reimbursed. In some instances, hospitals may offer treatment that is not reimbursed or is inadequately reimbursed on a limited basis. Engineered tissues also must allow companies to make a profit. Companies pay for product development according to regulatory requirements, production, marketing, and distribution of the product. Commercial success will depend on the number of patients that can benefit from the treatment, reimbursement for the product relative to the costs, and the risk of liability.

Clinical Value

Demonstrating that an engineered tissue is safe, effective, has desirable biologic and mechanical properties, and is cost-effective does not define its clinical value relative to other treatments or to the natural history of a disease or injury. The clinical value of tissue engineering treatments of musculoskeletal disorders can only be demonstrated by rigorous long-term prospective clinical studies that compare them with either the natural history of the treated condition or an accepted current treatment or both. Failure to include appropriate control or comparison groups and lack of long-term follow-up represent serious problems for interpretation of clinical trials.[20,21,66] Unfortunately, such studies are expensive, time consuming, and difficult to perform.

A prospective randomized trial comparing the outcomes of chondrocyte transplants with microfracture in 80 patients 18 to 45 years of age,[68] provides an example of this type of study. In this investigation, both treatments produced new tissue and improved joint function 2 years after the procedures. However, complications were slightly more common in patients with chondrocyte transplants. Chondrocyte transplants produced more hyaline-like cartilage, but the two approaches were not different with regard to formation of a new articular surface. The two treatments differ in terms of costs to patients, payers, physicians, and hospitals. Studies to compare growth factors and tissue engineered implants to other treatments that stimulate bone formation are needed. Furthermore, engineered tissues need to be compared with autografts and allografts, particularly for reconstruction of tendons and ligaments, and meniscus and segmental bone defects.

Some tissue engineering enthusiasts question why achieving clinical success for engineered tissue in orthopaedics should be difficult. They argue that development of a method for producing new tissue with properties that closely resemble normal tissue should lead to rapid clinical use. However, the natural history and severity of most musculoskeletal disorders along with the efficacy and safety of currently available orthopaedic treatments help explain the need to demonstrate clinical success for new treatments. Few musculoskeletal disorders are life threatening or limb threatening. Therefore, the outcome measures do not include the end point of preventing death and rarely include preserving a limb. This leaves less well-defined measures of outcome including pain relief, time to healing or return to work or activity, and improved function. Furthermore, most musculoskeletal disorders can be effectively treated with available interventions, making demonstration of the relative value of a new treatment more difficult.

Comparing tissue engineering treatments of a traumatic transection of the spinal cord and a 3-cm diameter full-thickness traumatic defect of the medial femoral condyle illustrates these points. The functional deficit caused by transection of the spinal cord is easily measured as it is uniform in each patient and among patients. It will not change with time and there is no current treatment. A treatment that produced a 50% improvement in 50% of patients, even with complications, would be immediately accepted as a great clinical success. Few people would suggest that a prospective randomized clinical trial was needed before additional research, development, or investment. In contrast, the functional deficit caused by a femoral condylar articular surface defect is difficult to measure. This defect varies among patients and can vary in the same person over time. Some articular surface defects lead to joint degeneration. However, the natural history is poorly understood and there are multiple potential treatments. A new treatment of femoral condylar defects that produced 50% improvement in 50% of patients would be considered a failure by most patients, physicians, and payers.

Achieving Early Clinical Success

Given the difficulties in making a tissue engineering approach clinically successful for orthopaedic patients, it is appropriate to consider strategies to accelerate the evolution of tissue engineering. The first step in such an effort is to identify tissue engineering approaches that are most likely to become clinically successful in the near future and develop these methods for clinical use. At the same time, approaches that may produce greater benefits, but take longer to develop, will continue to be evaluated. Promoting the natural repair response in vivo is generally less complex, has fewer risks of complications, and fewer problems with producing material with necessary biologic and mechanical characteristics than fabricating new tissue in vitro for implantation. Furthermore, an in vivo approach may be more rapidly accepted by clinicians and patients. This approach is also expected to become cost effective and profitable more quickly than the in vitro development of implants. Bone and vascularized dense fibrous tissues have generally effective repair and regenerative responses. In contrast, mature articular cartilage and intervertebral disk do not have effective healing mechanisms.[25,26] Thus, reconstruction of bone and dense fibrous tissue defects that are unlikely to heal or that have failed to heal offers the best opportunity to demonstrate the benefits of tissue engineering in the near future. The use of minimally invasive techniques such as injection of cells, growth factors, or matrices might be combined with noninvasive techniques such as controlled loading and motion, ultrasound or electromagnetic fields. The noninvasive methods of applying forces and energy may be of particular value in promoting the incorporation and remodeling of engineered tissues.[25,26] If in vivo tissue engineering safely and effectively reconstructs bone and fibrous tissue defects, perhaps greater interest in the development of other applications will be generated. Other applications include reconstruction of degenerated articular surfaces and intervertebral disks using tissues and structures fabricated in vitro.[66]

Summary and Conclusions

Applied musculoskeletal bioengineering, use of materials, forces, and energy to restore the structure and function of the musculoskeletal system have played a major role in the advancement of orthopaedics. These techniques provide orthopaedists with the ability to replace joints, reduce and stabilize fractures, and correct deformities. If tissue engineering can be used to replace lost or degenerated tissue with new living tissue, then orthopaedics will enter a new and exciting era. It is estimated that more than a half a million people in the United States per year could benefit from engineered bone, articular cartilage, intervertebral disk, dense fibrous tissue, and/or meniscus. The number of patients who could benefit from such approaches will increase rapidly with the aging of the population. Success in patients with the most apparent clinical needs will lead surgeons to extend the indications for use of engineered tissues to treatment of musculoskeletal disorders that are currently treated by other methods. Tissue engineering proponents need to appreciate that developing a method that predictably forms new functional tissue will not have a rapid stunning clinical impact comparable to the discovery of penicillin treatment of bacterial infections, a treatment that did not need prospective clinical trials to prove its value and whose clinical use was limited only by the rate of production of penicillin.[69,70] Instead, discovering a method of producing new functional tissue for the treatment of musculoskeletal disorders is only the first step in achieving clinical success. It is then necessary to critically evaluate the clinical needs that are most likely to be met by the engineered tissue and show that it meets rigorous standards for clinical safety, efficacy, and cost effectiveness. An engineered tissue will need to be profitable to achieve widespread clinical use. Because of the benign natural history and available treatments of most musculoskeletal disorders, the clinical value of treatments based on engineered tissues must ultimately be determined by prospective clinical studies.

References

1. Bulstrode C, Buckwalter JA, Carr A, et al (eds): *Oxford Textbook of Orthopaedics and Traumatology*. Oxford, England, Oxford University Press, 2002.

2. Young BH, Peng H, Huard J: Muscle-based gene therapy and tissue engineering to improve bone healing. *Clin Orthop* 2002;403:S243-S251.

3. Iwata H, Sakano S, Itoh T, Bauer TW: Demineralized bone matrix and native bone morphogenetic protein in orthopaedic surgery. *Clin Orthop* 2002;395:99-109.

4. Luyten FP, Dell'Accio F, De Bari C: Skeletal tissue engineering: Opportunities and challenges. *Best Pract Res Clin Rheumatol* 2001;15:759-769.

5. Musgrave DS, Fu FH, Huard J: Gene therapy and tissue engineering in orthopaedic surgery. *J Am Acad Orthop Surg* 2002;10:6-15.

6. Laurencin CT, Ambrosio AM, Borden MD, Cooper JA Jr: Tissue engineering: Orthopedic applications. *Annu Rev Biomed Eng* 1999;1:19-46.

7. Martinek V, Fu FH, Lee CW, Huard J: Treatment of osteochondral injuries: Genetic engineering. *Clin Sports Med* 2001;20:403-416.

8. Hardouin P, Anselme K, Flautre B, Bianchi F, Bascoulenguet G, Bouxin B: Tissue engineering and skeletal diseases. *J Bone Spine* 2000;67:419-424.

9. Temenoff JS, Mikos AG: Injectable biodegradable materials for orthopedic tissue engineering. *Biomaterials* 2000;21:2405-2412.

10. Goomer RS, Maris TM, Gelberman R, Boyer M, Silva M, Amiel D: Nonviral in vivo gene therapy for tissue engineering of articular cartilage and tendon repair. *Clin Orthop* 2000;379:S189-S200.

11. Mason JM, Breitbart AS, Barcia M, Porti D, Pergolizzi RG, Grande DA: Cartilage and bone regeneration using gene-enhanced tissue engineering. *Clin Orthop* 2000;379:S171-S178.

12. Vacanti CA, Bonassar LJ: An overview of tissue engineered bone. *Clin Orthop* 1999;367:S375-S381.

13. Jackson DW, Simon TM: Tissue engineering principles in orthopaedic surgery. *Clin Orthop* 1999;367:S31-S45.

14. Bonassar LJ, Vacanti CA: Tissue engineering: The first decade and beyond. *J Cell Biochem Suppl* 1998;30:297-303.

15. Brittberg M, Lindahl A, Nisson A, Ohlsson C, Isaksson O, Peterson L: Treatment of deep cartilage defects in the knee with autologous chondrocyte transplantation. *N Engl J Med* 1994;331:889-895.

16. Buckwalter JA, Hunziker EB: Orthopaedics: Healing of bones, cartilages, tendons, and ligaments. A new era. *Lancet* 1996;348:S2-18.

17. Payumo FC, Kim HD, Sherling MA, et al: Tissue engineering skeletal muscle for orthopaedic applications. *Clin Orthop* 2002;403:S228-S242.

18. Praemer AP, Furner S, Rice DP: *Musculoskeletal Conditions in the United States*. Rosemont, IL, American Academy of Orthopaedic Surgeons,1999, p 182.

19. Silva MJ, Sandell LJ: What's new in orthopaedic research. *J Bone Joint Surg Am* 2002;84:1490-1496.

20. Buckwalter JA: Advancing the science and art of orthopaedics: Lessons from history. *J Bone Joint Surg Am* 2000;82:1782-1803.

21. Buckwalter JA: Integration of science into orthopaedic practice: Implications for solving the problem of articular cartilage repair. *J Bone Joint Surg Am* 2003;85:1-7.

22. Porter R: *The Greatest Benefit to Mankind*. New York, NY, Norton, 1998, p 831.

23. LeVay D: *The History of Orthopaedics*. Park Ridge, NJ, Parthenon, 1990, p 693.

24. Woo SY, Buckwalter JA (eds): *Injury and Repair of the Musculoskeletal Soft Tissues*. Park Ridge, IL, American Academy of Orthopaedic Surgeons, 1988, p 548.

25. Buckwalter JA: Tendon, ligament, meniscus and skeletal muscle healing, in Rockwood CA, Green DP, Bucholz RW, Heckman JD (eds): *Fractures*. Philadelphia, PA, Lippincott, 2001, pp 273-284.

26. Buckwalter JA, Einhorn TA, Marsh JL: Bone and joint healing, in Rockwood CA, Green DP, Bucholz RW, Heckman JD (eds): *Fractures*. Philadelphia, PA, Lippincott, 2001, pp 245-271.

27. Buckwalter JA: Effects of early motion on healing of musculoskeletal tissues. *Hand Clin* 1996;12:13-24.

28. Buckwalter JA: Joint distraction for osteoarthritis. *Lancet* 1996;347:279-280.

29. Buckwalter JA, Grodzinsky AJ: Loading of healing bone, fibrous tissue, and muscle: Implications for orthopaedic practice. *J Am Acad Orthop Surg* 1999;7:291-299.

30. Salter RB: *Continous Passive Motion: A Biological Concept for the Healing and Regeneration of Articular Cartilage, Ligaments and Tendons*. Baltimore, MD, Williams and Wilkins, 1993, p 419.

31. Aldegheri R, Trivella G, Saleh M: Articulated distraction of the hip. *Clin Orthop* 1994;301:94-101.

32. van-Valburg AA, van-Roermund PM, Lammens J, Melkebeek J, Verbout AJ, Lafeber FPJG, Bijlsma JWJ: Can Ilizarov joint distraction delay the need for an arthrodesis of the ankle? A preliminary report. *J Bone Joint Surg Am* 1995;77:720-725

33. van-Valburg AA, van-Roermund PM, Lammens J, et al: Promising results of Ilizarov joint distraction in the treatment of ankle osteoarthritis. *Trans Orthop Res Soc* 1997;22:271.

34. van-Valburg AA, van-Roermund PM: Hlam v-R, Verbout AJ, Lafeber FPJG, Bijlsma JWJ: Repair of cartilage by Ilizarov joint distraction tested in the Pond-Nuki model for osteoarthritis. *Trans Orthop Res Soc* 1997;22:494.

35. Aynaci O, Onder C, Piskin A, Ozoran Y: The effect of ultrasound on the healing of muscle-pediculated bone graft in spinal fusion. *Spine* 2002;27:1531-1535.

36. Gebauer GP, Lin SS, Beam HA, Vieira P, Parsons JR: Low-intensity pulsed ultrasound increases the fracture callus strength in diabetic BB Wistar rats but does not affect cellular proliferation. *J Orthop Res* 2002;20:587-592.

37. Cook SD, Salkeld SL, Popich-Patron LS, Ryaby JP, Jones DG, Barrack RL: Improved cartilage repair after treatment with low-intensity pulsed ultrasound. *Clin Orthop* 2001;391:S231-S243.

38. Nolte PA, van der Krans A, Patka P, Janssen IM, Ryaby JP, Albers GH: Low-intensity pulsed ultrasound in the treatment of nonunions. *J Trauma* 2001;51:693-703.

39. Eck JC, Hodges SD, Humphreys SC: Techniques for stimulating spinal fusion: Efficacy of electricity, ultrasound, and biologic factors in achieving fusion. *Am J Orthop* 2001;30:535-541.

40. Cook SD, Ryaby JP, McCabe J, Frey JJ, Heckman JD, Kristiansen TK: Acceleration of tibia and distal radius fracture healing in patients who smoke. *Clin Orthop* 1997;337:198-207.

41. Heckman JD, Ryaby JP, McCabe J, Frey JJ, Kilcoyne RF: Acceleration of tibial fracture-healing by non-invasive, low-intensity pulsed ultrasound. *J Bone Joint Surg Am* 1994;76:26-34.

42. Linovitz RJ, Pathria M, Bernhardt M, et al: Combined magnetic fields accelerate and increase spine fusion: A double-blind, randomized, placebo controlled study. *Spine* 2002;27:1383-1389, discussion 1389.

43. Inoue N, Ohnishi I, Chen D, Deitz LW, Schwardt JD, Chao EY: Effect of pulsed electromagnetic fields (PEMF) on late-phase osteotomy gap healing in a canine tibial model. *J Orthop Res* 2002;20:1106-1114.

44. Ryaby JT: Clinical effects of electromagnetic and electric fields on fracture healing. *Clin Orthop* 1998;355:S205-S215.

45. Bassett CA: Beneficial effects of electromagnetic fields. *J Cell Biochem* 1993;51:387-393.

46. Sharrard WJ: A double-blind trial of pulsed electromagnetic fields for delayed union of tibial fractures. *J Bone Joint Surg Br* 1990;72:347-355.

47. Goldring MB: Anticytokine therapy for osteoarthritis. *Expert Opin Biol Ther.* 2001;5:817-829.

48. Heckman JD, Sarasohn-Kahn J: The economics of treating tibia fractures: The cost of delayed unions. *Bull Hosp Jt Dis* 1997;56:63-72.

49. Buckwalter JA, Mankin HJ: Articular cartilage II: Degeneration and osteoarthrosis, repair, regeneration and transplantation. *J Bone Joint Surg Am* 1997;79:612-632.

50. Buckwalter JA: Aging and degeneration of the human intervertebral disc. *Spine* 1995;20:1307-1314.

51. Buckwalter JA, Goldberg V, Woo SL-Y (eds): *Musculoskeletal Soft-Tissue Aging: Impact on Mobility*. Rosemont, IL, American Academy of Orthopaedic Surgeons, 1993, p 423.

52. Buckwalter JA, Heckman JD, Petrie D: Aging of the North American population: New challenges for orthopaedics. *J Bone Joint Surg Am* 2003;85:748-758.

53. Buckwalter JA, Woo SL-Y, Goldberg VM, et al: Soft tissue aging and musculoskeletal function. *J Bone Joint Surg Am* 1993;75:1533-1548.

54. Goldberg VM, Buckwalter JA, Hayes WC, Koval KJ: Orthopaedic challenges in an aging population. *Instr Course Lect* 1997;46:417-422.

55. Buckwalter JA, Martin JA, Mankin HJ: Synovial joint degeneration and the syndrome of osteoarthritis. *Instr Course Lect* 2000;49:481-489.

56. Martin JA, Buckwalter JA: Telomere erosion and senescence in human articular cartilage chondrocytes. *J Gerontol A Biol Sc Med Sci* 2001;56A:B172-B179.

57. Martin JA, Buckwalter JA: Human chondrocyte senescence and osteoarthritis. *Biorheology* 2002;39:145-152.

58. Klapach AS, Callaghan JJ, Goetz DD, Olejniczak JP, Johnston RC: Charnley total hip arthroplasty with use of improved cementing techniques: A minimum twenty-year follow-up study. *J Bone Joint Surg Am* 2001;83:1840-1848.

59. Callaghan JJ, Squire MW, Goetz DD, Sullivan PM, Johnston RC: Cemented rotating-platform total knee replacement: A nine to twelve-year follow-up study. *J Bone Joint Surg Am* 2000;82:705-711.

60. Callaghan JJ, Albright JC, Goetz DD, Olejniczak JP, Johnston RC: Charnley total hip arthroplasty with cement: Minimum twenty-five-year follow-up. *J Bone Joint Surg Am* 2000;82:487-497.

61. Jubb RW: Oral and intra-articular remedies: Review of papers published from March 2001 to February 2002. *Curr Opin Rheumatol* 2002;14:597-602.

62. Beary JF III: Joint structure modification in osteoarthritis: Development of SMOAD drugs. *Curr Rheumatol Rep* 2001;3:506-512.

63. Polisson R: Innovative therapies in osteoarthritis. *Curr Rheumatol Rep* 2001;3:489-495.

64. De Nanteuil G, Portevin B, Benoist A: Disease-modifying anti-osteoarthritic drugs: current therapies and new prospects around protease inhibition. *Farmaco* 2001;56:107-112.

65. Buckwalter JA, Stanish WD, Rosier RN, Schenck RC Jr, Dennis DA, Coutts RD: The increasing need for nonoperative treatment of patients with osteoarthritis. *Clin Orthop* 2001;385:36-45.

66. Buckwalter JA: Repair and regeneration of articular cartilage: Potential applications in osteoarthritis and evaluation of outcomes. *Osteoarthritis Cartilage* 2002;10A:S15.

67. Buckwalter JA: Do intervertebral discs deserve their bad reputation? *Iowa Orthop J* 1998;18:1-11.

68. Knutsen G, Engebretsen L, Ludvigsen TC, et al: Autologus chondrocyte implanation versus microfracture: A prospective randomized Norwegian multicenter trial. *Osteoarthritis Cartilage* 2002;10A:S28.

69. Fleming A: History and development of penicillin, in Fleming A (ed): *Penicillin: Its Practical Application.* London, England, Butterworth, 1946, pp 1-23.

70. Fleming A (ed): *Penicillin: Its Practical Application.* London, England, Butterworth, 1946, p 380.

Chapter 2

The Clinical Market for Tissue Engineered Products in Orthopaedics

Anthony Ratcliffe, PhD

Abstract

Tissue engineering products must be evaluated according to regulatory requirements and developed in a manner that will ensure that they can be approved for marketing. Tissue engineered products for orthopaedic indications may facilitate repair, induce repair, or serve as a functional replacement. Each of these functions is associated with increasing levels of difficulty in development, manufacturing, and use. In the United States, the regulations for a 510(k) device represent the most rapid pathway to market and would be expected to take 2 to 3 years. The regulatory pathway for some tissue engineered products may require 8 years or more. Although the cost to develop each product will vary substantially, it is likely that a product cleared for market as a 510(k) device will cost $5 to $20 million to develop. The cost of developing a class III tissue engineered device is likely to range from $50 to $200 million and the cost to bring a biologic tissue engineered product to market is likely to be $50 to $300 million or more. The development of tissue engineered products that are successful both clinically and commercially will require use of an appropriate business model, realistic timelines, and effective methods for research and development.

Introduction

If laboratory and translational research is to lead to improved orthopaedic clinical practice, new products will need to be successful in the treatment of a clinical problem and be commercially viable. Products must be evaluated according to regulatory requirements and developed in a manner that will ensure that they can be approved for marketing. Several substantial hurdles, independent of the research activities, need to be overcome to bring new technologies to the clinic. It is critical that these issues be addressed early in the development of a new product. In addition, product development and business plans should fully consider these issues to set appropriate expectations and maximize the opportunity for success. This chapter outlines some of the challenges of commercialization of new products and provides some guidance for the successful development of tissue engineered products for orthopaedic indications.

The Current Status of the Tissue Engineering Industry

Although the field of tissue engineering is relatively new, tissue engineered products are already on the market. These products have not been profitable to date, but commercial success of some of these products is expected within the next few years. Within wound care, several dermal-based products are indicated for use in the treatment of burns and chronic wounds.[1-6] In orthopaedics, an autologous chondrocyte product for the treatment of articular cartilage defects (Carticel, Genzyme Biosurgery, Cambridge, MA) has been on the market for several years and clinical data continue to show that is has clinical benefit. The growth factor bone morphogenetic protein-2 within a collagen sponge and cage has been approved for spinal fusion procedures,[7,8] and osteogenic protein-1 has been approved for bone repair in specific indications of long bone fractures.[9]

Tissue Engineered Products Mechanisms of Action

Tissue engineered products can participate in repair of tissues and organs in several ways. Facilitating repair, inducing repair, or serving as a functional replacement are functions associated with increasing levels of difficulty in development, manufacturing, and use. The most simple use of a tissue engineered product is to facilitate repair of a damaged tissue, essentially providing an improved environment for the body to repair itself. These tissue engineered products tend to be acellular and composed of extracellular matrix, either assembled from individual matrix components[10] or grown using cells that deposit the matrix but are subsequently killed during processing.[11] These products have usually been used to treat second- and third-degree burns, with significant success.

Tissue engineered products used to induce repair contain an extracellular matrix with viable cells[2,5] that actively participate in the repair process. The cells participate, at least in part, by secreting growth factors that participate in different stages of the wound repair cascade, including angiogenesis.[12-15] Both autologous and allogeneic tissues can be used and there are data indicating that allogeneic cells can survive for a long time within a wound site.[1,16] These products have been used in the repair of chronic wounds such as diabetic foot ulcers and venous ulcers.

The tissue engineered products that serve as a tissue replacement that functions at the time of implantation or soon after are the most difficult to develop. To date, only feasibility studies of tissue replacement products have been performed. Animal studies indicate that these products have some potential to repair defects in cartilage and bone;[17,18] however, these have not yet been used in clinical trials.

Regulatory Pathways, Timelines, and Cost

In the United States, the Food and Drug Administration (FDA) regulates the development of new products through the Center for Devices and Radiological Health, Center for Drug Evaluation and Research, or Center for

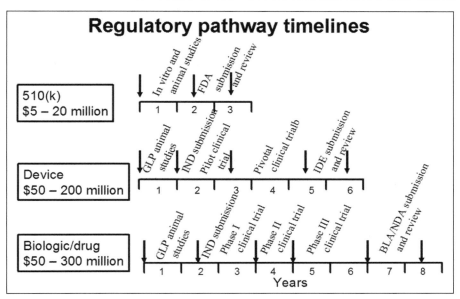

Figure 1 Approximate timeline and cost for the development of tissue engineered products, when regulated as a 510(k) device, a device, or a biologic/drug. These timelines assume that all feasibility studies have been completed, and that the pre-clinical and clinical studies progress without major interruption. IDE = Investigational Device Exemption; IND = Investigational New Drug Application; BLA = Biologics License Application; NDA = New Drug Application.

Biologics Evaluation and Research (CBER). Tissue engineered products have historically been regulated as devices, although their obvious biologic nature has meant that there is increased interest in these products being regulated through CBER. In reality, many tissue engineered products do not fit well into either center. The best regulatory approach might be to establish a center specifically designed to regulate tissue engineered products. Regulatory systems outside the United States are usually specific for each country and thus need to be considered individually. Regardless of the country, regulatory requirements for tissue engineered products will require rigorous studies to help ensure that products are effective and have an excellent safety profile.

Regulatory requirements have an enormous impact on the development timeline for a tissue engineered product (Figure 1). In the United States, the regulations for a 510(k) device represent the most rapid pathway to market. To be regulated as a 510(k) device, the new product must have be substantial equivalence to other devices approved before 1976. In reality, this also means that the product must be acellular. The completion of the necessary studies and obtaining clearance for the product to be marketed can be expected to take 2 to 3 years with this regulatory process. A substantial length of time may be required to develop a product concept to the stage where it is regarded as appropriate for study in animal models according to good laboratory practice (GLP). Resources are also needed for subsequent clinical studies and development of processes to manufacture the product. If the product is regulated as a class III device, as with other tissue engineered products approved for marketing, animal studies followed by pilot and pivotal clinical studies will be required to demonstrate safety and effectiveness. For tissue engineered products for orthopaedic indications, it seems likely that preclinical and clinical studies will be at least one year in duration. Thus the combined time for the development and approval of a product could be more than 6 years, with 4 years dedicated to completing clinical trials. If the product is regulated as a biologic, then additional animal studies of toxicology and dosing will be required in addition to phase I, II, and III clinical studies. These regulatory requirements may result in a development time of 7 or more years. Although the actual time to develop a particular product is unique and can be modified and potentially compressed, past experience with tissue engineered and related orthopaedic products suggests that these timelines are realistic.

The cost of developing these products is associated with the required regulatory pathway. The cost of developing each individual product will vary substantially, however, it is likely that a product cleared for market as a 510(k) device will cost $5 to $20 million to develop, assuming modest manufacturing capital costs and completion of one animal study. The cost of developing a class III tissue engineered device is likely to range from $50 to $200 million. This estimate is based on previously approved products and assumes that the development pathway proceeds without major interruption. The cost to bring a biologic tissue engineered product to market is likely to be $50 to $300 million or more. These costs are substantial and a business model should be robust. The revenue generated from the sale of these products should more than offset the investment. A useful approach would be to ensure that multiple indications can eventually be identified and approved for one product, therefore effectively spreading the development cost and increasing the revenue from an individual product.

Orthopaedic Tissue Engineered Product Pipeline and Potential Market Size

A potential listing of product areas, potential regulatory pathways, and times to market, is provided below.

Short term (2 to 4 years), 510(k) clearance
   –Consisting of extracellular matrix with or without scaffold, acellular
   –Repair device for ligaments, tendons, meniscus, bone, cartilage

Medium term (4 to 8 years), devices
   –Consisting of extracellular matrix with or without scaffold, acellular
   –Repair device for ligaments, tendons, meniscus, bone, cartilage

Long term (> 8 years), devices, biologics
  –Consisting of extracellular matrix and/or scaffold, cells and/or active agents
  –Repair construct using extracellular matrix, cells, and/or active agents
  –Replacement construct using extracellular matrix, cells and/or active agents for cartilage, bone, intervertebral disk, ligaments, and tendons

Although this is not intended as a complete list, it seems likely that most product concepts being developed today will fit within the categories suggested.

For each indication, there is a potential market size and a realistic market size. The annual number of procedures for soft connective tissues in the United States is approximately 400,000 to 600,000, including procedures for meniscus, anterior cruciate ligament, rotator cuff, and articular cartilage. Spinal fusion is the most common procedure for bone repair, but repair of nonunions and other bone graft procedures are also commonly performed. For each indication, factors that might influence the use of the product within this patient population should be considered. These factors include patient factors, availability of competitive products, cost, and anticipated performance of tissue engineered products. A realistic analysis of these factors will help determine the actual potential market. Finally, after the product is approved, there should be a consideration of the time needed to reach peak market sales as this may take 5 to 10 years to achieve.

Ultimately the success of a product will be measured not only by its clinical performance, but also by its financial performance. The costs of manufacturing, research, and development need to be considered along with estimates of how much can reasonably be charged for the product. It seems likely that revenues generated by tissue engineered products are unlikely to exceed $100 million annually in the near future. By some standards this is modest revenue, although by others this revenue can be regarded as good. Estimations of net present value may or may not indicate that this is a good investment. However, it seems that products regarded as successful both clinically and commercially can be developed using a robust and realistic business model, realistic timelines, and effective research and development efforts.

Summary

In addition to research challenges, there are hurdles in the development of tissue engineered products for use in orthopaedic practice. There are a number of product concepts that can reasonably be expected to be effective clinically. Development pathways to result in commercially successful products are also in place. Plans should have realistic timelines, budgets, market projections, and expectations of performance. The FDA plays a critical role in the development of tissue engineered products and should be involved early and often, hopefully in a collaborative mode. If these guidelines are followed, then the development of clinically and commercially successful products can be achieved, and the tissue engineered products can be successfully incorporated into clinical practice.

Acknowledgments

This work was supported by grants AR46133 and DE13437 from the NIH, and NIST ATP award 70NANB0H3001.

References

1. Mansbridge J: Tissue-engineered skin substitutes. *Expert Opin Biol Ther* 2002;2: 25-34.

2. Naughton G, Mansbridge J, Gentzkow G: A metabolically active human dermal replacement for the treatment of diabetic foot ulcers. *Artif Organs* 1997;21:1203-1210.

3. Griffith LG, Naughton G: Tissue engineering: Current challenges and expanding opportunities. *Science* 2002;8:1009-1014.

4. Parenteau N: Skin: The first tissue-engineered products. *Sci Am* 1999;280:83-84.

5. Eaglstein WH, Alvarez OM, Auletta M, et al: Acute excisional wounds treated with a tissue-engineered skin (Apligraf). *Dermatol Surg* 1999;25:195-201.

6. Falanga V, Sabolinski M: A bilayered living skin construct (APLIGRAF) accelerates complete closure of hard-to-heal venous ulcers. *Wound Repair Regen* 1999;7:201-207.

7. Boden SD, Kang J, Sandhu H, Heller JG: Use of recombinant human bone morphogenetic protein-2 to achieve posterolateral lumbar spine fusion in humans: A prospective, randomized clinical pilot trial: 2002 Volvo Award in clinical studies. *Spine* 2002;27:2662-2673.

8. Burkus JK, Transfeldt EE, Kitchel SH, Watkins RG, Balderson RA: Clinical and radiographic outcomes of anterior lumbar interbody fusion using recombinant human bone morphogenetic protein-2. *Spine* 2002;27:2396-2408.

9. Friedlaender GE, Perry CR, Cole JD, et al: Osteogenic protein-1 (bone morphogenetic protein-7) in the treatment of tibial nonunions. *J Bone Joint Surg Am* 2001;83:S151-S158.

10. Stern R, McPherson M, Longaker MT: Histologic study of artificial skin used in the treatment of full-thickness thermal injury. *J Burn Care Rehabil* 1990;11:7-13.

11. Noordenbos J, Dore C, Hansbrough JF: Safety and efficacy of TransCyte for the treatment of partial-thickness burns. *J Burn Care Rehabil* 1999;20:275-281.

12. Mansbridge JN, Liu K, Pinney RE, Patch R, Ratcliffe A, Naughton GK: Growth factors secreted by fibroblasts: role in healing diabetic foot ulcers. *Diabetes Obes Metab* 1999;5:265-279.

13. Mansbridge J, Liu K, Patch R, Symons K, Pinney E: Three-dimensional fibroblast culture implant for the treatment of diabetic foot ulcers: metabolic activity and therapeutic range. *Tissue Eng* 1998;4:403-414.

14. Pinney E, Liu K, Sheeman B, Mansbridge J: Human three-dimensional fibroblast cultures express angiogenic activity. *J Cell Physiol* 2000;183:74-82.

15. Jiang WG, Harding KG: Enhancement of wound tissue expansion and angiogenesis by matrix-embedded fibroblast (dermagraft): A role of hepatocyte growth factor/scatter factor. *Int J Mol Med* 1998;2:203-210.

16. Kern A, Liu K, Mansbridge J: Modification of fibroblast gamma-interferon responses by extracellular matrix. *J Invest Dermatol* 2001;117:112-118.

17. Schreiber RE, Kerby BM, Dunkelman NS, Symons KT, Rekettye LM, Willoughby J, Ratcliffe A: Repair of osteochondral defects with allogeneic tissue engineered cartilage implants. *Clin Orthop* 1999;367:S382-S395.

18. Ishaug SL, Crane GM, Miller MJ, Yasko AW, Yaszemski MJ, Mikos AG: Bone formation by three-dimensional stromal osteoblast culture in biodegradable polymer scaffolds. *J Biomed Mater Res* 1997;36:17-28.

Chapter 3

Bringing Orthopaedic Tissue Engineering From Bench to Bedside: Clinical Research Challenges

David W. Levine, MD, MPH

Abstract

Advances in cell biology and the creation of biomaterials may help to address limitations in several therapeutic areas including orthopaedics. Tissue engineering techniques offer the potential for repairing and regenerating tissue that has the characteristics of normal healthy tissue. Products developed through tissue engineering often have characteristics of both drugs and medical devices and have been called "surgiceuticals." The successful commercialization of tissue engineering products requires the development of new regulatory requirements. However, applying the regulations used for pharmaceutical agents is not appropriate for tissue engineering products that are surgically implanted and would hinder advances in the use of tissue engineering products. Corporations, regulatory authorities, third party payers, and clinicians must work together to create a new framework for clinical development of tissue engineering products. The potential risks and benefits to patients as well as issues pertaining to clinical trial methodology must be considered in the development of new regulatory pathways for tissue engineered products.

Introduction

Tissue engineering products have unique characteristics distinct from traditional pharmaceutical agents or medical devices. The successful commercialization of tissue engineering products requires that corporations, regulatory authorities, third party payers, and clinicians recognize these unique characteristics and create a new framework for the clinical development of these products. This framework should address:

- Results—What are the appropriate preclinical models? What are the end points for preclinical and clinical research? What types of clinical trials of surgically implanted products are ethical, practical, and scientifically sound?

- Regulation—What is the appropriate regimen to protect patient safety, establish efficacy, and allow innovation? Does the regulatory pathway support a return on the investment?
- Reimbursement—How should new technologies be assessed and financed?

This chapter focuses on the implications of the different requirements of clinical trials of the surgical delivery of tissue engineered products compared with those of pharmaceutical agents and medical devices.

Background

Traditional pharmaceutical agents are delivered to the patient through relatively noninvasive routes of administration, including oral, intramuscular, or intravenous administration. Standardized dosing, blinding, and termination of treatments in clinical studies of drugs are often straightforward. Although the drug development process is lengthy and expensive, the regulatory requirements and clinical trial methods are, for the most part, well established.

Traditional medical devices are usually surgically implanted devices or instruments used in a surgical procedure. The regulatory pathway is usually well defined, and in some cases extensive clinical trials might not be required.

Tissue engineering products have some characteristics of drugs and some characteristics of medical devices. Similar to drugs, they have biologically active components that may have dose-dependent effects. As with devices, tissue engineered products are surgically implanted and may have biomaterial components. The regulatory pathways for tissue engineered products are evolving. I call these products with hybrid drug-device characteristics "surgiceuticals."

"Surgiceuticals" offer great promise to address unmet needs in surgical therapy. Traditional surgical modalities and devices have enabled great advances in patient care, but have limitations. These limitations include prostheses that may not function as well or for as long as the original organ/tissue. In addition, prosthetics are not available for many indications and some procedures are palliative but not curative. For surgical treatments that require donated allogeneic tissue or organs, shortages of donor tissue/organs, immunogenicity, and disease transmission are critical issues. Advances in cell biology and the creation of new biomaterials may help to address these limitations in several therapeutic areas including orthopaedics. These advances create the potential for repairing and regenerating tissue that has the characteristics of normal healthy tissue.

A significant challenge in realizing the potential of tissue engineering is the design and conduct of clinical trials. Countries differ greatly in the development of regulations for marketing approval and policies for reimbursement for tissue engineered products. However, regulatory and/or reimbursement authorities are expected to require that tissue engineering products be subject to more stringent controlled clinical trials than are traditionally used for studies of surgical techniques. The need to conduct controlled clinical trials is not in question, but the design of clinical trials that are ethical, practical, clinically relevant, and scientifically rigorous requires careful consideration. There is a tendency to apply clinical study methods used to develop pharmaceutical agents to the evaluation of surgical tissue engineering techniques. It is the thesis of this chapter that the application of clinical trial methods and regulations used to develop drugs to the study of surgically implanted tissue engineered products is not appropriate and would hinder the clinical advancement of tissue engineering. Instead, trial design framework that recognizes the specific issues with surgical approaches and strikes the appropriate balance between scientific rigor, ethical values, practical implementation, and clinical relevance should be implemented. This proposed framework stems from an analysis of the state of surgical clinical research and the differences between pharmaceutical agents, medical devices, and surgically implanted tissue engineered products.

Current State of Clinical Research in Surgery

Randomized controlled trials (RCTs), or studies with controls of any kind, for the evaluation of surgical techniques are rare. When randomized trials are conducted, they are often inadequately designed or executed.

In a review of RCTs in the general surgery literature, Solomon and associates[1,2] observed that 3% to 7% of reported studies were randomized trials. Patient preferences, uncommon conditions, and a lack of surgical community equipoise were the most common reasons that trials of surgical interventions were not randomized. In addition, the methods used in trials of surgical techniques were determined to be of low quality. Solomon and McLeod[2] also proposed an algorithm to determine which types of surgical studies could be conducted as RCTs in an ideal clinical situation defined by conditions such as unlimited funding, adequate expertise, willingness of investigators to participate, and other factors. They concluded that only 38.8% of the surgical studies evaluated could have been done as an RCT according to their criteria.

Evaluations focused on the orthopaedic literature provided similar results in terms of frequency of RCTs and quality of the methods used. Rudicell and Esdaile[3] conducted a Medline search of RCTs in orthopaedics during a 5-year period and found only four orthopaedic trials that included a surgical cohort. Two studies compared two surgical procedures and two compared a surgical to a nonsurgical treatment. In a review of the quality of the statistical analyses of 103 original articles in the 1984 *Journal of Bone and Joint Surgery* (British), Morris[4] found five randomized controlled trials. Freedman and associates[5] reviewed the sample size and statistical power of studies in the 1997 volumes of the British and American *Journal of Bone and Joint Surgery*. They found that 33 studies (4.6%) of 717 were RCTs and only three of these 33 studies described sample size calculations. Using a post-hoc method to calculate power assessments, they concluded that 25 studies did not have adequate sample sizes to detect small differences and the studies used only 10%, on average, of the required sample size.

Bhandari and associates[6] reviewed the quality of RCTs reported in the American volume of *Journal of Bone and Joint Surgery* between 1988 through 2000. They observed that only 72 studies (2.9%) met their criteria for RCTs. They concluded that, "Few studies published in the *Journal of Bone and Joint Surgery* were randomized trials. More than half of the trials were limited by a lack of concealed randomization, lack of blinding of outcome assessors, or failure to report reasons for excluding patients." They also reported that trials comparing drug treatments had higher quality scores than trials comparing surgical interventions.

Not only are RCTs of orthopaedic interventions infrequent, but those performed tend to assess adjuvant therapy or technique modification, rather than fundamentally different treatment options. In an analysis of studies in the 1998 volumes of the *Journal of Bone and Joint Surgery,* seven of the American (6%) and 11 of the British articles (10%) were RCTs.[7] Of the seven RCTs in the American volume, five compared adjuvant therapy or modification of technique and two compared different treatment interventions. In the British volume, nine of the 11 RCTs compared adjuvant therapy or techniques and only two compared two fundamentally different treatment options. Few studies were comparative. The predominate study designs were case series and case reports.

Why so few randomized trials in orthopaedics and what are the implications for conducting clinical studies for tissue engineered products for orthopaedic indications? There are many reasons for both the dearth of RCTs and problems with methods used in trials of orthopaedic interventions. These include time, cost, and complexity of RCTs and the lack of training in clinical research for orthopaedic surgeons.

Differences Between Pharmaceuticals and "Surgiceuticals"–Implications for Trials of Tissue Engineering Products

There are several differences between drug therapy and surgical therapy that present challenges in designing well-controlled trials of surgical interventions. These include issues related to reversibility or termination of therapy, patient preferences, standardization of dosing, blinding, variables associated with administration of therapy, and evolution of surgical techniques.

Reversibility or termination of therapy is one of the key issues and relates to the broader issue of patient and physician equipoise for participation in clinical trials. Unlike participation in a trial of a drug, a patient's decision to enter into a trial with one or more surgical arms is not reversible. There is what I term a "Rubicon effect" in that it is more difficult to enroll patients in trials of surgical interventions in comparison to trials of drug treatments.

In order for a trial to be ethical and practical, participating patients and physicians must clearly understand treatment alternatives. Comparing the risks and benefits of alternative treatments is a critical issue for both patients and investigators. It is a dramatically different proposition for a patient to enter a study with two treatment arms, each involving administration of a

Table 1 Framework for Feasibility Assessment

Patient equipoise	Willing to try anything	Will consider alternatives	Clear treatment preferences
Surgeon equipoise	Able to perform either treatment, no clear preference	Procedure preference and different skills	Unwilling or unable to perform one of the treatment choices
Disease characteristics*	Life threatening, no effective treatment	Serious, disabling, limited current treatment options	Nonserious, elective treatment, multiple options
Procedure characteristics/ study questions*	Adjuvant to surgery technique	Modification of existing technique	Choice between two fundamentally different treatments
Short- versus long-term morbidity*	No significant differences in short-term morbidity	Some differences in short-term morbidity	Significant differences in short-term morbidity
Screening/patient identification	Easy to identify patients preoperatively	Some difficulty with preoperative screening	Diagnosis usuallymade intraoperatively
Number of eligible patients	Large	Medium	Small

*These factors relate to both equipoise and logistical issues

drug, compared with a study that has two treatment groups in which the patient will be treated by either a minimally invasive surgery or a considerably more invasive surgery. Patients have distinct perceptions regarding acceptable risk to benefit ratios. These preferences can limit their willingness to participate in RCTs of surgical interventions as is reflected by relatively high rates of refusal to participate in such trials.

A Framework for Controlled Trials of "Surgiceutical" Tissue Engineering Products

Table 1 provides a systematic framework to assess the feasibility of a surgical RCT for a specific question, product, or procedure comparison.

The first column of Table 1 includes issues that should be assessed to determine the feasibility of conducting an RCT for orthopaedic indications. A potential study that has all of the characteristics listed in the second column of Table 1 would be a very feasible RCT. A randomized controlled study design would not be feasible for a potential study that has the characteristics identified in the fourth column. Studies with characteristics in the third column or a mix of columns may or may not be able to be conducted with as an RCT.

Trials of surgical interventions should use the most rigorous controlled trial design that is ethical and feasible. An RCT is the first choice and the characteristics listed in Table 1 could be used to help determine the feasibility of conducting the study as an RCT. If it is not feasible to use a randomized controlled study design, then alternative controlled study designs, such as a well-designed prospective cohort study or within patient control designs, should be considered. All study designs, including the RCT, have strengths and limitations. The unique strength of an RCT is to control for unknown potential confounders. There are also limitations to RCTs. These include the risk of bias in the patient population selection if the percentage of eligible patients declining to participate is high. Bias in the studied population will limit the ability to generalize the study results to the overall patient population. When an adequately powered RCT is not ethical or feasible, several alternate study designs can be used to provide controlled evaluations. It is beyond the scope of this chapter to assess all of these alternatives and their strengths and limitations for evaluation of tissue engineering products. It is critical to emphasize that rote application of a study design specific for the evaluation of drugs is not appropriate and will hinder the clinical evaluation and approval of tissue engineering products.

Patients, clinicians, companies, regulators, and third party payers all have a vested interest in appropriate design of trials of tissue engineered products. Issues of patient risk and benefit are of critical importance. Patients participating in clinical trials face the risks of new technology, but may also benefit if the new intervention is efficacious. Trials must be rigorously designed to minimize risk and ensure that the study will provide scientifically valid data to justify any inherent risks. Clinicians have a strong interest in advancing new products that will meet unmet medical needs and develop new treatment algorithms. For companies, clinical trials are a very critical and expensive component of product development. The choice of study design and end points can have significant impact on the likelihood of regulatory approval, reimbursement, and clinical acceptance.

Clinical Trials Designed to Move Tissue Engineering Products to the Operating Room

A new framework for designing controlled clinical trials should be applied to tissue engineered products. This framework involves a systematic evaluation of individual product, disease, and procedure characteristics to design trials that have the most scientifically rigorous design that is ethical and feasible for a particular product. This could significantly enhance the quality of controlled clinical research in orthopaedic surgery and provide regulatory authorities, third party payers, and clinicians high quality clinical information.

The framework for clinician participation in these clinical trials will be significantly different than current norms for orthopaedic clinical research. Current research efforts are predominantly single center, case series or reports, and often retrospective. Tissue engineering clinical research will be large multicenter efforts conducted under Good Clinical Practice guidelines and regulations. The studies will be prospective, controlled, and properly powered.

Regulatory authorities could use this framework to tailor clinical trial requirements for product approval to the unique characteristics of tissue engineering products.

Third party payers should reassess how they evaluate and pay for new technologies. The costs of many interventions routinely used in clinical practice are often reimbursed although these interventions have not been thoroughly evaluated in controlled clinical trials. The time, expense, and rigor of studies required for regulatory approval should, in general, be sufficient to secure reimbursement for approved products. This is especially true when traditional treatments that have not undergone similar evaluation are routinely reimbursed.

Summary

The successful commercialization of tissue engineering products requires the development of new regulatory requirements. However, applying the regulations used for pharmaceutical agents is not appropriate for tissue engineering products that are surgically implanted and would hinder advances in the use of tissue engineering products. Patients, clinicians, companies, regulators, and third party payers all have an interest in appropriate design of trials of tissue engineered products. Issues of patient risk and benefit are of critical importance. Interested parties must work together to create a new framework for clinical development of products developed through tissue engineering.

References

1. Solomon M, Laxamana A, Devore L, McLeod R: Randomized control trials in surgery. *Surgery* 1994;115:707-712.

2. Solomon M, McLeod R: Should we be performing more randomized control trials evaluating surgical operations? *Surgery* 1995;118:459-467.

3. Rudicell S, Esdaile J: The randomized clinical trial in orthopaedics: Obligation or option? *J Bone Joint Surgery Am* 1985;67:1284-1293.

4. Morris R: A statistical study of papers in the Journal of Bone and Joint Surgery. *J Bone Joint Surgery Br* 1988;70:242-2464.

5. Freedman K, Bak S, Bernstein J: Sample size and statistical power of randomized, controlled trials in orthopaedics. *J Bone Joint Surg Br* 2001;83:397-402.

6. Bhandari M, Richards R, Sprague S, Schemitsch E: The quality of reporting of randomized trials in the Journal of Bone and Joint Surgery from 1988 through 2000. *J Bone Joint Surg Am* 2002;84:388-396.

7. Levine D: "Unique challenges of surgical versus pharmaceutical trials," in AAOS Orthopaedic Clinical Trials: Design, Implementation and Regulation Issues. Washington, DC, December, 2000.

Chapter 4

The Clinician's Prospective on Tissue Engineering

William J. Maloney, MD

Abstract

Orthopaedic surgery is driven by new technology and orthopaedic surgeons are accustomed to evaluating and implementing new technologies. New strategies are being developed to repair damage to articular cartilage. For this discussion, a focal chondral or osteochondral lesion in the weight-bearing region of the distal femur in which the surrounding articular cartilage is normal or near normal is used as an example. Currently available therapies for such a lesion include marrow stimulation techniques performed arthroscopically (abrasion, drilling, microfracture technique), osteochondral autografting (mosaic plasty), and autologous chondrocyte implantation. As new techniques are being developed, they will need to be assessed in terms of safety and efficacy. New treatments should be compared with currently available therapies for many aspects including difficulty of procedures, invasiveness of the technique, length of rehabilitation, and clinical outcomes. The cost effectiveness of new treatments must also be considered. Tissue engineering techniques have several potential advantages over these procedures, including eliminating donor site morbidity. It is important that scientists developing new technologies have a clear understanding of the clinical problem from the perspective of both the surgeon and patients. As new treatments are marketed, orthopaedic surgeons will need to continually review new information and advise patients as to the appropriate indications for these treatments.

Introduction

Orthopaedic surgery is uniquely situated to be the leader in bringing tissue engineering technology to the clinic. Orthopaedic surgeons are accustomed to dealing with new technologies. Over the past two decades, orthopaedic surgeons haves implemented new technologies including porous coated implants in hip and knee arthroplasty, alternative weight-bearing surfaces for hip and knee implants, bone graft substitutes, new instrumentation for correction of spinal deformity, and use of bioactive compounds such as bone morphogenetic protein-2 for spinal fusion. New procedures are continuously being developed to address clinical problems in a less invasive manner. Spinal surgery can now be performed thoracoscopically in some patients. Arthroscopic techniques have been developed and refined to the extent that the majority of knee and shoulder procedures minimize trauma to soft tissues and accelerate rehabilitation. Less invasive surgical techniques for fracture fixation have been developed, and efforts are now being directed toward less invasive methods for hip and knee arthroplasty.

Orthopaedic surgery is also well suited for tissue engineered products because of the nature of the clinical problems encountered. Unlike many specialties who evaluate and treat systemic disease, orthopaedic surgeons frequently address localized problems that are either degenerative or traumatic. Thus, local treatments can be used, thereby eliminating some of the potential hazards associated with systemic therapy.

The orthopaedic surgeon will obviously need to be an intermediary between new techniques and the patient. It will be important that surgeons be fully educated about new techniques and be able to advise the patients as to the appropriate indications and potential risks.

Safety

It is essential that both the safety and efficacy of engineered tissues be thoroughly evaluated. Patients are increasingly more aware of therapeutic options and are concerned about both medical and surgical complications related to interventions. The increased availability of information, especially on the Internet, is both a benefit and a hindrance. Because the information obtained is often erroneous or misinterpreted, the orthopaedic surgeon will need to be well versed in these technologies to accurately discuss these treatments with patients.

The safety concerns for implanting living cells into a patient are the same as those associated with blood transfusions or the use of allograft tissue. The concerns can be broadly categorized into two areas, transmission of viral disease and host rejection. The orthopaedic community as well as patients will have to be convinced that the cells are free of viral and bacterial contamination. These safety issues will be addressed by adherence to good manufacturing and good laboratory practices. Despite these regulations, the potential for mistakes must be recognized and regulatory agencies must be prepared to monitor new technologies appropriately as they move toward incorporation into clinic practice.

Efficacy

The indications for each technology must be clearly defined and the efficacy of the technology for each clinical problem should be determined. The efficacy of new technologies will be compared with existing technologies with regard to outcome and the factors needed to achieve the best possible clinical outcome. There are several important questions that must be answered. Does the new technology require a difficult, complicated surgical intervention? How does the new technology compare with existing technologies in terms of technical difficulty? Can the procedures be performed using minimally invasive approaches such as an arthroscopic procedure or does it require an open procedure such as a joint arthrotomy? Does the technology require a single procedure or does it require several procedures? Is the rehabilitation complicated or simple? Can rehabilitation be done on an outpatient basis or is an inpatient hospital stay required? Does the patient require a prolonged period of non–weight-bearing activity or can they quickly resume activities? These important questions will have to be defined for the surgeon before the technology can be implemented in their patients. It is important that investigators involved in product development have a clear understanding of the clinical problem, the current treatments of those problems, and the implications from the perspective of surgeons and patients.

Evaluation of Efficacy: Currently Available Therapies

An analysis of currently available treatments of cartilage repair is instructive as it helps define expectations for new therapies. Damage to and subsequent degeneration of articular cartilage as well as its relationship to osteoarthritis is a significant health problem in the United States. According to a national health survey, approximately 70% of the population aged 65 years and older is affected in some way by osteoarthritis.[1] A significant number of the patients with osteoarthritis are not currently being treated by physicians.[2] Many patients are untreated and others use alternative therapies. It has been estimated that the annual total direct medical costs related to articular cartilage disease is in the range of $8 billion in the United States.[3] Currently, there is no cure for osteoarthritis and no treatments that have been proven to retard disease progression.

For this discussion, currently available therapies directed at chondral and osteochondral defects will be presented as an example of how a clinician might evaluate a tissue engineered therapy. An important first step is to clearly define the disease process that is being evaluated. It is often difficult to clearly demarcate focal degenerative processes from an osteoarthritic joint in patients who have articular cartilage damage and degeneration. This discussion focuses on a focal chondral or osteochondral lesion in the weight-bearing region of the distal femur in which the surrounding articular cartilage is normal or near normal. Current therapies for this type of lesion include marrow stimulation techniques performed arthroscopically (abrasion, drilling, microfracture technique), osteochondral autografting (mosaic plasty), and autologous chondrocyte implantation.

Marrow Stimulation Procedures

Marrow stimulation procedures are designed to permit pluripotential marrow cells access to the damaged region.[4] These cells are believed to have chondrogenic potential and will lead to tissue repair under the right mechanical and biologic environments. The surgical procedure is straightforward, minimally invasive, and uses an arthroscopic approach. The subchondral bone is violated using a variety of techniques such as abrasion arthroplasty with power burr, subchondral drilling, and the microfracture technique. The goal is to stimulate the formation of a repair tissue that is fibrocartilaginous. The procedure can be performed by a general orthopaedic surgeon as an outpatient procedure. It requires no special training and it is relatively inexpensive as there are no implantables. This is a low-risk procedure for the patient.

The rehabilitation process is relatively inconvenient for the patient as it is generally recommended that the patient be toe-touch weight bearing for 6 to 8 weeks. Outcomes have been variable and probably are related, in part, to the nature of the cartilage lesions treated, the surgical technique, and patient compliance with rehabilitation. In general, this procedure is thought to improve symptoms in approximately 70% of patients.

Autologous Osteochondral Transplantation

Autologous osteochondral transplantation involves harvesting osteochondral plugs from the periphery of the patellofemoral articulation and transplanting these osteochondral plugs into the damaged area of the weight-bearing surface.[5] This is a more complicated surgical procedure than the marrow stimulation technique. It can be performed either with an open arthrotomy, a mini-arthrotomy, or arthroscopically depending on the experience of the surgeon, accessibility to the lesion, and the ability to obtain sufficient donor plugs. This is a slightly more extensive operation compared with that required for the marrow stimulation techniques and there are some concerns related to donor site morbidity. The rehabilitation processes are similar. Patients must be non–weight-bearing or protected weight bearing for approximately 6 weeks. The advantage of this procedure over marrow stimulation techniques is that it transplants hyaline cartilage; however, there are concerns about long-term donor site morbidity. Predominantly reported by a single investigator, results have been good to excellent at intermediate-term follow-up in approximately 90% of the patients.[6]

Autologous Chondrocyte Implantation

Autologous chondrocyte implantation involves arthroscopic assessment and biopsy to obtain a full-thickness cartilage specimen over an area 5 mm wide and 10 mm long.[7-9] A typical biopsy site is the superior intercondylar notch proximal to the sulcus terminalis. The biopsy specimen is then sterilely transported to Genzyme Biosurgery (Cambridge, MA) where cell culturing and cryopreservation is performed. A second surgical procedure is then performed a minimum of 6 weeks after the biopsy. At the second operation, periosteum is harvested and is sutured in place over the defect through an arthrotomy. The cultured autologous chondrocytes are then implanted by injecting them beneath the periosteal patch, which has been sutured in place and sealed with fibrin glue. Postoperative rehabilitation includes a 6-week period of protected weight bearing followed by progression of full weight bearing by 4 months, as well as continuous passive motion for 6 to 8 hours per day. This procedure is technically more complex than autologous osteochondral transplantation. Both procedures can be performed on an outpatient basis. Because it involves two operations, autologous chondrocyte implantation is associated with a slightly greater risk to the patient. Autologous chondrocyte implantation is more expensive because of the need for two procedures and the cost of the preparation of the autologous chondrocytes. Improvement has been reported in 80% to 90% of patients based on intermediate length follow-up data.[7-9] Tissue biopsies have determined that the resulting repair tissue is a combination of hyaline-like cartilage and fibrocartilage.

Assessment of Tissue Engineered Products

Studies of the treatment of articular cartilage lesions provide some insight as to expectations for a tissue engineered product. The procedures used with tissue engineered products are relatively safe with a low rate of complications. Although there are a few prospective studies and an almost complete lack of comparative studies,[10] the more complicated and expensive procedures may be justified as they can address a wide range of pathologies and possibly result in better outcomes. A tissue engineered product would have to at least match these results providing intermediate-term pain relief in 80% to 90% of patients. In addition, the technology would have to be associated with little or no additional risk to the patient.

Clinicians may not require rigorous comparative studies before widespread incorporation of a new therapy into clinical practice. New technologies often are used on a fairly widespread basis after regulatory approval even if there is a lack of data on long-term clinical outcomes. Clinicians depend on the regulatory process to ensure product safety and that efficacy claims are appropriate. Conversely, clinicians often quickly abandon a therapy if he or she feels that it is not beneficial.

Although repaired tissue and tissue regeneration can be distinguished, the clinician's main concern is patient outcome. Patient outcome is determined by evaluating pain relief and return to function, rather than by histologic criteria. Although the goal is often regeneration of normal tissue, this may be difficult to achieve in the majority of patients and may, in fact, be unnecessary if the clinical outcome experienced by the patient is satisfactory.

The clinician will also need to evaluate the potential advantages of any tissue engineered product compared with currently available therapies. Returning to the example of the articular cartilage lesion, tissue engineering techniques have several potential advantages. First, there should be no donor site morbidity as donor tissue should not be required. Both the mosaic plasty technique and the articular chondrocyte implantation require harvesting of normal articular cartilage from "noncritical" areas. Not only is there a limit to the amount of tissue that can be harvested, but the question of morbidity associated with harvesting of the osteochondral plugs remains unanswered. The use of tissue engineered products would eliminate the initial surgical procedure required for autologous chondrocyte implantation. The degree of difficulty of the procedure, the need for new training, and the required surgical expertise are also of interest. Tissue engineered products should be no more difficult to perform than the mosaic plasty or articular cartilage implantation and may actually be less challenging. The clinician should consider the implications of the technology as it relates to patient rehabilitation. When used for the treatment of articular cartilage lesions, it is expected that tissue engineered products would require rehabilitation processes similar to those of current therapies.

Finally, the clinician should have an objective way to assess outcome. Radiographs often suffice for the evaluation of bone healing; however, radiographs are inadequate for the assessment of the repair of soft tissues such as articular cartilage or tendon. More sensitive imaging techniques such as MRI will be required not only to assess issues such as incorporation of engineered tissues, but to evaluate patients who are not doing well clinically.

Summary

The field of orthopaedic surgery is ideal for bringing the principles of tissue engineering from the laboratory bench to the bedside. Orthopaedic surgeons have a long history of introducing new technologies and orthopaedic surgical problems tend to be focal and not systemic, making them ideal targets for tissue engineering strategies. It is important from the perspective of orthopaedic surgeons that tissue engineering products have specific indications. Ideally, new methods should be relatively easily to implement and use minimally invasive techniques with no increased risk to the patient. From the patient's viewpoint, the outcomes must be at least equivalent and hopefully superior to currently available treatments and involve a reasonable rehabilitation process. Finally, the cost effectiveness of new technologies must be considered.

References

1. Katz SI: Department of Health and Human Services, National Institutes of Health, Statement of the Director, Washington, DC, National Institute of Arthritis and Musculoskeletal and Skin Diseases. 1998, House Appropriations Subcommittee Hearings-Testimony. http://www.dhhs.gov/progorg/asl/testify/b970305a.txt.

2. Jackson DW, Simon TM: Tissue engineering principles in orthopaedic surgery. *Clin Orthop* 1999;367:S31-S45.

3. Lockshin M: Acting Director, National Institutes of Health, Department of Health and Human Services, National Institute of Arthritis and Musculoskeletal and Skin Diseases-Testimony, Washington, DC, House Appropriations Labor, Health and Human Services, and Education FY96 Labor Health and Human Services Appropriations, March 21, 1995. http://www.dhhs.gov/progorg/asl/testify/b950321h.txt.

4. Johnson LL: Clinical methods of cartilage repair: Arthroscopic abrasion arthroplasty. A review. *Clin Orthop* 2001;391:S306-S317.

5. Hangody L, Kish G, Karpati Z, et al: Mosaic plasty for the treatment of articular cartilage defects: Application in clinical practice. *Orthopedics* 1998;21:751-758.

6. Hangody L, Feczko P, Bartha L, Bodo G, Kish G: Mosaic plasty for the treatment of articular defects of the knee and ankle. *Clin Orthop* 2001;391:S328-S336.

7. Brittberg M, Lindahl A, Nilsson A, et al: Treatment of deep cartilage defects in the knee with autologous chondrocyte transplantation. *N Engl J Med* 1994;331:889-895.

8. Brittberg MB, Tallheden T, Sjogren-Jansson E, Lindahl A, Peterson L: Autologous chondrocytes used for articular cartilage repair: An update. *Clin Orthop* 2001;391:S337-S348.

9. Minas T: Autologous chondrocyte implantation for focal chondral defects of the knee. *Clin Orthop* 2001;391:S349-S361.

10. Hangody L, Kish G, Karpati Z: Arthroscopic autogenous osteochondral mosaic plasty: A multicentric, comparative, prospective study. *Index Trauma Sport* 1998;5:3-9.

Chapter 5

Cell-based Therapies: Toward an Integrated United States-European Vision?

Frank P. Luyten, MD, PhD

Abstract

The repair of damaged tissues or organs represents a major challenge for our aging society. Cell-based therapies are emerging as a potential strategy for tissue repair. However, there are many concerns that must be addressed before these therapies can be used in the clinical setting. Key issues include the need for standard methods to characterize cell populations. The safety, efficacy, and tolerability of these therapies have yet to be fully explored in preclinical and clinical studies. In addition, companies who develop these techniques must address regulatory aspects, including the development of manufacturing processes, designing studies to support product claims, and obtaining funding from investors. It is hoped that regulatory agencies will work together to establish standards to facilitate the development of cellular therapies that may ultimately offer patients an effective method to repair tissues.

Introduction

The increasing age of our population presents an enormous challenge for society in general and for the health care system in particular. Good health is frequently jeopardized by organ failure, particularly of the heart, liver, kidneys, or lungs, but the locomotor system is of equal importance for the preservation of quality of life. The repair of damaged tissues or organs is a major challenge given the shortage of donor tissues and organs, the unpredictable outcome of tissue or organ transplants, and the morbidity associated with prevention of transplant rejection. Thus "regenerative medicine" has recently gained attention. The regeneration of damaged tissues can, in theory, be achieved by targeted activation and/or modulation of signaling pathways to stimulate reparative processes, cell-based therapies, and the use of bioengineered implants.

Cell-based therapies have recently drawn particular attention as both mature and somatic pluripotential stem cells harvested postnatally are easily obtainable and expandable. Although cell-based therapy has shown promise in preclinical and clinical studies, major hurdles need to be addressed before these therapies can be used in the clinical setting. This review of the challenges in the development of these therapies is intended to be thought provoking and, hopefully, discussions of the issues presented will contribute to the development of successful cell-based therapies.

Defining The Cellular Product

It is unthinkable to expect reproducibility and consistency in studies if the cellular product is not well characterized, standardized, and made with a well-controlled manufacturing process. At present, this is one of the major challenges for the development of cell-based therapies. It is well known that distinct pharmaceutical compounds from the same family (for instance, the bisphosphonates or nonsteroidal anti-inflammatory drugs) often exhibit different clinical efficacy and toxicology because of differences in pharmacokinetic and pharmacodynamic properties. However, it appears that insufficient attention is being paid to the characterization of cellular products. Furthermore, it is often assumed that expanded cell populations derived from articular cartilage are always the "same" chondrocytes, or that stem cells derived from bone marrow, fat, muscle, skin or any other source are all very comparable. Data indicate that every organ or tissue contains multipotent stem cells and that these cells probably all contribute to local tissue repair. It is still unclear, however, whether tissue-derived stem cells have advantages compared with blood-derived or bone marrow-derived stem cells. A better characterization of different cell populations derived from a variety of tissues is needed to help define expectations for the various cell-based therapies. Advances in characterization of cells by the presence or absence of cell surface markers have been made based on work in the field of hematology. It would appear to be appropriate to define cell populations by the presence of certain markers (positive markers) and the absence of others (negative markers). The use of cell surface markers to characterize cells is facilitated by the ability to use enrichment procedures. Although it may seem obvious that the assays for cell surface markers need to be standardized and performed according to good manufacturing practice conditions, this has yet to be achieved. Reports of the characterization of cell populations indicate that the cell preparations are heterogeneous, although they are sometimes enriched with a certain subpopulation of cells. Thus there is a need to develop and evaluate the biological properties of more homogeneous populations. The functions and properties of specific subpopulations need to be defined, especially with regard to the contribution of a given subpopulation of cells to tissue repair. For example, myoblasts may contribute to certain repair processes of muscle but another cell subpopulation, the satellite cells, may also be important for tissue repair.[1]

The distinction between differentiated cell populations and stem cell preparations may be a matter of relative amount or enrichment in a given type of cell. Most tissues contain cells at different stages of differentiation including some very early precursors, partially committed cells, and terminally differentiated cells. The heterogeneity of the degree of differentiation is clearly demonstrated by the unexpected finding of multipotent precursor cells in a preparation of articular cartilage derived cells, even before expansion. This tissue is considered to contain primarily one cell population of highly specialized chondrocytes.[2] Therefore, every tissue-derived cell product is likely to be a mixture of both mature and precursor cells. The use of mixed cell preparations may have important clinical implications for both efficacy and toxicology.

Ideally, the characterization of cellular and combination products should be linked to specific biological functions appropriate to the desired clinical application. Preferably the identifying characteristics, including molecular markers, should include both positive and negative markers. Challenges in characterizing cell products are exemplified by the characterization of proper "hyaline chondrocytes" for use in repair of cartilage. Promising results with autologous chondrocyte implantation for the repair of symptomatic cartilage defects have led to the development of commercially available cellular products. However, these products have yet to be fully characterized in a standardized way. Classical markers such as collagen type II, the production of highly sulfated proteoglycans, anchorage independent growth, and the lack of expression of bone markers, all help define these highly specialized cells. However, many known or unknown functional aspects of the chondrocyte are not assessed with these markers. Therefore, it is proposed that a proper chondrocyte cell population for the repair of articular cartilage defects be defined as a cell pool capable of organizing, under challenging in vivo conditions, stable hyaline-like cartilage that is resistant to vascular invasion and mineralization or replacement by bone.[3] These functional characteristics are critically important as the repair tissue needs to resist the progression of the underlying bone front. Another consideration in long-term repair is the tendency of articular cartilage to become "endochondral" in patients with osteoarthritis. In this situation, early bone markers appear, indicating loss of the stable chondrocyte phenotype and the beginning of bone differentiation. Therefore, it may be clinically relevant to implant a cell population able to resist these events. Knowing the biological behaviors of a cell population would be valuable for patients at risk for osteoarthritis or for other disorders. In addition, it is important to realize that the behavior of a cell population is also affected by its environment. We recently demonstrated that cell populations derived from articular cartilage may contribute to the in vivo formation of stable cartilage, muscle, or bone depending on their phenotypic stability and the microenvironment in which they are transplanted.[2]

Preclinical Studies

Contribution of the Cellular Product to Tissue Repair

Cell Tracking In studies of cell-based therapies, it is critical to determine the relative contribution of the cellular supplementation to the observed tissue repair, especially when cells are delivered as a suspension. This is certainly relevant for studies of cell-based therapies for repair of musculoskeletal tissues. Tracking the cells in relevant in vivo models is one means to determine the contribution of the cell-based therapy. Cell tracking can be accomplished in several ways, including the use of male-derived cells in female animal models and tracking cells with the Y chromosome. Another method is the use of cells from different species, such as human cells in nude (nu/nu, immunoincompetent) mouse models. The human specific alu sequences can be detected by in situ hybridization or by using polymerase chain reaction probes specific for human and/or mouse sequences.[3] Cells can also be labeled with fluorescent dyes[4] or magnetic particles. Although the use of most fluorescent dyes does not appear to affect the biological behavior of the cells, the dye is only detectable for approximately 3 months in vivo. These dyes usually bind to the lipid layers of the cell membrane and they will quickly become undetectable when the implanted cells proliferate rapidly. Labeling of long-term duration may be achieved by using genetically engineered cell populations, for instance green fluorescent protein (GFP) expressing cells in mice.[1] To obtain cells that chronically express GFP in other mammalian species, retroviral infection of the studied cell populations is a possible approach. However, the biological functions of viral infected cells may be affected and the results may not accurately reflect the in vivo outcome. Therefore, it is imperative that cell labeling procedures are validated in terms of the impact on the biological behavior of the cell population and for inflammatory responses.

Quantification of the Engraftment One of the strengths of cell tracking is the ability to quantify the extent that a cellular product contributes to tissue repair. An example of this approach was recently published.[1] Adult bone marrow-derived cells (BMDC) contribute to muscle tissue in a progressive manner. Following irradiation-induced damage, transplanted GFP-labeled BMDC become satellite cells. The stress-induced progression of BMDC to muscle satellite cells to muscle fibers contributed to as many as 3.5% of muscle fibers. Although establishing an important proof of principle, these results demonstrate the challenges to improving engraftment procedures. For muscle repair, the contribution of implanted cells to repair processes should probably approach 10% to 15% to be clinically relevant.

Quality of the Repair Tissue The ultimate goal of tissue repair is to obtain durable, high quality tissue repair, or restitutio ad integrum, rather than scar tissue. The optimal repaired tissue would eventually be indistinguishable from the original neighboring tissue and would be fully integrated functionally. Functional integration of the grafted material in the original tissue and development of the proper phenotype to achieve long lasting effects would enhance the cost effectiveness of cell-based treatments that are expensive. For joint surface defects, a durable effect is crucial to prevent the progression to osteoarthritis. Cell labeling techniques allowed us to determine that properly expanded and quality controlled chondrocytes resided in the joint surface defect in a goat model.[4] More importantly, after 10 weeks the implanted cells expressed collagen type II, indicating that the cells contributed to hyaline-like cartilage repair.[4] Results after 1 year indicated the potential of a reasonable tissue regeneration with proper tissue organization in this animal model (unpublished data, 2003). These studies provide evidence that proper characterization of biological function of the expanded cell population may reasonably predict proper tissue repair in preclinical models and possibly in clinical studies.

Toxicology

Precautions to preempt the possibility of tragic setbacks of the sort that has damaged other promising fields are imperative. Toxicology issues include verification of cell fusion (tetraploid-diploid), "cell leaking," and proper tissue differentiation. The functional integration of properly differentiated cells is crucial. For instance, improper repair of cardiac muscle using cell-based therapies may lead to side effects including potentially life-threatening rhythmic disturbances. Therefore, toxicology studies must evaluate proper tissue differentiation and potential improper tissue formation. We used expanded articular cartilage-derived cell populations to convincingly demonstrate that the phenotypic stability and the microenvironment determine the tissue formation processes.[2] In a damaged muscle environment, for example, stable articular chondrocytes still organize cartilage tissue and contribute only minimally to muscle repair. However, cells from the same donor that were extensively passaged in vitro and thus "dedifferentiated" lost the potential to make any stable cartilage but contributed significantly to in vivo muscle repair. The latter experiment demonstrates that the signaling environment in a damaged muscle overrides other signaling networks and that cells with some phenotypic plasticity are recruited into the muscle lineage to the extent possible. It can be expected that the environmental signaling in a full-depth cartilage defect, where there is direct contact with the underlying bone, is strongly influenced by "bone/bone marrow" signaling because there is limited signaling of the nearby cartilage.

Labeled cell populations also provide the opportunity to look for cells distant from the site of engraftment. Cells injected systemically can be tracked as they travel to various sites within the body, thus providing information about their activities. The use of cell populations with limited phenotypic stability and high plasticity, such as multipotent stem cells, may reveal their contribution to different cell compartments of the body and tissues. However, pluripotent cells carry the risk of uncontrolled events such as inappropriate tissue formation or even tumorigenesis.

A particularly worrisome and yet unaddressed issue relates to the potential toxicity of the engrafted cell population in a pathologic environment. For example, joint surface defects could be repaired in an otherwise normal joint using allogeneic mesenchymal stem cells. In most patients who are typically young and with otherwise normal joints, the signaling environment in the joint is one of physiologic homeostasis. This signaling environment may dramatically change if an inflammatory joint disease subsequently develops in the same patient. Rheumatoid arthritis or an inflammatory flare in more severe osteoarthritis could lead to increased levels of cytokines in the joint fluid, inducing expression of major histocompatibility complexes on the allogeneic cells and possible rejection of the allogeneic mesenchymal stem cells.

Clinical Studies

Proof of principal studies are needed to confirm the promise of cell-based therapies and tissue engineering approaches in the clinical setting. It is

Table 1 Overview of Classification of Cell and Tissue Therapeutic Products in Europe

Product Type	Country	Classification		
		Biologic/ Pharmaceutical	Unregulated	Other
Autologous chondrocyte implant with cultured chondrocytes	Belgium			Tissue bank
	Germany	Pharmaceutical		
	France			Tissue bank but only research, market approval on hold
	Finland		X	
	Ireland		X	
	Italy		X	
	Netherlands		X	
	Austria	Pharmaceutical		
	Spain			Tissue bank
	United Kingdom			Medical Device Act
	Sweden	Pharmaceutical		
	United States	Biological		

increasingly important to validate treatment protocols in prospective, randomized clinical trials, including multicenter studies. To establish treatment algorithms for specific clinical conditions, the evidence-based medicine approach is the only viable or defensible paradigm. A report on the lack of efficacy of débridement and lavage as a treatment for gonarthrosis has once again drawn attention to the fact that "eminence-based medicine" is not reliable.[5] The design of clinical trials of cell-based therapies is challenging. Studies should be properly randomized and evaluations should be performed in a blinded manner. Internationally recognized outcome measurements need to be developed and validated. In addition, it is important to provide education regarding the use and interpretation of these measures to the medical surgical community, patients, and industry personnel.

Regulatory Aspects: United States and Europe

A major hurdle to achieve the goals discussed is the lack of an international regulatory framework for both development and market authorization for new cellular and tissue engineered products. In the United States, the Center for Biologics Evaluation and Research of the Food and Drug Administration (FDA) has demonstrated a willingness to spend energy and time in develop-

Table 2 Some Examples of Combination Products 'Potential' Product Designation

Product Type	Region/ Country	Status	Regulation
Autologous cultured chondrocytes	United States	Biologic	21CFR1271 CBER
	European Union Countries	No Pan–European Union wide legislation	
		From unregulated to biologic to tissue bank service	Individual country and from 2004 forward?
Autologous cultured chondrocytes with synthetic membrane	United States	Biologic + device (Combination product)	21CFR1271 CBER+CDRH Joint Committee
	European Union Countries	Medical devices incorporating 'products' from human origin specifically excluded from Medical Device Directive, see above	No biologic either
Autologous cultured chondrocytes with cytokines	United States	Biologic	21CFR1271
	European Union	Biologic	EMEA Central approval

CFR=Code of Federal Regulations (of the FDA)
CDRH=Center for Devices and Radiologic Health
EMEA=European Agency for the Evaluation of Medicinal Products

ing new regulations for this very challenging type of product. However, regulations in Europe are still very nebulous and largely dependent on legislation in individual countries. Although recent developments by the European Commission indicate that initiatives are being taken, harmonization of legislation on a global level is lacking. In other areas of biotechnology, enormous effort has been expended on international harmonization through the International Conference on Harmonization of Technical Requirements for Registration of Pharmaceuticals for Human Use process.

An example of the divergence in regulation in the European market for autologous chondrocytes is presented in Table 1. Situations in which the "same" cellular product ends up in very different regulatory tracks depending on its combination with other products (Table 2) should be avoided if the ultimate clinical efficacy and therapeutic claims are the same. For instance, a cultured cell population capable of forming stable cartilage in vivo, can either be a tissue bank service, a medicinal product if delivered in a collagen gel solution, a device when grown on a membrane or a biological, or a pharmaceutical when cultured in serum-free conditions depending on the country (United States or European Union). The regulatory tracks for these products are completely divergent and this puts a tremendous strain on the mostly young industrial partners. This situation can lead to loss of interest by investors and difficulty in obtaining the funds needed to complete expensive but essential clinical studies. Consolidation of the regulatory procedures regarding these novel treatments into one department in one agency is an absolute requirement. International agreements on regulatory requirements will be necessary if these new therapeutics are to be developed within the context of evidence-based medicine.

Summary

Although cell-based therapies are a potentially effective strategy to repair tissues, a number of issues must be addressed before these therapies will be available for use in the clinical setting. There is a need to standardize the markers and methods to characterize cell populations, which appear to contain a mixture of cell types. It is also essential to determine the relative contribution of the cellular supplementation to the observed tissue repair using cell tracking or other methods. Both the quantity and quality of the repaired tissue must be determined. Another major issue is the toxicity of cell-based therapies. The proper differentiation and functional incorporation of the implanted cells need to be assessed. The safety, efficacy, and tolerability of these therapies must be thoroughly evaluated in both preclinical and clinical studies. In addition, companies who develop these techniques must address regulatory and other aspects, including the development of manufacturing processes, designing studies to support product claims, and obtaining funding from investors. It is hoped that regulatory agencies will work together to establish standards to facilitate the development of cellular therapies and combination products that may ultimately offer patients an effective method to repair tissues.

References

1. La Barge MA, Blau HM: Biological progression from adult bone marrow to mononucleate muscle stem cell to multinucleate muscle fiber in response to injury. *Cell* 2002;111:589-601.

2. Dell'Accio F, De Bari C, Luyten FP: Microenvironment and phenotypic stability specify tissue formation by human articular cartilage-derived cells in vivo. *Exp Cell Res* 2003;287:16-27.

3. Dell'Accio F, De Bari C, Luyten FP: Molecular markers predictive of the capacity of expanded human articular chondrocytes to form stable cartilage in vivo. *Arthritis Rheum* 2001;44:1608-1619.

4. Dell'Accio F, Vanlauwe J, Bellemans J, Neys J, De Bari C, Luyten FP: Expanded phenotypically stable chondrocytes persist in the repair tissue and contribute to cartilage matrix formation and structural integration in a goat model of autologous chondrocyte implantation. *J Orthop Res* 2003;21:41-48.

5. Moseley JB, O'Malley K, Petersen NJ, et al: A controlled trial of arthroscopic surgery for osteoarthritis of the knee. *N Engl J Med* 2002;347:81-88.

References With Respect to Regulation

1. Points to Consider on the Manufacture and Quality Control of Human Somatic Cell Therapy Medicinal Products, CPMP/BWP/41450/98, 31 May 2001.

2. Proposal for a directive on setting standards of quality and safety for the donation, procurement, testing, processing, storage and distribution of human tissues and cells, Directorate General SANCO, Doc 2002/0128.

3. Directive 2001/20/EC on the Conduct of Clinical Trials, 4 April 2001.

4. Directive 2001/83/EC on medicinal products for human use, 6 November 2001.

5. Council Directive 93/42/EEC of 14 June 1993 concerning medical devices.

6. European Commission, DG Enterprise, MEDDEV 2.4/1 Rev.8, July 2001.

7. Current Good Tissue Practice for Manufacturers of Human Cellular and Tissue-based Products; Inspection and Enforcement, 66 Federal Register 1508; 8 January 2001

8. Human Cells, Tissues and Cellular and Tissue based products; Establishment Registration and Listing, 66 Federal Register 5447; 19 January 2001.

9. Regulatory Challenges of Tissue-Engineered Products, Ronald S. Warren, RAPS, Regulatory Affairs Focus, Feb 2002.

10. Overview of the Regulation of Human Cellular and Tissue-based Products, Ann Price, RAPS, Regulatory Affairs Focus, July 2002.

11. Combination Products in Europe: a case study, Connie Garrison, RAPS, Regulatory Affairs Focus, March 2002.

Chapter 6

An Industry Perspective on the Development of Osteogenic Agents Using Recombinant Human Bone Morphogenetic Protein-2 (rhBMP-2) as an Example

Howard Seeherman PhD, VMD
Cristina Csimma, RPh, MHP
Alexandre Valentin-Opran, MD
John Wozney, PhD

Abstract

Bringing a therapeutic agent to market for orthopaedic indications is a lengthy, complex procedure that involves basic science research, manufacturing, preclinical studies of pharmacology and safety, marketing, clinical development, registration, and sales. Osteogenic agents have recently been developed for clinical use for the repair of bone. The challenges faced by companies developing therapeutic agents for orthopaedic indications include the design of preclinical and clinical studies to document the safety, tolerability, and efficacy of the new product. In addition, several practical issues to ensure that the agent can be manufactured in a cost-effective manner must be considered. The development of rhBMP-2 delivered with an absorbable collagen sponge (ACS) for treatment of open tibial fractures is described as an example.

Introduction

Fractures represent a significant public health issue worldwide. The incidence of fractures in the United States has been estimated to be approximately 7 million per year.[1,2] A recent 3-year prospective study indicated that the lifetime risk for sustaining a fracture (1 in 2 persons) is approximately equivalent to the risk for coronary artery disease.[3] Thus, therapeutic agents to accelerate the rate of fracture healing and/or increase the assurance of fracture healing would have a major impact on quality of life. Bringing a therapeutic agent to market for orthopaedic indications is a lengthy, complex procedure that involves basic science research, manufacturing, preclinical studies of pharmacology and safety, marketing, clinical development, registration, and sales. The Pharmaceutical Research and Manufacturers Association estimates the average costs of research and development for a new drug to be $802 million and the average development time to be 10 to 15 years.[4] The purpose of this paper is to provide an industry perspective on the development of therapeutic agents for orthopaedic indications using rhBMP-2 (dibotermin alfa) as an example. Emphasis will be placed on the clinical development of rhBMP-2 delivered with an ACS for treatment of open tibial fractures. More recent efforts to expand the indications for rhBMP-2 to both closed and open fractures using a new injectable and implantable calcium phosphate matrix (CPM) paste carrier will also be discussed. Parallel programs being pursued using rhBMP-2/ACS for spinal fusion are reviewed elsewhere, as are reports reviewing the progress with other osteogenic agents.

The Development of the rhBMP-2/ACS

Basic Science and Manufacturing

The development of rhBMP-2/ACS began with basic scientific studies showing that rhBMP-2 is a differentiation factor with bone inducing properties.[5] The osteoinductive properties of rhBMP-2 were first demonstrated after implantation at ectopic (non-bony) sites in rodents.[6] Subsequent studies determined that rhBMP-2 is a soluble, local-acting signaling molecule that binds to specific cell surface receptors.[7] These receptors are involved in signaling within the cell through Smad proteins that translocate to the nucleus. Once in the nucleus, they combine with other factors to initiate transcription of a defined set of genes responsible for the differentiation of stem cells into cartilage and bone cells. Although rhBMP-2 alone is sufficient for osteoinductive activity, optimal bone induction is achieved when rhBMP-2 is used in combination with a delivery matrix.[8,9] The matrix helps to maintain the concentration of rhBMP-2 at the repair site long enough for bone-forming cells to migrate to the site of repair, proliferate, and differentiate in response to rhBMP-2. Some carrier systems provide an osteoconductive matrix on which bone cells can lay down bone. The matrix can also impart physical properties required for injection or implantation at the repair site.

Large-scale manufacturing based on recombinant technology using cultures of genetically engineered mammalian cells paved the way for further development of rhBMP-2.[10] Genetically engineered mammalian cells were created by introducing an expression vector containing the cDNA of the human BMP-2 gene into a parental cell line. A clone that demonstrated DNA integration along with adequate rhBMP-2 expression and genetic stability was selected. Master and working cell banks then were established for consistent clinical and commercial manufacturing. The purification consists of chromatographic methods and membrane ultrafiltration resulting in active rhBMP-2 in a formulation buffer.

Preclinical Pharmacology and Safety

Preclinical studies are performed to establish safety and to demonstrate potential for efficacy in subsequent human clinical trials. Although the majority of the studies of preclinical safety of pharmaceutical products are well defined by regulatory agencies, the requirements for preclinical efficacy studies are more variable. In general, the benefits of a comprehensive preclinical program are weighed against the resources needed to complete an extensive evaluation. In particular, the relevancy of animal models to clinical practice must be considered. Thus, the continuum of programs evaluating efficacy in preclinical models ranges from studies only performed in rodents to those that include large animal and nonhuman primate data.

Preclinical studies have documented the efficacy of rhBMP-2/ACS to induce bone in numerous animal models. In particular, rhBMP-2/ACS has been shown to accelerate osteotomy repair by 33% in a rabbit ulna osteotomy model[11] and to accelerate tibia osteotomy healing in goats.[12] The combination of rhBMP-2/ACS has also been used to bridge critical-sized defects and accelerate allograft/host bone union in dogs,[13,14] and to generate spinal fusions in rabbits, dogs, sheep, and nonhuman primates.[9] Bone induction has also been demonstrated in animal models for oral/maxillofacial indications.[15] Bone formation in all of the animal models was limited to the site of local application, and the newly formed bone underwent normal remodeling. Both the local and systemic safety of rhBMP-2/ACS has also been verified in animal models. Some pharmacologic data are available. The predicted half-life of rhBMP-2 at the site of ACS implantation was 3.76 ± 1.39 days in the rabbit ulna osteotomy model.[11] Systemic half-life of released rhBMP-2 into the circulation was only several minutes following intravenous injection in rats.[15]

Clinical Development

The clinical development program involves the design of human clinical trials to adequately determine safety and efficacy of the experimental agent for the appropriate orthopaedic indications. Conducting large clinical trials for orthopaedic indications is particularly challenging. Standards of care that can be used at multiple clinical sites for a particular indication are needed. Defining appropriate objective outcome assessments, such as fracture healing, can be particularly challenging. Clinical assessment of the time of fracture healing and radiographic confirmation of fracture healing are often inconsistent. Other outcome assessments, such as the need for additional interventions, need to be defined. The criteria for when these interventions are to be used must also be determined. In trials that do not use placebo controls, the potential for bias needs to be evaluated.

Marketing Studies and Literature Review

The clinical development program for rhBMP-2/ACS began with marketing studies combined with an extensive literature review and meetings with thought-leaders in trauma surgery to confirm that there was a need for new treatments of open tibial fractures that are often associated with high rates of reported delayed union and nonunion.[15,16] The rates of delayed union in tibial shaft fractures ranged from 16% to 60% for less severe fractures (Gustilo-Anderson type I, II, and IIIA) and from 43% to 100% for more severe fractures (Gustilo IIIB and IIIC).

Observational Study

An observational clinical study was then performed to establish baseline information on the incidence, treatment, morbidity, and outcomes in patients with tibial fractures requiring surgical treatment.[15] Eighty-two evaluable patients were enrolled from 10 Level I trauma centers in the United States. The study confirmed that a high percentage of fractures resulted in delayed union and secondary intervention. Twenty-five of 53 patients (47%) with open fractures required a second surgical intervention for delayed union compared with six of 29 patients (21%) with closed fractures. This study validated the use of reduction in the rate of second interventions as an appropriate efficacy end point. Outcome assessments were also evaluated and the sample size required to achieve statistically significant differences in efficacy in a large pivotal study was estimated. The mean cost of secondary interventions ($1,995) was also determined for pharmacoeconomic evaluation of rhBMP-2/ACS as a potential therapeutic agent. The observational clinical study identified significant issues to be considered in the design of a large multinational clinical study. These issues included establishing a standard of care that could be adhered to at multiple institutions, assessing the frequency of visits after treatment, tabulating the drop-out rate, and determining the length of the study evaluation period. The results of this study also demonstrated the challenges of using radiographic end points as the sole criteria for determining fracture healing.

Study Designs

Based on the results of the above evaluation, several phase I/II pilot clinical studies confirmed the safety and feasibility of implanting rhBMP-2/ ACS in open tibial fractures.[15] A prospective, multinational, randomized, controlled, single-blind (patients unaware of treatment received) phase III pivotal study was then performed. This trial included 450 patients with open tibial fractures enrolled at 49 sites in 11 countries.[16] The pivotal study was designed to show superiority of rhBMP-2/ACS to standard of care defined as fracture fixation with the use of an intramedullary nail (reamed or unreamed) within 14 days after injury. Patients were randomized to: standard of care (intramedullary nail fixation and routine soft-tissue management), and standard of care plus 0.75 mg/mL rhBMP-2/ACS, or 1.50 mg/mL rhBMP-2/ACS. Randomization was also stratified by fracture severity using the Gustilo classification. Patients with Gustilo type I, II, and IIIA were assigned to stratum A and patients with Gustilo type IIIB were assigned to stratum B. Using this study design, differences between standard of care plus rhBMP-2/ACS and standard of care are not only required to be statistically significant but also clinically relevant for a successful outcome. This design is in contrast to the study design used to evaluate equivalency of replacement of autograft with rhBMP-2/ACS in combination with a metal fusion cage for spinal fusion. In the equivalency study design, statistical equivalence of two treatments is required for a successful outcome. As with all appropriately designed studies, the sample sizes must be sufficiently large to ensure acceptable power of the statistical analyses.

This study did not include a group given placebo (ACS alone) as a control based on the lack of any beneficial effect in preclinical studies and the perceived risk of placing a foreign body into potentially contaminated open tibial fractures. Inclusion of this type of placebo group would not allow for a double-blinded study design (both patients and investigators blinded) because the treatment groups would still have to be compared with standard of care. However, because the treatments were not administered by clinicians in a double-blinded manner, there is the potential for bias in the clinical assessment of the study end points. To address the potential for bias in the clinical assessment, an independent panel blinded to treatment evaluated all of the radiographs. In addition, the time to secondary intervention was compared in patients treated with rhBMP-2 and the patients receiving standard of care.

The primary efficacy end point was the proportion of patients who required a secondary intervention resulting from delayed union or nonunion within 12 months. Secondary interventions were defined as any procedure or event that would promote fracture healing. Secondary interventions were classified as most invasive (autogenous bone grafting, exchange nailing and plate fixation, fibular osteotomy, or bone transport), less invasive (intentional nail dynamization, unintentional nail dynamization due to screw breakage, or exchange to a functional brace), and noninvasive (ultrasound, magnetic field, or electrical stimulation). Clinical investigator assessments of fracture healing were used as a secondary end point. This assessment was based on radiographic evidence of healing noted on radiographs taken after surgery, at 6, 10, 14, 20, 26, 39, and 52 weeks after treatment, full weight bearing, and lack of tenderness at the fracture site on palpation. An independent panel, blinded to the treatment and patient data, evaluated all radiographs to determine time to fracture union, defined as bridging of at least three of four cortices. Monitoring of adverse events included recording reports of local inflammation, infection, hardware failure, pain, complications of wound healing, and antibody responses to bovine or human type-1 collagen or rhBMP-2.

A total of 400 centers in 14 countries were initially screened to determine the potential for compliance with the defined standard of care and to determine if each site had sufficient resources to successfully complete the clinical trial. A total of 49 centers participated in the study with 26 sites recruiting greater than 80% of the patients. A total of 421 patients (94%) completed the 12-month follow-up. There was a concentration dependent decrease in the number of secondary interventions in the patients receiving rhBMP-2 (26% of patients in the group receiving 1.50 mg/mL rhBMP-2 and 37% of patients in the group receiving 0.75 mg/mL rhBMP-2 required secondary interventions) compared with the group that received standard care (46%, $P = 0.0004$). The group receiving 1.50 mg/mL rhBMP-2 had a 44% reduction in the risk of secondary intervention (relative risk = 0.56; 95% CI = 0.40

to 0.78; $P = 0.0005$), significantly fewer interventions ($P = 0.0326$), and less invasive interventions ($P = 0.0265$) compared with the standard care group. The reduction in secondary interventions between the 1.5-mg/mL rhBMP-2 group compared with the standard of care group was also significant for both strata of wound severity (Gustilo type I, II, and IIIA and Gustilo type IIIB). There was also a significant difference among treatment groups favoring the 1.5-mg/mL rhBMP-ACS ($P = 0.0013$) when the nail type (reamed or unreamed) was considered for all primary and secondary efficacy end points tested (Cochran-Mantel-Haenszel and Fisher's exact tests and Logistic Regression Analysis). This result indicates efficacy of rhBMP-2 in patients treated with reamed as well as unreamed intramedullary nailing. In the subpopulation of smokers, there was an overall increase in risk of secondary interventions (52%) compared with nonsmokers (39%). However, patients treated with 1.5 mg/mL rhBMP-2 who were smokers had a reduced rate of secondary interventions compared with the smokers in the standard of care group (30% versus 52%, $P = 0.0138$).

Clinical assessment of fracture healing indicated that 50% of the patients were healed at 145, 187, and 184 days in the groups treated with 1.5 mg/mL rhBMP-2, 0.75 mg/mL rhBMP-2, and standard care, respectively. Significantly more patients who received 1.50 mg/mL rhBMP-2 were healed at 10 weeks ($P < 0.05$) and from 14 weeks to 12 months ($P < 0.003$) compared with the standard care group. At 6 months, 58% of patients in the 1.5-mg/mL rhBMP-2 cohort were determined to be clinically healed compared with 38% of patients in the standard care cohort ($P = 0.0008$). The clinical assessment of fracture healing occurred sooner than that of the panel reading the radiographs, (50% united at 6 months versus 9 months). However, the clinical assessments of success (healing) or failure (secondary intervention or not healed by 12 months) were similar to results determined by the radiographic assessment of fracture union in 384 of the patients (91%).

Although the overall rate of fracture site infection was similar in all groups, there were significantly less fracture site infections in patients with Gustilo type IIIA and B injuries who were treated with 1.5 mg/mL rhBMP-2 (24%) compared with the standard care group (44%, $P = 0.012$). In addition, there were significantly fewer hardware failures ($P = 0.017$) and faster wound healing (83% versus 65% at 6 weeks; $P = 0.001$) in 1.50-mg/mL rhBMP-2 group compared with the patients treated with standard care. Antibodies to BMP-2 were detected in 1%, 2%, and 6% of patients in the standard care, 0.75-mg/mL, and 1.5-mg/mL rhBMP-2 cohorts, respectively. Antibodies to bovine type I collagen developed in 6%, 15%, and 20% of patients treated with standard care, 0.75 mg/mL, and 1.5 mg/mL rhBMP-2, respectively. There was no apparent relationship between the presence of antibodies and clinical outcome or adverse events.

The mean times to secondary intervention were 104.5 days for the standard care group and 105 days and 107 for the 0.75-mg/mL rhBMP-2 and 1.5-mg/mL rhBMP-2 groups, respectively. There were no statistical differences between these groups with regard to mean time to secondary intervention, suggesting that there was no bias in deciding when secondary interventions were required. A lack of bias was also indicated by the correlation between clinical assessment of healing and the independent evaluation of radiographs as well as the concentration-dependent decrease in hardware failure.

This study represents the largest prospective, controlled, multicenter clinical trial of the use of BMPs in the initial surgical management of acute long bone fractures. Based on these results, rhBMP-2/ACS was approved in the European Union for treatment of patients with severe open tibial fractures (http://www.eudra.org/humandocs/Humans/EPAR/inductos/inductos.htm). Approval for marketing in the United States is pending Food and Drug Administration consideration of a panel decision recommending the use of rhBMP-2/ACS in severe open long bone fractures when deemed appropriate by the treating clinician. The combination of rhBMP-2/ACS has also received marketing approval in the United States and Canada for use in spinal fusion in combination with an interbody cage (http://www.fda.gov/cdrh/mda/docs/p000058.pdf).[17]

The Development of rhBMP-2/CPM

Efforts have recently shifted from a focus on rhBMP-2/ACS to the development of an injectable/implantable CPM paste formulation that would expand the indications for rhBMP-2 to both closed and open fractures compared with the surgically implanted rhBMP-2/ACS formulation. The CPM delivery vehicle is an endothermically setting paste designed to form amorphous hydroxyapatite granules containing rhBMP-2 when fully reacted.[8,9] A prolonged release of rhBMP-2 (predicted half-life = 8.32 ± 2.71 days) is achieved through resorption of the CPM paste by osteoclasts and giant cells.[18] In addition, the CPM paste promotes bone formation by osteoconduction.

Preclinical results in rat, rabbit, canine, and nonhuman primate models have demonstrated a dramatic acceleration of diaphyseal bone healing ranging from 30% to 50% after a single injection of rhBMP-2/CPM paste.[18] Ongoing preclinical studies also indicate rapid bone induction in metaphyseal locations in response to rhBMP-2/CPM. Clinical trials for both diaphyseal and metaphyseal fracture indications are anticipated.

Summary

Although the clinical development of osteogenic factors for orthopaedic indications is complex, the programs outlined above demonstrate that these new therapeutics can be successfully taken from the laboratory bench to the clinic. In addition to the studies involving fractures, BMPs are also being evaluated to increase bone density locally in anatomic locations such as the hip, wrist, and spine, which are at increased risk for fracture in patients with osteoporosis. Dramatic increases in bone density have been demonstrated in animal models following local treatment with rhBMP-2. Preliminary studies are also being initiated to evaluate the use of rhBMP-12 for tendon and ligament repair, treatment of tendinopathy, and for tendon to bone reattachment surgeries, including anterior cruciate ligament repairs and rotator cuff surgery. The role of growth and differentiation factor-8, another member of the BMP family of molecules, is being investigated for treatment of muscle disorders such as muscular dystrophy and chronic muscle wasting conditions. Strategies to increase muscle strength may also play a role in preventing osteoporotic fractures by reducing the incidence of falling. Taken as a whole, the programs to develop BMPs for clinical use demonstrate tremendous potential for improving patient care and their long-term well-being.

References

1. Praemer A, Furner S, Rice DP: *Musculoskeletal Conditions in the United States.* Rosemont, IL, American Academy of Orthopaedic Surgeons, 1999.

2. *Ortho FactBookTM US.* Chagrin Falls, OH, Knowledge Enterprises, 2000.

3. Brinker MR, O'Conner MS: Trauma Care Delivery: The Effect of EMTALA Regulations. Paper presented at: Annual Meeting of the Orthopedic Trauma Association, 2002, Toronto, Ontario, Canada.

4. The Myth of "Rising Drug Prices" Exposed, 2002.

5. Sakou T: Bone morphogenetic proteins: From basic studies to clinical approaches. *Bone* 1998;22:591-603.

6. Urist MR: Bone formation by autoinduction. *Science* 1965;150:893-899.

7. Yamashita H, Ten Dijke P, Heldin CH, Miyazono K: Bone morphogenetic protein receptors. *Bone* 1996;19:569-574.

8. Seeherman H: The influence of delivery vehicles and their properties on the repair of segmental defects and fractures with osteogenic factors. *J Bone Joint Surg Am* 2001;83:S79-S81.

9. Seeherman H, Wozney J, Li R: Bone morphogenetic protein delivery systems. *Spine* 2002;27:S16-S23.

10. Wozney JM, Rosen V, Celese AJ, et al: Novel regulators of bone formation: Molecular clones and activities. *Science* 1988;242:1528-1534.

11. Bouxsein ML, Turek TJ, Blake CA, et al: Recombinant human bone morphogenetic protein-2 accelerates healing in a rabbit ulnar osteotomy model. *J Bone Joint Surg Am* 2001;83:1219-1230.

12. Welch RD, Jones AL, Bucholz RW, et al: Effect of recombinant human bone morphogenetic protein-2 on fracture healing in a goat tibial fracture model. *J Bone Miner Res* 1998;13:1483-1490.

13. Sciadini MF, Johnson KD: Evaluation of recombinant human bone morphogenetic protein-2 as a bone-graft substitute in a canine segmental defect model. *J Orthop Res* 2000;18:289-302.

14. Pluhar GE, Manley PA, Heiner JP Jr, Seeherman HJ, Markel MD: The effect of recombinant human bone morphogenetic protein-2 on femoral reconstruction with an intercalary allograft in a dog model. *J Orthop Res* 2001;19:308-317.

15. Valentin-Opran A, Wozney J, Csimma C, Lilly L, Riedel GE: Clinical evaluation of recombinant human bone morphogenetic protein-2. *Clin Orthop* 2002;395:110-120.

16. Govender S, Csimma C, Genant HK, Valentin-Opran A: BMP-2 Evaluation in surgery for tibial trauma (BESTT) study group: Recombinant human bone morphogenetic

protein-2 for treatment of open tibial fractures. *J Bone Joint Surg Am* 2002;84:2123-2134.

17. Boden SD, Zdeblick TA, Sandhu HS, Heim SE: The use of rhBMP-2 in interbody fusion cages: Definitive evidence of osteoinduction in humans: A preliminary report. *Spine* 2000;25:376-381.

18. Seeherman H, Li R, Blake C, et al: A single injection of rhBMP-2/calcium phosphate paste given one week after surgery accelerates osteotomy healing by 50% in a non-human primate fibula osteotomy model. Paper presented at: Transactions of the 48th Annual Meeting of the Orthopaedic Research Society, Dallas, TX, 2002.

Chapter 7

Commercial Approval of rhBMP-2 in Spinal Fusions: Bringing the Product to the Market

Bill McKay, ME

Abstract

For years, orthopaedic surgeons have been anxiously awaiting a commercially available recombinant human bone morphogenetic protein (rhBMP) for their bone grafting needs. They had been hearing about the promising preclinical results of BMPs since the isolation and production of recombinant BMPs in the mid 1980s. It was not until approximately 18 years and several hundreds of millions of dollars later that the first rhBMP (INFUSE Bone Graft/LT-CAGE Lumbar Tapered Fusion Device [Medtronic Sofamor Danek]) was finally approved for marketing by the Food and Drug Administration (FDA, on July 2, 2002) as being equivalent to the use of autogenous bone graft. Even more remarkable is the fact that it has been almost 40 years (1965) since Dr. Marshal Urist made the discovery that demineralized bone matrix (DBM) had bone regenerative properties due to the presence of osteoinductive proteins. The amount of research and development required to bring such a technology to market is truly staggering. Numerous hurdles had to be overcome, including the development of laboratory methods and equipment capable of isolating the individual proteins from bone protein extracts, the recombinant production of a large molecular weight protein with proper three-dimensional conformation, identification of a suitable carrier, determination of the proper BMP concentration for human use, design of a clinical study that would show statistical equivalence to autograft, and finally FDA approval. In addition, due to FDA regulatory requirements, minor changes to the device are not possible once clinical trials have been initiated. Several years of basic research and development as well as clinical trial investigations would be lost if minor changes were needed after clinical trials had begun. Any miscalculations along the way would require starting over from the beginning. The FDA approval of rhBMP-2 with an absorbable collagen sponge carrier (INFUSE Bone Graft) was based on a clinical trial using INFUSE Bone Graft with the LT-CAGE Lumbar Tapered Fusion Device in an anterior lumbar interbody fusion procedure. This approval represents a landmark in orthopaedics, but it is only the first of many rhBMP-2 product approvals expected in the future. Many other clinical investigations are ongoing. These trials are evaluating different carriers, concentrations of BMP, and grafting techniques. As with any new technology, it is expected that rhBMP-2 products will continue to evolve to achieve optimal safety and efficacy for different grafting procedures. Recombinant human BMPs have the potential to be one of those unique technologies that could revolutionize surgery by changing the way bone grafting procedures are performed.

Introduction

Spinal fusion procedures typically require greater amounts of bone graft than other orthopaedic procedures because bone formation is desired in areas where bone usually does not exist, such as in intervertebral disk spaces and along the muscle bed of intertransverse processes. Autogenous bone graft has been considered the gold standard for conducting spinal fusion procedures; however, material for bone grafts can only be harvested in limited quantities, and the procedures are associated with morbidity. Allograft bone, including DBM, is sometimes used to supplement autograft bone when autograft bone is in short supply. Both autograft and DBM have bone regenerative properties attributed to the presence of BMPs. These osteoinductive proteins are present only in trace amounts (ie, nanogram quantities) in bone, which may explain some of the variability in bone forming ability observed with bone grafts.[1] Due to the low levels of BMPs in bone, it is not possible to develop a commercially viable BMP product by isolating and concentrating human BMP. To overcome this limitation, the individual BMPs from bone have been isolated and produced recombinantly.[2-5] Once recombinant proteins were available, it was possible to develop a bone graft substitute with the potential to replace and possibly outperform autogenous bone graft by providing greater and more consistent concentrations of BMP.

Wozney[4,5] first produced recombinant BMP-2 15 years ago. For years after this discovery, surgeons read about the dramatic effects of rhBMP-2 and its superiority to autogenous bone in preclinical studies. They could not understand why it was taking so long for a recombinant human BMP to reach the market. There was much anticipation surrounding the use of rhBMP-2 because it appeared that this new unique technology had the potential to revolutionize orthopaedic and spinal surgery. In the vast majority of orthopaedic and spinal operations, surgeons attempt to unite broken bones or fuse motion segments. A product such as rhBMP-2 would be beneficial by resulting in healing equal to or better than autograft and by eliminating the need for harvest of autogenous bone or use of allograft derived bone.

Unfortunately, obtaining clearance of rhBMP-2 for marketing in the United States required the hard work of literally hundreds of people and hundreds of millions of dollars. The research and development of rhBMP-2 into a commercial product (INFUSE Bone Graft) began in 1984 and involved the following major milestones.

1. 1984-1988: Discovery and recombinant production of rhBMP-2

2. 1988-1993: Product development process
3. 1993-1996: Preclinical evaluation process
4. 1995: Medtronic Sofamor Danek licensed BMP rights for spinal applications from Genetics Institute
5. 1996-2000: Clinical trial evaluations
6. 2000-2002: Regulatory approval process

The discovery, recombinant production, preclinical evaluation, pilot and pivotal clinical studies, and regulatory approval processes that culminated in approval for marketing of rhBMP-2 are reviewed.

The Discovery and Recombinant Production of rhBMP-2 (1984 to 1988)

Scientists at Genetics Institute Inc (Andover, MD; now Wyeth) were intrigued with the work of Dr. Urist[2] at University of California-Los Angeles in which he demonstrated that DBM had bone regenerative properties. Dr. Urist[3] determined that there were unidentified morphogenetic proteins in DBM that were responsible for this bone formation. He called these proteins BMPs. In 1984, Genetics Institute began a research program to identify the specific proteins in the DBM that were responsible for initiating bone formation.[4,5] The program involved testing different fractions of DBM protein extracts in an in vivo rat subcutaneous ectopic implant model. Proteins with osteoinductive properties would result in the formation of a bone nodule in this assay. Approximately 2 years later, in 1986, an entire family of BMPs was isolated. Ten kilograms of bone was used as the starting material, yielding only 10 mg of BMPs. This yield of BMPs from bone was approximately one part per billion and illustrates the difficulty in the isolation of BMPs and the need for sophisticated biochemical analytical equipment. The proteins were numbered chronologically as they were isolated. The first protein, BMP-1, was determined to be a procollagen protein and not osteoinductive. The next protein, BMP-2, was shown to be inductive and was among the more potent BMPs identified. Therefore, BMP was the lead molecule to be investigated for possible orthopaedic uses. There are now nearly 20 known BMPs.

Extraction of a specific protein from bone was not a viable option for a commercial product due to the limited supply of human bone. Therefore, to ensure a large consistent supply of BMP-2 for research purposes, it had to be produced recombinantly. Consisting of two chains of 114 and 131 amino acids each, rhBMP-2 is a glycosolated homodimer protein with a specific three-dimensional, folded conformation. This three-dimensional conformation is required for proper binding of BMP-2 to cell receptors and for osteoinductive activity. Molecular biologists needed to splice the human BMP-2 gene into a cell line that would produce the human BMP-2 correctly and efficiently. A common mammalian cell line (Chinese hamster ovary [CHO] cells) was chosen over the bacteria cell line *Escherichia coli*. Genetically engineered CHO cells are grown in large stainless steel tanks and excrete the BMP-2 protein across their cell membranes into the culture medium, eliminating the need to break open the CHO cells to obtain the rhBMP-2 and avoiding contamination with cellular debris. The rhBMP-2 is then purified from the culture medium with the use of chromatography. A buffered solution to facilitate the freeze-drying process and ensure stability of the rhBMP-2 at room temperature was developed. The manufacturing of rhBMP-2 using this process is done under aseptic conditions, making the process labor intensive and expensive.

The Product Development Process (1988 to 1993)

After several years of work to isolate BMP-2 and to develop methods for production of the human recombinant protein, the actual product development process began. Although rhBMP-2 could be produced, little was known about its clinical utility. The optimal method of delivery, concentration, or dose was not known. Therefore, investigations were conducted using the same rat subcutaneous ectopic implant model that originally led to the discovery of BMPs. It was quickly determined that a larger, more consistent bone nodule could be formed when the BMP was delivered on a carrier matrix. A carrier matrix serves several functions including providing space maintenance where new bone is desired, maintaining the BMP concentration above a threshold concentration to initiate bone formation, and acting as a scaffold for cell attachment and early osteoid deposition. Approximately 100 different carrier materials were evaluated over many years. The selection of an appropriate carrier material requires careful consideration. Carrier materials should be biocompatible, bind BMP, and degrade over time so that ultimately only new bone is present. Some of the carrier materials evaluated included collagen, hyaluronic acid, other natural polymers, calcium phosphates (ie, hydroxyapatite and tricalcium phosphate), calcium sulfates, synthetic degradable polymers, DBM, and blood. It was concluded that bone formation could be achieved if the carrier was capable of retaining and delivering the BMP over a period of 2 to 3 weeks. Longer delivery times did not appear to be necessary. An open porous structure allowed rapid influx of cells and new bone formation. Binding of BMP to the carrier is important to reduce protein loss from the carrier during surgical implantation and to help retain the BMP in the carrier postimplantation to prevent unwanted bone formation in other areas. After several years, type I collagen sponges and calcium phosphate ceramics were identified as two of the better carriers. This finding was not surprising given that BMP is primarily found in bone tissue, which is composed of type I collagen and calcium phosphate. Once the carriers were identified, studies to completely characterize the carrier material and its interactions with rhBMP-2 were performed according to FDA requirements. Characterizations studied included in vitro binding of rhBMP-2 to the carrier, pharmacokinetic properties of the release of rhBMP-2 from the carrier in vivo, and standard tripartite biocompatability studies for the carrier and the carrier plus rhBMP-2 combination.

The collagen sponge selected was a type I bovine collagen sponge (HELISTAT, Integra LifeSciences) that had been used clinically for more than 20 years as a hemostatic sponge. Significant binding of rhBMP-2 to the collagen sponge occurs after 15 minutes of wait time after application of the rhBMP-2 solution to the collagen sponge. The in vivo half life of the rhBMP-2 from the collagen sponge was approximately 3 days as determined with the use of radiolabeled rhBMP-2. The remaining rhBMP-2 is undetectable at the site of implantation 3 weeks postimplantation.[6] The collagen sponge is relatively strong and very cohesive but is somewhat compressible by soft tissues such as muscle. Therefore, the collagen sponge is an excellent BMP carrier in applications where the sponge is not subjected to compression by overlying muscle. The first FDA-approved indication is an example of this type of application in which the rhBMP-2 soaked sponge is placed into a metallic interbody spinal fusion cage (LT-CAGE Lumbar Tapered Fusion Device, Medtronic Sofamor Danek) protecting it from compressive forces.

Delivery of BMP to locations that are subject to soft-tissue compression, such as posterolateral intertransverse process fusions, requires a different carrier that is able to offer some compression resistance and maintain space for new bone formation. Resorbable calcium phosphate ceramic materials were found to be ideally suited for this purpose.[7-9] Calcium phosphates selected and evaluated include hydroxyapatite (HA) and tricalcium phosphate (TCP). Biphasic compositions of HA/TCP were found to be optimal because TCP undergoes too rapid of resorption and HA too slow of resorption. Therefore, biphasic compositions of 60% HA/40% TCP and 15% HA/85% TCP are being investigated to determine if one of these combinations has advantages over the others in the clinical setting. Composites of ceramic and collagen sponges offer improved handling characteristics relative to loose ceramic granules and have proved to be highly effective in preclinical models of spinal fusion.[10] Figure 1 shows the three different carriers currently in clinical trials. Other carriers continue to be investigated as they become available to ensure that delivery of BMP is optimal and that the best handling properties are achieved.

The Preclinical Evaluation Process (1993 to 1996)

Any new device must be able to clearly demonstrate both its safety and efficacy in a preclinical (animal) model that is relevant to the ultimately desired intended clinical indication. Medtronic Sofamor Danek began collaborating with Genetics Institute to study the use of rhBMP-2 for spinal fusion. Back in the early 1990s, posterolateral fusion was the predominate spinal fusion technique, so rhBMP-2 was first evaluated in animal models of this fusion technique. Studies of the use of rhBMP-2 for spinal fusion were first published by Schimandle and associates[11] and Sandhu and associates.[12] Both of these investigators achieved 100% fusion rates with the use of rhBMP-2 in rabbit and canine posterolateral fusions using the collagen sponge carrier. Many additional preclinical posterolateral fusion studies were conducted and they consistently resulted in fusion rates equal to and frequently significantly greater than those of autograft.[13-17]

The utility of rhBMP-2 in spinal fusions became clear and Medtronic Sofamor Danek obtained exclusive rights to BMPs in North America for

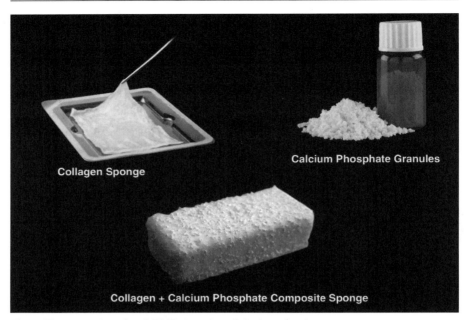

Fig. 1 Three carriers for rhBMP-2 under clinical evaluation.

spinal applications from Genetics Institute in 1995. Before proceeding to clinical investigations, studies in nonhuman primate studies were initiated because it was known that higher order species required greater concentrations of BMP to achieve the same successful results observed in animal models of species of lower order. This is believed to be due to slower bone formation processes in higher order species and possibly due to decreased availability of osteoprogenitor responding cells. Typical concentrations of BMP required in rats, rabbits, monkeys, and humans are 0.1, 0.43, 1.0, and 1.5 mg/mL, respectively. Use of rhBMP-2 on the plain collagen sponge was less successful in nonhuman primate posterolateral fusion studies, unless some type of bulking agent (autograft, allograft, or ceramic granules) was added to the collagen sponge to help it maintain its volume.[18] The slower bone formation capacity of higher order species requires longer residing carriers. This may explain why bulking agents were required to get the desired bone formation and fusion in nonhuman primates.

As interbody spinal fusion cages began to become more popular in the mid 1990s, several preclinical studies were conducted with rhBMP-2 using the collagen sponge carrier in interbody applications. The collagen sponge carrier proved to be an ideal carrier for use inside cages because the cage protected the sponge from any compression. The rhBMP-2 was shown to be superior to the use of iliac crest autogenous bone in several preclinical studies, including a study of nonhuman primates.[19-22] Figure 2 shows the superi-

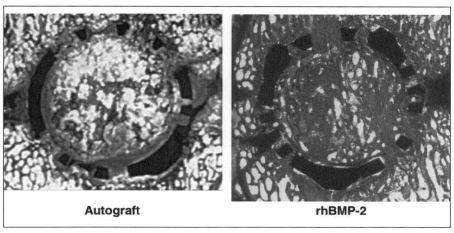

Autograft   **rhBMP-2**

Fig. 2 Histologic sections from a sheep study at 6 months, of a cage filled with iliac crest autograft or rhBMP-2 on a collagen sponge carrier.

ority in the quality of bone formation in a cage filled with rhBMP-2 compared to autograft as determined by histology.

In addition to the preclinical efficacy studies discussed, extensive safety and pharmacology studies were required to fully characterize the adsorption, distribution, metabolism, and excretion of the rhBMP-2 in animals.[23] Radiolabeled rhBMP-2 was used to track the amount of rhBMP-2 entering the blood stream from a local implant site, its distribution throughout the body, its hepatic metabolism, and renal excretion. These studies showed a pattern of normal protein metabolism. Finally, toxicity studies evaluated large doses of rhBMP-2, up to 1,000 times therapeutic doses, implanted either on a carrier at a bony site or injected directly into the blood stream.[23,24] No adverse effects were observed. These studies helped to define the safety profile of rhBMP-2 and supported the decision to evaluate rhBMP-2 in humans.

Clinical Trial Evaluations (1996 to 2000)

Medtronic Sofamor Danek had plans to evaluate rhBMP-2 (INFUSE Bone Graft) in both posterolateral and interbody fusion applications, but the required preclinical safety and efficacy data were first available for the interbody fusion cage application. Thus the initial clinical trial involved the use of the LT-CAGE Lumbar Tapered Fusion Device titanium cage, in both open and laparoscopic anterior lumbar single level interbody fusion surgical procedures. Per FDA requirements, this first study of a BMP in a spinal application was only a pilot clinical trial primarily designed to assess safety. The rhBMP-2 solution was applied onto the absorbable collagen sponge at the time of surgery, and this was placed inside the interbody cages.

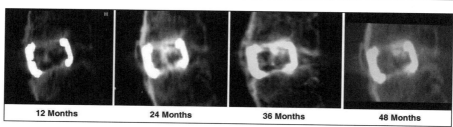

Fig. 3 Serial CT scans of a patient treated with INFUSE Bone Graft in an LT-CAGE Lumbar Tapered cage at 1, 2, 3, and 4 years.

Fourteen patients received the LT CAGEs and were randomized to two treatment groups as to the type of bone graft used to fill them, rhBMP-2 or iliac crest autogenous bone graft using a 3:1 scheme.[25] Eleven patients were treated with rhBMP-2 and three were treated with autograft. All patients treated with rhBMP-2, and two of the three patients in the autograft group were considered fused at 6 months. However, the FDA required a minimum of 1 year of follow-up of all patients in this pilot study before allowing a pivotal trial to begin. There were no reports of adverse events that raised any safety issues. Thus after appropriate follow-up, Medtronic Sofamor Danek was permitted to proceed with a larger, pivotal clinical trial. Figure 3 shows an example of a patient treated with rhBMP-2/INFUSE Bone Graft from the pilot study with 4 years follow-up.

The pivotal clinical trial was a multicenter (16 sites) prospective randomized clinical trial comparing INFUSE Bone Graft to iliac crest autogenous bone graft in the LT-CAGE.[26] A total of 143 patients were assigned to the INFUSE group and 136 were assigned to the autograft group. Both groups were similar with regard to demographic characteristics. Clinical outcome and fusion rates were statistically equivalent for both groups. There was, however, a statistically shorter length of surgery for the patients treated with INFUSE (1.6 hours versus 2.0 hours, $P < 0.0001$) with less blood loss (109.8 mL versus 153.1 mL, $P < 0.0173$). The INFUSE treated patients obviously had no iliac crest harvest site pain, in contrast to 32% of the patients in the autograft group reporting some harvest site pain 2 years after the surgery. INFUSE Bone Graft had a radiographic fusion rate based on radiographs and CT of 100% versus 97.1% for the autograft control. When adding in revisions as fusion failures, per the clinical protocol and irrespective of their radiographic status, the fusion rates were 94.5% versus 88.7% for INFUSE Bone Graft and autograft, respectively.

There was also an arm of the study involving a laparoscopic approach that included 136 patients treated with INFUSE.[27] The efficacy results for these patients were equivalent to groups treated with an open surgical approach. However, their length of hospital stay was statistically shorter (1.2 days versus 3.1 and 3.3 days for the investigational and control open surgical approach, respectively, $P < 0.0001$).

Pilot studies with the 60/40 BCP calcium phosphate carrier and rhBMP-2 in posterolateral fusion studies have been very promising. Luque[28] reported that 14 of 15 patients had successful posterolateral intertransverse process fusion. Boden and associates[29] reported on a pilot study in the United States that resulted in a fusion rate of 100%. Pivotal studies with this carrier as well as with a composite collagen/ceramic sponge carrier are ongoing.

The Regulatory Approval Process (2000 to 2002)

The rhBMP-2 and its carrier was officially classified as a "Device" as opposed to "Drug" by the FDA in 1993. It was specifically classified as a combination "Biologic Device," so it would be reviewed by both the CDRH (Center for Devices and Radiological Health) and CBER (Center for Biologics Evaluation and Research) of the FDA, with CDRH taking the lead review role. Once all the clinical data were collected and analyzed, a PMAA (PreMarket Approval Application) is written and submitted to the FDA. The PMAA for the INFUSE Bone Graft/LT-CAGE Lumbar Tapered Fusion Device was given "expedited" review status by the FDA. This status gave the PMAA for INFUSE PMAA a top priority for review. To facilitate the approval process, Medtronic Sofamor Danek also took advantage of another system in place at the FDA, a modular PMAA. In a modular PMAA, a company is permitted to submit components of the PMAA, ie, preclinical safety and efficacy data, several months prior to the submission of the clinical data. This format allows the FDA to review any available information, prior to completion of analysis and submission of the clinical trial data, that would be reviewed as part of the overall approval process. Historically, the review of PMAAs leading to approval have averaged 2.5 years. The approval for the INFUSE Bone Graft/LT-CAGE Lumbar Tapered Fusion Device took 18 months from the submission of the final module.

As the FDA reaches completion of their review of the PMAA, inspections are conducted at the company, the manufacturing sites, and clinical study sites. Any outstanding issues must be addressed prior to final approval by the FDA. The FDA approved the INFUSE Bone Graft/LT-CAGE Lumbar Tapered Fusion Device for marketing on July 2, 2002.

Summary

The process from Urist's early research on demineralized bone in the 1960s and 1970s to FDA approval of the first recombinant human BMP was long and arduous. Many people contributed in various capacities and significant funds were invested to achieve a commercially available BMP product. rhBMP-2 represents one of the few tissue engineered products that has made it out of the laboratory and is having an impact on surgeons' practices and benefiting patients. Review of the research and development steps should shed light on the particular challenges that had to be overcome to develop this product. An understanding of these challenges may help in the development of other tissue engineered products. Studies of additional carriers, rhBMP-2 concentrations, and clinical indications are being pursued by

Medtronic Sofamor Danek. It is anticipated that these additional investigations will lead to optimal use of rhBMP-2 for broader clinical bone grafting applications.

References

1. Jaw RYY: Amer Assoc Tissue Banks Annual Meeting, 2001.

2. Urist MR: Bone: Formation by autoinduction. *Science* 1965;150:893-899.

3. Urist MR, Mikulski A, Lietze A: Solubilized and insolubilized bone morphogenetic protein. *Proc Natl Acad Sci USA* 1979;76:1828-1832.

4. Wozney JM, Rosen V, Celeste AJ, et al: Novel regulators of bone formation: molecular clones and activities. *Science* 1988;242:1528-1534.

5. Wozney JM: Overview of bone morphogenetic proteins. *Spine* 2002;27:S2-S8.

6. Seeherman H, Wozney J, Li R: Bone morphogenetic protein delivery systems. *Spine* 2002;27:S16-S23.

7. Martin GJ, Boden SD, Morone MA, Moskovitz PA: Posterolateral intertransverse process spinal arthrodesis with rhBMP-2 in a non-human primate: Important lessons learned regarding dose, carrier, and safety. *J Spinal Disord Tech* 1999;12:179-186.

8. McKay B, Sandhu H: Use of recombinant human bone morphogenetic protein-2 in spinal fusion applications. *Spine* 2002;27:S66-S85.

9. Boden S, Martin G, Morone MA, Ugbo J, Moskoviz P: Posterolateral lumbar intertransverse process spine arthrodesis with recombinant human bone morphogenetic protein-2/hydroxyapatite-tricalcium phosphate after laminectomy in the nonhuman primate. *Spine* 1999;24:1179-1185.

10. Suh DY, Boden SD, Ugbo JL, et al: Evaluation of rhBMP-2 with various ceramic/collagen sponge carriers in posterolateral spinal fusion in the rabbit and non-human primate. AAOS Annual Meeting, 88, 2002.

11. Schimandle JH, Boden SD, Hutton WC: Experimental spinal fusion with recombinant human bone morphogenetic protein-2. *Spine* 1995;20:1326-1337.

12. Sandhu HS, Kanim LEA, Kabo JM, et al: Evaluation of rhBMP-2 with an OPLA carrier in a canine posterolateral (transeverese process) spinal fusion model. *Spine* 1995;20:2669-2682.

13. Hollinger EH, Trawick RH, Boden SD, Hutton WC: Morphology of the lumbar intertransverse process fusion mass in the rabbit model: A comparison between two bone graft materials–rhBMP-2 and autograft. *J Spinal Disord Tech* 1996;9:125-128.

14. Sandhu HS, Kanim LEA, Kabo JM, et al: Effective doses of recombinant human bone morphogenetic protein-2 in experimental spinal fusion. *Spine* 1996;21:2115-2122.

15. Fischgrund JS, James SB, Chabot MC, et al: Augmentation of autograft using rhBMP-2 and different carrier media in the canine spinal fusion model. *J Spinal Disord Tech* 1997;10:467-472.

16. Helm GA, Sheehan M, Sheehan JP, et al: Utilization of type I collagen gel, demineralized bone matrix, and bone morphogenetic protein-2 to enhance autologous bone lumbar spinal fusion. *J Neurosurg* 1997;86:93-100.

17. David SM, Gruber HE, Meyer RA, et al: Lumbar spinal fusion using recombinant human bone morphogenetic protein in the canine. *Spine* 1999;24:1973-1979.

18. Suh DY, Boden SD, Ugbo JL, et al: Use of rhBMP-2 supplemented with allograft bone chips in posterolateral fusions in non-human primates. AAOS Annual Meeting, 70, 2002.

19. Sandhu HS, Toth JM, Diwan AD, et al: Histological evaluation of the efficacy of rhBMP-2 compared to autograft bone in sheep spinal anterior interbody fusion. *Spine* 2002;27:567-575.

20. Zdeblick TA, Ghanayem AJ, Rapoff AJ, Swain C, Bassett T, Cooke ME: Cervical interbody fusion cage: An animal model with and without bone morphogenetic protein. *Spine* 1998;23:758-766.

21. Boden S, Martin G, Horton WC, Truss T, Sandhu H: Laparoscopic anterior spinal arthrodesis with rhBMP-2 in a titanium interbody threaded cage. *J Spinal Disord Tech* 1998;11:95-101.

22. Hecht BP, Fischgrund JS, Herkowitz HN, Penman L, Toth J, Shirkhoda A: The use of recombinant human bone morphogenetic protein-2 (rhBMP-2) to promote spinal fusion in a non-human primate anterior interbody fusion model. *Spine* 1999;24: 629-636.

23. Poynton AR, Lane JM: Safety profile for the clinical use of bone morphogenetic proteins in the spine. *Spine* 2002;27:S40-S48.

24. Meyer RA, Gruber HE, Howard BA, et al: Safety of recombinant human bone morphogenetic protein-2 after spinal laminectomy in the dog. *Spine* 1999;24:747-754.

25. Boden S, Zdebick TA, Sandhu H, Heim SE: The use of rhBMP-2 in interbody fusion cages. *Spine* 2000;25:376-381.

26. Gornet MF, Burkus K, Dickman CA, Zdeblick TA: rhBMP-2 with tapered cages: A prospective randomized lumbar fusion study. NASS, 2001.

27. Zdeblick TA, Heim SE, Kleeman TJ, et al: Laparoscopic approach with tapered metal cages: rhBMP-2 vs. autograft. NASS, 200, 2001.

28. Luque E: Latest clinical results using demineralized bone materials and rhBMP-2: The Mexican experience. Total Spine: Advanced Concepts and Constructs. Feb 2000.

29. Boden SD, Kang J, Sandhu H, Heller JG: Use of recombinant human bone morphogenetic protein-2 to achieve posterolateral lumbar spine fusion in humans. *Spine* 2002;27:2662-2673.

Chapter 8

Hyaluronan-based Scaffolds in the Treatment of Cartilage Defects of the Knee: Clinical Results

Alessandra Pavesio, BSc
Giovanni Abatangelo, MD
Anna Borrione, BSc
Domenico Brocchetta, MD
Claudio DeLuca, BSc
Anthony P. Hollander, PhD
Elisaveta Kon, MD
Francesca Torasso, BSc
Stefano Zanasi, MD
Maurilio Marcacci, MD

Abstract

The repair of damaged cartilage is particularly challenging because the intrinsic mechanisms to repair cartilage in adults are limited. The use of tissue engineering to combine isolated cells with polymer scaffolds to generate new tissues has emerged as a potential therapeutic option. Hyalograft C (Fidia Advanced Biopolymers, Abano Terme, Italy) is a hyaluronan-based scaffold for the delivery of cultured autologous chondrocytes. Results with the use of this material for the treatment of full-thickness cartilage defects have been promising. We retrospectively reviewed data from 111 patients treated with Hyalograft C for cartilage lesions of the knee who had a follow-up time postimplantation of at least 12 months. Our data indicate that the vast majority of patients had improvements in relief of symptoms, mobility, pain reduction, and quality of repaired cartilage tissue. When the size of lesions and time from implantation were taken into account, the results suggested that there was greater improvement with increasing lesion size or longer time of follow-up. These results indicate that the use of Hyalograft C for the delivery of cultured autologous chondrocytes represents a promising new treatment of repair of cartilage.

Introduction

Articular cartilage damage is one of the most common injuries seen in orthopaedic practice. A retrospective review of 31,516 knee arthroscopies showed that the incidence of chondral lesions was 63%, with an average of 2.7 lesions per knee. Full-thickness cartilage lesions with exposed bone were observed in almost 20% of patients, with 5% of these occurring in patients younger than age 40 years.[1]

The repair of cartilage lesions represents a great challenge for orthopaedic surgeons because adult cartilage lacks the intrinsic ability for repair and lesions may lead to osteoarthritis. Many surgical techniques have been proposed for the treatment of cartilage lesions. Options that address symptoms include arthroscopic lavage and débridement. Approaches aimed at promoting an effective repair and restoring the damaged articular cartilage include bone marrow stimulation techniques, periosteal and perichondrial grafts, osteochondral autograft, and allograft transplantation.[2] Although some treatment algorithms have been proposed, there is no consensus on a preferred approach to repair of full-thickness defects.[3,4]

Autologous chondrocyte implantation (ACI), first introduced in Sweden in 1987, is widely used for the treatment of full-thickness defects of the articular cartilage.[5] However, despite the promising long-term clinical results,[6] the use of ACI has several limitations essentially correlated with the complexity of the surgical procedure that involves the implantation of autologous chondrocytes under a periosteal flap. In addition, because the cells are grown in a liquid suspension there is concern regarding the lack of chondrocyte differentiation and limited matrix production in vitro.

To overcome these limitations, tissue-engineering techniques that combine isolated cells with polymer scaffolds to generate new tissues or tissue equivalents are gaining attention for their potential to recreate three-dimensional tissues such as cartilage.[7,8] Three-dimensional scaffolds based on Hyaff 11-based (Fidia Advanced Biopolymers, Abano Terme, Italy) are able to support the in vitro ingrowth of chondrocytes, which remain highly viable and capable of expressing the original phenotype.[9] Hyaff 11 is obtained by chemical modification of hyaluronan, a naturally occurring, widely distributed, and highly conserved glycosaminoglycan. Hyaff 11 may be processed into stable configurations to produce a variety of biodegradable devices with different physical forms and in vivo residence times. Extensive biocompatibility studies have demonstrated the safety of biomaterials containing Hyaff 11.[10]

Dedifferentiated chondrocytes seeded into the Hyaff scaffold produce a characteristic extracellular matrix, rich in proteoglycans, and express typical markers of hyaline cartilage, such as collagen II and aggrecan.[11,12] When implanted in full-thickness defects of the femoral condyle in rabbits, chondrocytes cultured in the Hyaff matrix regenerated a cartilage-like tissue.[13,14] It has been also hypothesized that Hyaff 11, when used in conjunction with mesenchymal cells, can lead to an in situ differentiation of these progenitor cells to a chondrocyte phenotype.[14]

These preclinical findings supported the use of Hyaff 11 scaffolds for autologous chondrocyte implantation in a clinical setting and led to the development of Hyalograft C.

Hyalograft C in Clinical Practice

Hyalograft C is a tissue-engineered graft consisting of autologous chondrocytes grown on a three-dimensional Hyaff 11 scaffold in a nonwoven configuration. Since its introduction to clinical practice in 1999 for the treatment of cartilage lesions of the knee, more than 1,600 patients have been grafted at more than 70 clinical centers in Italy and in Austria.

Autologous chondrocytes are isolated from a patient's cartilage harvested arthroscopically from a non–weight-bearing area in the knee. The chondrocytes are expanded in vitro as described previously[5] and are subsequently cultivated into the Hyaff 11 scaffold for 14 days.[15] The graft is implanted at the lesion site by a limited arthrotomy under regional or general anesthesia in a tourniquet-controlled bloodless field, or arthroscopically using a customized cannula as proposed by Marcacci and associates.[16] Articular cartilage lesions are prepared with curettage débriding to the best cartilage available. Border stabilization is performed, and subchondral perforation is limited to minimize bleeding at the site of implantation. Hyalograft C is trimmed, if necessary, to fit the defect and is then placed into the lesion. In the majority of patients, no graft fixation is required; however, depending on the size and location of the defect, fibrin glue and/or sutures may be used to keep the graft in place.[17] After the release of the tourniquet, the knee is mobilized to confirm adherence and stability of the graft before standard closure of the arthrotomy.

Postsurgical joint drainage and compressive bandaging are applied for a minimum of 24 hours postoperatively. Immobilization is recommended for 24 hours postgrafting, after which a postoperative rehabilitation program begins with passive motion and isometric exercises for the quadriceps. Articular weight bearing can be applied from the third week and gradually increased thereafter. Initial clinical reports, which focused on the safety profile and symptomatology of patients treated with Hyalograft C, were very encouraging.[15,17-22]

Hyalograft C Patient Registry

A patient registry has been established to provide a large database to evaluate the outcomes of Hyalograft C treatment in clinical practice.[23,24] The objective of this registry is to collect relevant clinical data to assess the outcome of the treated lesions and to monitor the long-term progress of patients. We present the updated findings in a cohort of 111 patients who were treated at eight Italian clinical centers.

Patients and Methods

Data from 111 patients treated with Hyalograft C for cartilage lesions of the knee and who had a postimplantation follow-up time of at least 12 months were reviewed retrospectively. Treatment outcomes were obtained prospectively. Data were collected with the use of a case report form that accounted for the most recent guidelines of the International Cartilage Repair Society.[25]

The average follow-up time for this series of patients was 18 months (standard deviation, SD = 5.9). There were 68 males and 43 females with a mean age of 37.4 years (SD = 12.9). Patients were affected by chondral defects of the knee caused by trauma (34.6%), osteochondritis dissecans (11.8%), and lesions that were microtraumatic and/or degenerative in nature (53.6%).

Most of the patients (76.6%) had a single lesion, and the remaining 23.4% had multiple lesions. There were a total of 138 defects with an average of 1.2 defects per patient. As expected, the majority of the defects were localized on condyle (74.6%), primarily on the medial femoral condyle (63.8%). The remainder of the defects were localized on the trochlea (8.7%), patella (8.7%), and tibial plateau (8.0%). The defects treated were graded Outerbridge IV in the vast majority of cases (85.4%), and the remaining defects were determined to be grade III. The mean surface area implanted per patient was 3.42 cm^2 (SD = 2.57) with a maximum value of 12.25 cm^2.

More than half of the patients (53.2%) had previously undergone surgical procedures to the affected knee. Remarkably, 36% of the total population had been subjected to cartilage surgery in the past, including débridement and bone marrow stimulation techniques, which eventually proved unsuccessful. A total of 29.7% of patients had previously undergone meniscus and/or ligament surgery. Osteotomy had been performed on 6.3% of the patients.

In our study, 28% of patients underwent one or more associated surgical procedures at the time of Hyalograft C grafting. These procedures included high tibial valgus osteotomy and/or patellar realignment osteotomy (7.2% of patients), and meniscus and/or ligament surgery (24.3% of patients).

Hyalograft C was implanted without any coverage or fixation system in 58.6% of the patients, and fibrin glue and/or sutures were used in 34.2% of the patients. Hyalograft C was applied under a periosteal flap in only eight patients, at the very beginning of the clinical experience with this novel approach. Ten patients were implanted with the use of a minimally invasive arthroscopic technique.[5]

The functional outcome was obtained using four end points: patient's subjective evaluation of knee conditions and quality of life, surgeon's knee functional test, arthroscopic evaluation of cartilage repair, and histologic assessment of the grafted site. Patients were asked for a subjective evaluation of their knee symptoms and physical function using the International Knee Documentation Committee (IKDC) Subjective Knee Evaluation Form,[25] and of their quality of life using the EuroQol EQ-5D questionnaire.[26] Baseline data were obtained by asking patients to answer these questionnaires retrospectively for the conditions of their knees and their quality of

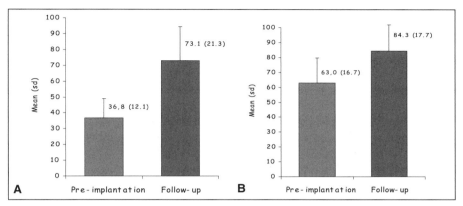

Figure 1 Patients' subjective assessment (N = 111) **(A)** IKDC Subjective Knee Evaluation: mean score at preimplantation and at last follow-up. Patients were asked to rate their knee symptomatology, functionality, and activity level. High scores represent a high level of function and low level of symptoms. Therefore, a score of 100 is interpreted to mean no limitations to daily living activities or sports and the absence of symptoms. Difference between preimplantation and follow-up values proved statistically significant ($P < 0.0001$, Student's t-test). **B,** EuroQol EQ-5D Visual Analog Scale (EQ VAS): mean score preimplantation and at last follow-up. EQ VAS was used for patient self-rating of the global health state. A value of 100 represents the best imaginable health state. Difference between baseline and post-treatment values proved statistically significant ($P < 0.0001$, Student's t-test).

life after injury but just before Hyalograft C implantation. Knee functional outcome was assessed for 61 patients according to the IKDC objective knee examination form.[25]

Thirty-seven patients underwent second-look arthroscopy, performed as a consequence of an adverse event in four patients and for investigative purposes in 33 patients. Quality of the repair tissue was classified according to the integration of the graft into the surrounding cartilage (0 to 4 points), the degree of defect fill (0 to 4 points), and the macroscopic appearance (0 to 4 points). The best possible defect repair score was 12 points.[6] Twenty-four of these patients consented to biopsy harvest in the area of graft implantation for histologic and immunohistochemical assessment of the repair tissue. Finally, the occurrence of adverse events in the postoperative period was recorded.

The entire set of data was entered into a single database. Statistical analyses were performed using the SAS system statistical software package (SAS Institute, Cary, North Carolina, USA).

Results

Patient's Assessment of Knee Conditions and Quality of Life

Results revealed that 92.7% of the patients experienced a subjective improvement in knee function and symptoms. Only 7 patients (6.4%) pre-

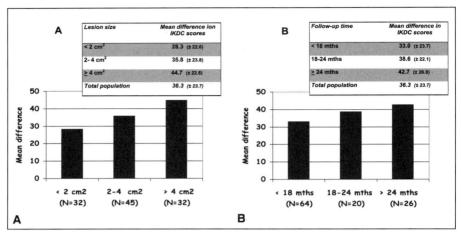

Figure 2 IKDC subjective knee evaluation analyzed by lesion size and follow-up times from implantation. **A,** Mean difference between follow-up and preimplantation IKDC subjective knee evaluation scores in patient subpopulations by different lesion sizes. **B,** Mean difference between follow-up and preimplantation IKDC subjective knee evaluation scores in patient subpopulations by different follow-up times. In both situations, a greater difference reflects a greater improvement in the subjective assessment of the patient's knee.

sented a worsening of their knee conditions at follow-up compared with their preoperative status.

The mean subjective IKDC score obtained at baseline was 36.8 (SD = 12.1) and 73.1 (SD = 21.3) at the follow-up control. The difference was statistically significant ($P < 0.0001$, Student's t-test, Figure 1, A).

A total of 89.9% of patients experienced an improvement in their quality of life, as assessed by the EQ-VAS. The situation was found to be unchanged in six patients and worsened in five patients, four of whom also experienced a worsening in the subjective knee evaluation scores.

The mean EQ-VAS score (Figure 1, B) changed from 61.0 (SD = 16.7) to 84.3 (SD = 17.1), and the difference in mean EQ-VAS score was statistically significant ($P < 0.0001$, Student's t-test). As expected, the majority of the improvements were related to mobility (63% of patients improved), usual activities (68%), and reduced pain/discomfort (83%). When the IKDC subjective outcome was analyzed according to lesion size and time from Hyalograft C implantation, a positive trend of subjective improvement was observed with increasing size and follow-up times (Figures 2, A and 2, B).

Surgeon's Knee Examination

Figure 3 shows the percentage of patients (N = 61) displaying the four possible IKDC knee group grades. At follow-up, final grade was normal in 59.0% of patients and nearly normal in 31.1% of patients. Thus 90.1% of patients displayed knee conditions within the two best categories. The knee

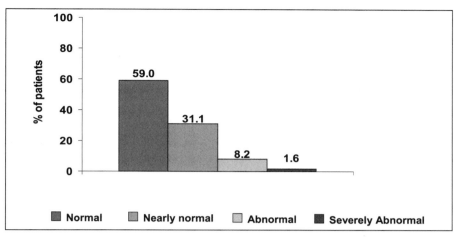

Figure 3 IKDC knee examination performed by the surgeon (N = 61). Patient distribution among the four knee functional group grades (normal, nearly normal, abnormal, or severely abnormal). The final functional grade was determined by the lowest ratings in effusion, passive motion deficit, and ligament examination sections of the IKDC Knee Examination Form.

rated as severely abnormal in only one patient and this may have been related to the occurrence of a marked fibroarthrosis.

Macroscopic Assessment of the Repair Tissue

The mean repair score for 37 patients was 10.7 (range, 3 to 12) at a mean arthroscopy time from implantation of 12.2 months. Cartilage repair was found to be normal (score = 12) in 43.2% of the patients (N =16), and nearly normal in 54.1% of the patients (N = 20, 9 of which scored 11). The cartilage repair was therefore rated as biologically acceptable in 36 (97.3%) of 37 patients. The cartilage repair in only one patient, associated with periosteal hypertrophy, was rated as severely abnormal, having a biologically unacceptable appearance with significant fissuring of the repair tissue and no integration with the surrounding cartilage.

Biopsy Analysis

Biopsy samples (N = 24) were analyzed in a blinded manner by two independent investigators who rated the regenerated cartilage as hyaline-like, fibrocartilage, or mixed tissue based on cellularity, cell distribution, matrix composition, and immunolocalization of collagen type I and II. Based on these criteria, the majority of the biopsies (15 of 24) were rated as hyaline-like, five as mixed tissue, and four as fibrocartilage. The average time of postimplantation second-look arthroscopy for this subset of patients was 14.2 months. An additional analysis of the same samples will use a recently developed technique for quantification of the relative proportion of collagen I and II.[27]

Adverse Events

A total of six adverse events were recorded (N = 111). Intra-articular adhesions were reported for two patients 12 months after implantation. One patient, treated for a lesion on the medial femoral condyle, underwent medial meniscus surgery and tibial osteotomy concurrently with Hyalograft C implantation. This patient was overweight. A marked fibroarthrosis was seen on arthroscopic examination. The second patient, treated for a patella lesion, had patella alignment concomitant to the biopsy for chondrocyte procurement. The patient did not comply with the recommended postoperative rehabilitation program and reported a marked stiffness. Symptoms resolved after arthroscopic adhesiolysis in both patients.

Periosteal hypertrophy was observed in two of the eight patients in this cohort in which a periosteal flap was used to secure Hyalograft C. One patient experienced significant knee catching and locking 12 months after implantation, which resolved after arthroscopic excision. The other patient was asymptomatic. In this patient, the hypertrophy was observed incidentally at the time of a revision for a traumatic anterior cruciate ligament rupture 24 months after implantation and the hypertrophic tissue was removed.

The two remaining reported adverse events were fever from the first to the fifth postoperative day and dehiscence of the osteotomy wound in a patient who had osteotomy concurrently with Hyalograft C application. No graft failures, defined as requiring a revision to remove the implant or reimplant it or a procedure that violates the subchondral bone to treat the defect, have been reported.

Summary

Hyalograft C, a hyaluronan-based scaffold for the delivery of cultured autologous chondrocytes, is showing promise for the treatment of full-thickness cartilage defects. Recently introduced into clinical practice, this product is receiving much attention in the orthopaedic community, and more than 1,600 patients have been treated to date. Advantages of Hyalograft C compared with the currently available ACI technique include ease of handling and the potential to be tailored to fit the defect. Hyalograft C can be delivered by a minimally invasive technique, including arthroscopy,[16] thus significantly reducing surgical time and morbidity. With this novel approach, no periosteal coverage is required to keep the graft in place.

We reviewed data from 111 patients treated with Hyalograft C who had a mean follow-up period of 18 months. Our results indicate that treatment with Hyalograft C was associated with improvements in relief of symptoms, mobility, pain reduction, and quality of repaired cartilage tissue in the majority of patients. Additional analyses that accounted for the size of lesions and time from implantation suggested that there was greater improvement with increasing size of the lesion or longer follow-up time. The safety profile of the treatment appears positive, with a limited number of adverse events reported.

The authors acknowledge that the basal and preimplantation data presented are from a retrospective review and no control group was used. Thus, this study lacks the statistical power of a randomized, blinded, prospective clinical trial. However, our study provides data that are valuable for the design of other clinical studies. Additional follow-up data are needed to provide better information about long-term outcomes.

With the above limitations in mind, we consider Hyalograft C to be an effective treatment option for large defects (greater than 2 cm^2), particularly for patients with high physical demands and for those who have not responded to alternative cartilage repair techniques, regardless of the size of the lesion and the patient's demand.

Participants in the Hyalograft C Study Group

Diego Ghinelli, MD
Donato Rosa, MD
Luigi Pederzini, MD
Fabrizio Pellacci, MD
Andrea Ricciardiello, MD

References

1. Curl WW, Krome J, Gordon S, Rushing J, Smith BP, Poehling GG: Cartilage injuries: A review of 31,516 knee arthroscopies. *Arthroscopy* 1997;13:456-460.

2. Farnworth L: Osteochondral defects of the knee. *Orthopedics* 2000;23:146-157.

3. Minas T, Nehrer S: Current concepts in the treatment of articular cartilage defects. *Orthopedics* 1997;20:525-538.

4. Alparslan L, Winalski SC, Boutin RD, Minas T: Postoperative magnetic resonance imaging of articular cartilage repair. *Semin Musculoskelet Radiol* 2001;5:345-363.

5. Brittberg M, Lindahl A, Nilsson A, Ohlsson C, Isaksson O, Peterson L: Treatment of deep cartilage defects in the knee with autologous chondrocyte transplantation. *N Engl J Med* 1994;331:889-895.

6. Peterson L, Minas T, Brittberg M, Nilsson A, Sjögren-Jansson E, Lindahl A: Two to 9-years outcome after autologous chondrocyte transplantation of the knee. *Clin Orthop* 2000;374:212-234.

7. Freed LE, Marquis JC, Nohria A, Emmanual J, Mikos AG, Langer R: Neocartilage formation in vitro and in vivo using cells cultured on synthetic biodegradable polymers. *J Biomed Mater Res* 1993;27:11-23.

8. van Susante JL, Buma P, van Osch GJ, et al: Culture of chondrocytes in alginate and collagen carriers gel. *Acta Orthop Scand* 1995;66:549-556.

9. Brun P, Abatangelo G, Radice M, et al: Chondrocyte aggregation and reorganization into three-dimensional scaffolds. *J Biomed Mater Res* 1999;46:337-346.

10. Campoccia D, Doherty P, Radice M, Brun P, Abatangelo G, Williams DF: Semisynthetic resorbable materials from hyaluronan esterification. *Biomaterials* 1998;19:2101-2127.

11. Aigner J, Tegeler J, Hutzler P, et al: Cartilage tissue engineering with novel nonwoven structured biomaterial based on hyaluronic acid benzyl ester. *J Biomed Mater Res* 1998;42:172-181.

12. Grigolo B, Lisignoli G, Piacentini A, et al: Evidence for redifferentiation of human chondrocytes grown on a hyaluronan-based biomaterial (Hyaff® 11): Molecular, immunohistochemical and ultrastructural analysis. *Biomaterials* 2002;23:1187-1195.

13. Grigolo B, Roseti L, Fiorini M, et al: Transplantation of chondrocytes seeded on a hyaluronan derivative (Hyaff® 11) into cartilage defects in rabbits. *Biomaterials* 2001;22:2417-2424.

14. Solchaga LA, Yoo JU, Lundberg M, et al: Hyaluronan-based polymers in the treatment of osteochondral defects. *J Orthop Res* 2000;18:773-780.

15. Scapinelli R, Aglietti P, Baldovin M, Giron F, Teitge R: Biological resurfacing of the patella: Current status. *Clin Sports Med* 2002;21:547-573.

16. Marcacci M, Zaffagnini S, Kon E, Visani A, Iacono F, Loreti I: Arthroscopic autologous chondrocyte transplantation: Technical note. *Knee Surg Sports Traumatol Arthrosc* 2002;10:154-159.

17. Zorzi C: Hyalograft C clinical study group: Tissue engineered cartilage grafting. Preliminary clinical data. ICRS Symposium, Gothenburg, Sweden 2000, 124.

18. Schatz KD: Erste Erfahrungen Mit Matrixassistierter Knorpel-Zelltransplantation (Summary experiences with autologous chondrocyte implantation). *Arthritis Rheum (Munch)* 2001;5:262-268.

19. Schatz KD: Treatment of cartilage lesions using expanded cartilage cells seeded on a three dimensional fibrous scaffold based on benzylic ester of hyaluronic acid: Preliminary clinical results. *Osteoarthritis Cartilage* 2001;9:PA32-S29.

20. Zanasi S, Ricciardiello A: Tissue-engineered cartilage in complex and salvage procedures. ICRS Symposium, Toronto, Canada 2002, 156.

21. Berruto M, Paresce E, Murgo A, Uderzo E, Odella S, Mazza M: Hyalograft C as salvage procedure in the treatment of traumatic chondral defects of patello-femoral joint: Clinical, histological and BRI 2 years follow-up study. ICRS Symposium, Toronto, Canada, 2002, 173.

22. Marlovits S, Striessnig G, Resinger C, Trattnig S, Vécsei V: Hyaluronan matrix-associated chondrocyte transplantation for the treatment of post-traumatic chondromalacia patella: Early clinical results of a pilot study. ICRS Symposium, Toronto, Canada, 2002, 181.

23. Pavesio A, Abatangelo G, Borrione A, et al: Hyaluronan-based scaffolds (Hyalograft C) in the treatment of knee cartilage defects: Preliminary clinical findings, in "Tissue engineering of cartilage and bone" Novartis Foundation Symposium No. 249, John Wiley and Sons, Chichester, UK, 2003, 203.

24. Kon E, Marcacci M, Zanasi S, Brocchetta D: Hyalograft C implantation in knee cartilage lesions: A long term follow-up project. ICRS Symposium, Toronto, Canada, 2002, 56.

25. Cartilage Injury Evaluation Package ICRS: 2000. http://www.cartilage.org/Evaluation_Package/ICRS_Evaluation.pdf

26. Rabin R, de Charro F: EQ-5D: A measure of health status from the EuroQol Group. *Ann Med* 2001;33:337-343.

27. Hollander AP, Dickinson SC, Sims TJ, Soranzo C, Pavesio A: Quantitative analysis of repair tissue biopsies following chondrocyte implantation, in "Tissue engineering of cartilage and bone" Novartis Foundation Symposium No. 249, John Wiley and Sons, Chichester, UK, 2003, 218.

Chapter 9

The Role of the National Institutes of Health in Extramural Clinical Studies and Trials

James S. Panagis, MD, MPH

Abstract

In the United States, the National Institutes of Health (NIH) is the principal federal agency that funds basic biomedical and clinical research. This overview of the role of the NIH in extramural clinical studies and trials discusses activities related to: (1) clinical research support and administration, (2) human subject protection, and (3) clinical research training and career development. Potential grant applicants are encouraged to look to the NIH for research support and training in the clinical and basic biomedical sciences.

Introduction

The National Institutes of Health is the principal federal agency that funds basic biomedical and clinical research. The NIH includes 27 separate institutes and centers and it is one of eight health agencies within the US Department of Health and Human Services. The broad focus on basic biomedical and clinical research allows the NIH to support research from the "bench top to the patient's bedside." This support is critical for the development of new or improved strategies for the prevention, diagnosis, treatment and cure of diseases. To accomplish this mission, the NIH supports a broad range of research and research training, both internally (intramural research) and externally (extramural research).

The purpose of this chapter is to review the role of the NIH in supporting clinical trials through its extramural research activities. In particular, it will focus on three goals/activities of this extramural support: (1) clinical research support and administration, (2) human subject protection, and (3) clinical research training and career development. In addition, particular sections will focus on specific activities of the National Institute of Arthritis and Musculoskeletal and Skin Diseases (NIAMS), the Government's focal point for research into the causes, treatment, and prevention of arthritis and musculoskeletal and skin diseases.

Human subject research is the study of living persons with whom an investigator directly interacts, intervenes, or obtains identifiable, private information. It includes both clinical research and basic research on human tissues, if you can identify the source. Clinical research is research conducted on human subjects or on material of human origin that is identifiable with the source person. Policies cover large and small-scale, exploratory, and observational studies. Examples of clinical research include: (1) studies of mechanisms of disease, therapeutic interventions, clinical trials, development of new technologies, (2) epidemiologic and behavioral studies, and (3) outcomes research and health services research. In the United States, studies of human subjects designed to answer questions about biomedical or behavioral interventions (eg, drugs, treatments, devices, or new ways of using known treatments, to determine whether they are safe and efficacious) are designated as Phase I through Phase III clinical trials. Although the general objectives and features of these trials have been defined, there may be exceptions for the evaluation of certain indications. It should be noted that the Food and Drug Administration (FDA) regulates trials that are part of a drug development program, although these trials are often sponsored by drug companies (rather than by the NIH).

A Phase I clinical trial evaluates a new biomedical or behavioral intervention in a small group of people (eg, 20 to 80) for the first time to evaluate its safety (eg, determine a safe dosage range and identify side effects). Data on various pharmacologic measures may also be collected during Phase I to Phase III trials. Phase II trials evaluate the intervention in a larger group of people, usually several hundred, to determine efficacy for the intended indication and to further evaluate safety. Phase III trials usually enroll several hundred to several thousand subjects and compare the experimental intervention to other standard or experimental interventions (or possibly placebo). These studies also monitor adverse effects and collect information that will allow the intervention to be used safely. An NIH-defined Phase III clinical trial is a broadly based, or prospective investigation, including community and other population-based trials. Often, the objective of a Phase III trial is to provide evidence for changing policy or standard of care. Phase III trials may investigate pharmacologic, nonpharmacologic, and behavioral interventions for disease prevention, prophylaxis, diagnosis, or therapy.

Clinical Research Support and Administration

In fiscal year 2001, the NIH spent approximately $6.4 billion on extramural clinical research. This represents about 32% of its total research spending. The Institutes and Centers (I/Cs) that spent the most on clinical research in 2001 were the National Cancer Institute (NCI), the National Heart, Lung and Blood Institute (NHLBI), and the National Institute of Mental Health (NIMH). Clinical research expenditures increased by 44% (adjusted for inflation) from 1997 to 2001. In 2001, 40% of the funding for clinical research supported grants to individual investigators (primarily investigator-initiated or R01 applications). The remainder of these funds went to support center grants, cooperative agreements, program projects, and contracts.

The NIH Grants Process

It is beyond the scope of this chapter to provide a detailed primer on the NIH grants process. Such information can be found at: http://grants.nih.gov/grants/oer.htm and http://www.niams.nih.gov/rtac/grantapps/index.htm. The I/Cs award grants using a wide variety of mechanisms to support clinical research, clinical training and career development, and scientific conferences. Most unsolicited applications for grant support need not be directed to a specific I/C because many areas of research are relevant to more than one I/C. Instead, applications are received by the Center for Scientific Review (CSR) and assigned to one or more I/Cs for possible funding. With the exception of National Research Service Award Individual Fellowships (F32s), NIH grants are awarded to institutions, not to individual investigators. In fact, an application is submitted by an institution, on behalf of an investigator. The signatures on the face page of an application certify that the applicant institution is prepared to take on the many responsibilities that come with the acceptance of an NIH grant. The institutional office that administers grants, which may be known as "Grants and Contracts" or "Sponsored Research," is therefore an essential partner for the investigator preparing an application. This office can often provide forms, instructions, and advice, including notice of any special requirements the institution may have for the approval and submission of applications. The standard format for most grant applications is form PHS 398 (rev. 4/98). Application kits, including detailed instructions, are available at most institutional offices of sponsored research. Forms may also be obtained from the:

Division of Extramural Outreach and Information Resources
National Institutes of Health
6701 Rockledge Drive, MSC 7910
Bethesda, MD 20892-7910
Telephone: 301/435-0714,
Email: Grantsinfo@nih.gov, or at
http://grants.nih.gov/grants/funding/phs398/phs398.html

For most applications, standard application deadlines apply, as indicated in the application kit. A list of receipt deadlines can be found at: http://grants.nih.gov/grants/funding/submissionschedule.htm

Potential applicants are encouraged to consult with colleagues and institutional officials who have experience with the NIH grants process before preparing an application. All applications are subjected to rigorous peer review. In most cases, peer review is conducted by the CSR, using one of more than one hundred standing review groups or study sections. Some special types of applications, such as career development awards, are reviewed by committees convened by the I/Cs. Under the recently implemented "streamlined" review process, most review groups discuss only the applications judged, by scientific merit, to be in the top half of the total received. All applications discussed by a committee receive numerical scores from which a percentile ranking relative to other applications may be calculated. The scores, percentiles, and critiques are then communicated to the assigned I/C(s) in a summary statement that includes detailed critiques of the proposal. A copy of the summary statement is also sent to the investigator. An advisory council, composed of both scientists and members of the public, provides a second level of review to evaluate both the adequacy of the initial review and the relevance of applications to the mission of the I/C. Finally, the staff and director of the I/C consider the scores, percentiles, critiques, and advisory council recommendations in selecting applications for awards. The overall process may require up to 10 months from the receipt of an application to the issuance of an award. Applications that are not funded may be revised to address the concerns of the reviewers and resubmitted.

Although the traditional investigator-initiated research project (R01) grant is the most common form of NIH support, a large and varied selection of more specialized mechanisms has evolved to address specific needs. I/Cs often target specific areas of research by issuing Program Announcements (PA) and Requests for Applications (RFA). Finally, NIH policies and procedures are frequently revised to meet the needs of the Government and the research community.

With the advent of the World Wide Web, electronic links have become the medium of choice for scientists trying to stay informed about opportunities for funding and current policies and procedures. The NIH web site provides access to information, including all releases of the NIH Guide for Grants and Contracts at: http://grants.nih.gov/grants/guide/index.html. All announcements of special initiatives and policy changes are published in this document. In addition, investigators are offered the opportunity to receive the guide on a weekly basis by e-mail.

From the NIH home page (http://www.nih.gov), the most important link for investigators is Grants and Funding Opportunities (http://grants.nih.gov/grants/index.cfm). Links from this page provide several important sources of information. The grants page contains links to pages labeled funding opportunities, grants policy, awards data, and receipt dates. In addition, there are links to the NIH Guide and to specialized pages addressing such topics as human and animal protection, research training, and small business grants. The funding opportunities page provides links to descriptions and guidelines for most of the mechanisms used by NIH. The grants policy page, while concerned mostly with fiscal and administrative matters, may occasionally serve to clarify a point of interest to an investigator. The most important feature of the awards data page is the system called CRISP (Computer Retrieval Information on Scientific Projects) (https://www-commons.cit.nih.gov/crisp). This is a searchable database of information on currently active NIH awards, including brief abstracts of each funded project. The receipt dates page summarizes the current dates for various mechanisms and states the policy that determines whether an application is considered "on time" or not. Exploring the CRISP database can be an effective way for an investigator to learn about which studies are being supported in a specific area of research.

The grants and funding opportunities page includes links to information on research training opportunities, peer review, and the home pages of the various I/Cs of the NIH. The pages of the I/Cs contain information on missions, research interests, and current activities. Most importantly, I/C home pages contain links to the NIH guide announcements describing current initiatives, such as program announcements and requests for applications. These initiatives often have specific requirements and receipt dates that differ from those for unsolicited applications. Many I/C home pages, such as http://www.niaid.nih.gov/ncn/tools/howto$.htm, provide helpful advice on the preparation of applications. There is a link at http://www.csr.nih.gov/ for the CSR home page that contains rosters of the standing study sections and schedules of their meetings. Finally, the CSR page provides summaries of policies and procedures related to peer review. For example, "A Straightforward Description of What Happens to Your Research Project Grant Application (R01/R21) After it is Received for Peer Review" can be found at: http://www.csr.nih.gov/review/peerrev.htm. A tutorial on "How to Write a Human Subjects Application" can be found at: http://www.niaid.nih.gov/ncn/clinical/humansubjects/hs_01.htm. Guidelines for Submission of Applications for Investigator-Initiated Clinical Trials to the NIAMS can be found at: http://www.niams.nih.gov/rtac/clinical/index.htm.

The Role of the NIH Program Officer
The NIH Program Officer is the extramural research community's contact for technical and program information regarding a particular area of science. Although not a requirement, a potential applicant is encouraged to speak with an NIH Program Officer regarding program interests and priorities as well as scientific, technical, and procedural questions. The Program Officer is the initial contact for applications with annual direct costs equaling or exceeding $500,000/year. Post award, the NIH Program Officer monitors the progress of the grant, and may interact with the Data Safety Monitoring Board, or Safety Officer.

Some Recent Policy Developments
Below are brief synopses of some recent NIH grants policy developments. It is important to view the full announcement (at the cited web site) in each case for additional details on policy implementation.

Modular Grant Applications (http://www.nih.gov/grants/guide/notice-files/not98-178.html)) In this format, required since the June 1, 1999 receipt date, applications request direct costs in $25,000 modules, up to a total direct cost request of $250,000 per year for all unsolicited new, revised, and competing continuation individual research project grants (R01), small grants (R03), and exploratory/developmental grants (R21). Applications requesting more than $250,000 in any year will be required to follow the traditional application instructions and applicable NIH policies. The modular grant initiative expands the existing streamlining and reinvention initiatives that are designed to concentrate the focus of investigators, their respective institutions, peer reviewers, and NIH staff on the science NIH supports, rather than on the details of budgets.

Change in Policy of Supporting New Investigators (http://grants.nih.gov/grants/policy/r29transition.htm) The NIH no longer accepts applications for R29 (First Independent Research Support and Transition) awards. These awards, also known as FIRST awards, were designed for scientists who have not previously served as the principal investigators (PIs) on any

Public Health Service-supported research project other than a small grant (R03), Academic Research Enhancement Award (R15), an exploratory/developmental grant (R21), or certain research career awards (K01, K08, and K12). New investigators now submit standard R01 applications, with no special restrictions on budget, duration, or effort. This change in policy will allow investigators maximum freedom in identifying the level and period of support needed for the work they are planning. Peer reviewers are instructed to give new investigator applications special consideration in scoring (see http://www.drg.nih.gov/guidelines/newinvestigator.htm). Applications may be identified as originating from a new investigator by checking the box on line 3 of the Form 398 face page (rev. 4/98). In making this change, NIH has committed to supporting at least the same number of new investigators and, as necessary, directing more resources to their support.

Applications Requesting Direct Costs in Excess of $500,000 (http://www.nih.gov/grants/guide/notice-files/not98-030.html) Any applicant planning to submit an application requesting $500,000 or more in direct costs for any year must contact I/C program staff (NIH Program Officer) before submitting the application, and obtain agreement that the I/C will accept the application. This "prior approval" should be completed 6 weeks before the planned receipt date.

Explicit Statements of Review Criteria (http://grants.nih.gov/grants/guide/notice-files/not97-010.html) Although they still assign a single global score, peer reviewers are now instructed to address specifically five discrete criteria in evaluating applications. These criteria are significance, approach, innovation, investigator, and environment. Written critiques are constructed to highlight the reviewer's comments in each of these areas. The application does not need to be strong in all categories to be judged likely to have major scientific impact and thus deserve a high priority score. For example, an investigator may propose to carry out important work that by its nature is not innovative but is essential to move a field forward.

Other NIH Policies Regarding Human Subjects NIH policy and guidelines on the inclusion of women and minorities as participants in research can be found at: http://grants.nih.gov/grants/guide/notice-files/NOT-OD-02-001.html. The NIH policy and guidelines on the inclusion of children as participants in research are at: http://grants.nih.gov/grants/guide/notice-files/not98-024.html. NIH policy requires education on the protection of human subject participants for all investigators submitting NIH proposals involving human subjects (http://grants.nih.gov/grants/guide/notice-files/NOT-OD-00-039.html).

Criteria for Federal Funding of Research on Human Embryonic Stem Cells (hESC) (http://grants1.nih.gov/grants/stem_cells.htm and http://grants. nih.gov/grants/guide/notice-files/NOT-OD-02-005.html) Only research using hESC lines that are registered in the NIH Human Embryonic Stem Cell Registry will be eligible for Federal funding. It is the responsibility of the applicant to provide the official NIH identifier(s) for the hESC line(s) to be used in the proposed research. Applications that do not provide this information will be returned without review.

General Clinical Research Centers

Through the National Center for Research Resources (NCRR), the NIH supports General Clinical Research Centers (GCRCs). The GCRCs are a national network of 78 centers that provide optimal settings for medical investigators to conduct safe, controlled, state-of-the-art, inpatient and outpatient studies of both children and adults. The GCRCs also provide infrastructure and resources that support several career development opportunities. Investigators who have research project funding from NIH and other peer-reviewed sources may use GCRCs. To request access to a GCRC facility, eligible investigators should initially contact a GCRC program director, listed in the clinical research resources directory. Because the GCRCs support a full spectrum of patient-oriented scientific inquiry, researchers who use these centers can benefit from collaborative, multidisciplinary research opportunities. To ensure research diversity at the CRCs, no single group of investigators at a center may utilize more than 33% of the resources. The individuality of each GCRC is determined by the research strengths and needs of its host institution. Highly trained research personnel, a core laboratory, a bioinformatics system, and a metabolic kitchen are examples of GCRC resources. The GCRC research staff, including nurses, dietitians, biostatisticians, skilled technicians, and administrative personnel, help investigators by facilitating the day-to-day research process and assisting research patients in a supportive and efficient environment. Funding for GCRCs increased from $153 million in 1997 to approximately $221 million in 2001. The NIH supported 78 GCRCs in 2001. Their mission has recently expanded to include telecommunications and telemedicine. Information on GCRCs can be found at: http://www.ncrr.nih.gov/clinical/cr_gcrc.asp.

NIAMS Clinical Trial Planning Grant

The purpose of the NIAMS Clinical Trial Planning Grant (R21) is to provide support for the organization of activities critical for the successful implementation of clinical trials in areas within the NIAMS mission. This planning grant is intended to: (1) allow for early peer review of the rationale and design for high-risk, complex, or large-scale clinical trials, (2) provide support for the development of a detailed clinical trial research plan, including a manual of operations and procedures, as a means of decreasing the start-up time needed for initiating trials after the award is made, and (3) provide support to refine critical components of a clinical trial, such as experimental design, analytic technique, recruitment strategies, data management, and collaborative arrangements. This award is for 1 year and up to a total direct cost request of $100,000. Additional information can be found at: http://www.niams.nih.gov/rtac/funding/grants/pa/par_02_119.pdf.

Human Subject Protection

Research with human subjects is an essential element of biomedical and behavioral research and bioethical considerations impact the design and conduct of such research. Since 1947, when guidelines for research with human subjects were promulgated, there has been an increasingly widespread recognition of the need for voluntary and informed consent and scientifically valid design of experiments involving human subjects. This recognition has resulted in formalized federal regulations and guidelines for the protection of human subjects in federally sponsored research. These regulations, revised November 13, 2001, can be found at: http://ohrp.osophs. dhhs.gov/humansubjects/guidance/45cfr46.htm. They require that all research protocols involving human subjects be reviewed by an Institutional Review Board (IRB). This review ensures that: (1) risks are minimized and reasonable in relation to anticipated benefits, (2) there is informed consent, and (3) the rights and welfare of the subjects are maintained. Prior to submission of an application, questions regarding human subjects issues can be addressed to both NIH program and review staff. After peer review, such questions should be posed only to program staff (NIH program officer). They can confirm the involvement of human subjects, review potential risks and proposed protection measures. If a concern regarding human subjects was noted, the NIH program officer will advise what documentation is needed for NIH approval.

A glossary of terms for human subject requirements can be found at: http://www.niaid.nih.gov/ncn/clinical/decisiontrees/hsgloss.htm.

A decision tree for human subjects can be found at: http://www.niaid.nih. gov/ncn/clinical/decisiontrees/humansub.htm.

A decision tree for protection of human subjects from research risk can be found at: http://www.niaid.nih.gov/ncn/clinical/decisiontrees/human.htm.

Reporting

NIH policy requires that certain information regarding all research that involves human subjects be collected annually (eg, demographics of study populations in the noncompetitive renewal application or annual report and annual IRB/Independent Ethics Committee (IEC) review. The requirements may differ for each funding component and thus it is always best to review requirements with the awarding I/C. The awardee must complete the Inclusion Enrollment Report showing cumulative accrual information for each study. Such information is required for further processing of the award. This submission should be submitted annually as part of each noncompeting renewal or annual progress report. It is also required for new studies. The format for the Inclusion Enrollment Report is located at: http://grants.nih. gov/grants/funding/phs398/enrollmentreport.pdf.

A decision tree for inclusion of women plan can be found at: http://www. niaid.nih.gov/ncn/clinical/decisiontrees/women1.htm.

A decision tree for inclusion of minorities plan can be found at: http:// www.niaid.nih.gov/ncn/clinical/decisiontrees/minority1.htm.

A decision tree for children inclusion plan can be found at: http://www. niaid.nih.gov/ncn/clinical/decisiontrees/children.htm.

Additional reporting requirements may include: recruitment progress, indices of quality control, and related operational features. These are usually reported to the NIH Program Officer. Annual and final reports are required as with any grant. In addition, the NIAMS requires that a clinical trial be registered on its clinical trial database within 21 days of funding. The database is required to be updated quarterly, or more often if the status of the trial changes.

Institutional Review Board or Independent Ethics Committee Approval

Annually, the awardee will submit to the NIH I/C documentation of continuing review and approval from the local IRB/IEC, including a copy of the current IRB/IEC approved consent document and the OHRP Federal Wide Assurance (FWA) number for the institution/site. When there are other institutions involved in the research (eg, a multicenter clinical trial or study), the protocol must be reviewed and approved by each institution's IRB/IEC. Initial and annual documentation of continuing review and approval, including the current approved informed consent document and FWA number from each institution, must also be provided to the awarding NIH I/C. For international sites, approval from National IRB/IEC, if applicable, may be required in addition to or in lieu of approval from the local IRB/IEC.

To help ensure the safety of participants enrolled in NIH-funded studies, the awardee must provide the awarding I/C with copies of documents related to major changes in the status of ongoing protocols, including:

1. all amendments or changes to the protocol, identified by protocol version number and/or date,
2. all changes in informed consent documents, identified by version number and/or date, and dates during which they were valid,
3. termination or temporary suspension of patient accrual,
4. termination or temporary suspension of the protocol,
5. any change in IRB/IEC approval(s), and
6. any other problems or issues that could affect the participants in the study.

Notification of any of the above changes must be made within a specified period of time (usually within 3 days) by email or fax, followed by a letter cosigned by the PI and the institutional business official, detailing notification of the change of status to the local IRB and a copy of any responses from the IRB/IEC. A decision tree for IRB waiver can be found at: http://www.niaid.nih.gov/ncn/clinical/decisiontrees/irbwaiver.htm.

Data and Safety Monitoring Requirements

Independent monitoring is essential for all clinical trials involving investigational drugs, devices or biologics, and other clinical research perceived to involve more than a minimal risk. A risk is minimal where the probability and magnitude of harm or discomfort anticipated in the proposed research are not greater, in and of themselves, than those ordinarily encountered in daily life or during the performance of routine physical or psychological examinations or tests. For example, the risk of drawing a small amount of blood from a healthy individual for research purposes is no greater than the risk of doing so as part of a routine physical examination (45CFR46.102I).

Monitoring plans must be included in any application or proposal that proposes research involving more than minimal risk. However, final decisions regarding the type of monitoring to be employed are usually made jointly by the applicant and the awarding NIH I/C prior to initiation of the study. Discussions with the responsible NIH I/C program officer are recommended. An improved monitoring plan must be in place before patient enrollment begins. This safety monitoring can be performed by an independent safety officer or by a Data and Safety Monitoring Board (DSMB). An independent safety office is a physician (or other appropriate expert) who is independent of the study and is available in real time to review and document appropriate action regarding adverse events and other safety issues. Additional information can be found at: http://www.niams.nih.gov/rtac/clinical/dsmb4.pdf. The DSMB is an independent committee charged with reviewing safety information and trial progress. When a safety officer or monitoring board is organized, a description of the safety officer/board, its charter and/or operating procedures (including proposed meeting schedule and plan for review of adverse events), and roster and curriculum vitae from all members must be submitted to and approved by the awarding NIH I/C prior to initiation of the study (http://www.niams.nih.gov/rtac/clinical/index.htm). This board can provide advice with respect to study continuation, modification, and/or termination. All Phase II clinical trials must be reviewed by a DSMB (http://www.niams.nih.gov/rtac/clinical/dsmb3.html). Other trials may require DSMB oversight as well.

Investigational New Drug/Investigational Device Exemption (IND/IDE) Requirements

In the United States, clinical research involving the use of investigational therapeutics, vaccines, or other medical interventions (including licensed products/devices for a purpose other than the one(s) for which they were licensed) must be performed under an FDA Investigational New Drug/Investigational Device Exemption (IND/IDE) (http://www.fda.gov/cdrh/ode/idepolcy.html). Exemptions must be granted in writing by the FDA. If the proposed clinical trial will be performed under an IND/IDE, the awardee must provide the awarding NIH I/C with the name and institution of the IND/IDE sponsor, the date the IND/IDE was filed with FDA, the FDA IND/IDE number, any written comments from the FDA, and the written response to those comments. Under the IND/IDE, the sponsor is required to provide the FDA with safety reports of serious adverse events. Under the terms and conditions of the award, the awarding NIH I/C may require that the awardee submit copies to the responsible Program/Project officer some or all of the following:

1. Expedited Safety Report of unexpected or life-threatening experience or death: A copy of any report of unexpected or life-threatening experience or death associated with the use of an IND drug, must be reported to the FDA by telephone or fax as soon as possible, but no later than 7 days after the sponsor's receipt of the information. This

information usually must be submitted to the awarding NIH I/C Program/Project Officer within 24 hours of FDA notification.

2. Expedited Safety Reports of serious and unexpected adverse experiences: A copy of any report of unexpected and serious adverse experience associated with the use of an IND drug or any finding from tests in laboratory animals that suggests a significant risk for human subjects, which must be reported to the FDA in writing as soon as possible, but no later than 15 days after the IND sponsor's receipt of the information, must be submitted to the awarding NIH I/C Program/Project Officer within 24 hours of FDA notification.

3. IDE Reports of Unanticipated Adverse Device Effect: A copy of any reports of unanticipated adverse device effect submitted to the FDA must also be submitted to the awarding NIH I/C Program/Project Officer within 24 hours of FDA notification.

4. Other adverse events documented during the course of the trial should be included in the annual IND/IDE report and reported to the awarding NIH I/C annually.

Specific NIH I/Cs may have other reporting requirements, and these will be provided in the notice of grant award. In addition, final decisions regarding ongoing safety reporting requirements for research not performed under an IND/IDE are usually made jointly by the awardee and the awarding NIH I/C.

Certificates of Confidentiality

Certificates of confidentiality constitute an important tool to protect the privacy of research study participants. They are issued by the NIH to protect identifiable research information from forced disclosure. They allow the investigator and others who have access to research records to refuse to disclose identifying information on research participants in any civil, criminal, administrative, legislative, or other proceeding, whether at the Federal, state, or local level. Certificates of confidentiality may be granted for studies collecting information that if disclosed could have adverse consequences for subjects or damage their financial standing, employability, insurability, or reputation. By protecting researchers and institutions from being compelled to disclose information that would identify research subjects, certificates of confidentiality help achieve the research objectives and promote participation in studies assuring confidentiality and privacy to participants. Additional information can be found at: http://grants1.nih.gov/grants/policy/coc/index.htm and http://grants1.nih.gov/grants/policy/coc/faqs.htm.

Clinical Research Training

NRSA Institutional Training Grants

These 5-year institutional (renewable) grants enable institutions to offer funding for predoctoral and postdoctoral training of selected individuals. Applications for institutional research training grants may include a request for short-term predoctoral positions reserved specifically to train medical or other health professional on a full-time basis during the summer or other "off-quarter" periods. Short-term appointments are intended to provide health professional students with opportunities to participate in biomedical and/or behavioral research in an effort to attract these individuals into research careers. Predoctoral Training: 1. Candidates for PhD or equivalent degree, 2. Duration of up to 5 years of support, 3. Commitment: Full-time research traineeship, 4. Provisions: Stipend $16,500; tuition and fees: 100% of first $3,000; 60% of remainder; trainee travel: up to $1,000/year. Training-related expenses: $2,000/year. Postdoctoral Training: 1. MD or PhD or equivalent, 2. Duration of up to 3 years of support, 3. Commitment: Full-time research traineeship, 4. Provisions: Stipend $28,260 to $44,412 per year; tuition and fees: 100% of the first $3,000; 60% of remainder; trainee travel: up to $1,000/year; training-related expenses: $2,500/year, per postdoctoral trainee. Additional information on the NRSA Institutional training Grants can be found at: http://www.niams.nih.gov/rtac/funding/grants/pa/pa_00_103.pdf.

Career Development

Mentored Clinical Scientist Development Investigator Award (K08)

The purpose of the Mentored Clinical Scientist Development Investigator Award is to provide support to clinicians who need an intensive period of mentored research experience. Duration of support is up to 5 years. Support is provided for salary up to $75,000 (75% level of effort), fringe benefits, and other research expenses up to $20,000. Additional information on this mechanism can be found at: http://www.niams.nih.gov/rtac/funding/grants/pa/pa_00_003.pdf.

Mentored Clinical Scientist Development Program Award (K12)

In 2001, NCI, NCRR, NIA, NIDA, and NIDCR announced a new career development award, the Mentored Clinical Research Scholar (CRS) Program. This award provides support to institutions to establish career development programs for physicians and dentists so they, in turn, may develop the research skills necessary to become independent, patient-oriented, clinical investigators. The award is for 5 years with the trainees performing at a 75% level of effort. Candidates are provided a maximum of $90,000 for salary support for each year commensurate with the applicant institution's salary structure for persons of equivalent qualifications, experience, and rank. A director and an advisory committee will oversee the integrated, didactic, and mentored CRS research program. The program must include activities that will provide candidates with a comprehensive understanding of clinical research approaches that are fundamental and not necessarily disease-specific. The lead mentor should be an established clinical researcher who holds a faculty position. The mentor will work closely with the candidate to develop a tailored career development plan. Candidates for the program will be selected from among recently trained physicians and dentists. Funds provided by this award may be used to support full or partial completion of an advanced degree such as an MS or PhD in clinical investigation or an MPH. The maximum duration of support for a candidate cannot exceed

5 years. Candidates selected for the CRS program may be eligible for the NIH Loan Repayment Program for clinical investigators, which provides for repayment of the educational loan debt of physicians and dentists (http://grants.nih.gov/grants/guide/pa-files/PAR-00-140.html).

Career Transition Development Award (K22)

The overall goals of the NIAMS Career Transition Award program are to enable outstanding individuals to obtain research training experience in the NIAMS Intramural Research Program (IRP), and to facilitate their successful transition to an extramural environment as independent researchers. The award will provide 2 to 3 years of support for research training in a NIAMS intramural lab (Phase I), followed by 2 to 3 years of support for an independent research project in an extramural institution (Phase II). The combined duration of this award cannot exceed 5 years. It is anticipated that awardees will subsequently obtain research project grants such as the R01 (investigator-initiated research project) to support the continuation of their work.

Eligible candidates must be US citizens, noncitizen nationals, or lawfully admitted permanent residents. Former PIs of NIH research awards (R01), FIRST awards (R29), SBIR/STTR awards, subprojects of Program Projects (P01) or Center Grants (P50), K08 awards, or the equivalent, are not eligible. Total direct costs for all years, based on the individual's experience, cannot exceed $150,000 (salary up to $75,000). During Phase II, the candidate must spend a minimum of 75% of his/her full-time professional work conducting research and research career development. Additional information the NIAMS K22 can be found at: http://www.niams.nih.gov/rtac/funding/grants/pa/par_02_056.pdf. Additional information on the NIAMS IRP can be found at: http://www.irp.niams.nih.gov/NIAMS2/IRPHome.jsp. Other NIH Institutes/Centers may have different policies and procedures. It is suggested that interested applicants contact them directly.

Mentored Patient-Oriented Research Career Development Award (K23)

The purpose of the K23 award is to support investigators who are committed to conducting patient-oriented research under the supervision of a mentor (http://www.niams.nih.gov/rtac/funding/grants/pa/pa_00_004.pdf). This award is for 3 to 5 years and is nonrenewable. Candidates must give 75% minimum effort toward research career development and clinical research. Description: To support supervised study and research for clinically trained professionals who have the potential to develop into productive, clinical investigators focusing on patient-oriented research. Candidates must have completed their specialty and, if applicable, subspecialty training prior to receiving an award. Duration of this award is 3 to 5 years and is not renewable. Support is provided for salary up to $75,000 for a minimum 75% effort and up to $25,000 in selected research expenses.

Midcareer Investigator Award in Patient-Oriented Research (K24)

The K24 award provides support for protected time for clinicians to allow them time for patient-oriented research and to act as mentors for beginning clinical investigators. Candidates must have completed their specialty training within 15 years of submitting the application. Candidates must be working in a research environment, conducting patient-oriented research, and have independent research support. Duration of support is 3 to 5 years, and it is renewable. Awardees receive up to 50% of the NIH salary cap ($141,300 in 2000) in addition to fringe benefits. They also receive $25,000 per year for research support and must give from a minimum of 25% to a maximum of 50% effort toward mentoring and clinical research. Additional information on this mechanism can be found at: http://www.niams.nih.gov/rtac/funding/grants/pa/pa_00_005.pdf.

Clinical Research Curriculum Award (K30)

The NIH developed this program to attract talented individuals to the challenges of clinical research and to provide them with the critical skills that are needed to develop hypotheses and conduct sound research. The Clinical Research Curriculum Award (CRCA) is an award to institutions. It supports the goal of the NIH to improve the quality of training in clinical research. This award is intended to support the development of new didactic programs in clinical research at institutions that do not currently offer them, or to support the improvement or expansion of programs at institutions with existing programs. There is a core of knowledge and skills common to all areas of clinical research that should form the foundation of the well-trained, independent clinical researcher. Formal course work includes the design of clinical research projects, hypothesis development, biostatistics, epidemiology, and legal, ethical, and regulatory issues related to clinical research.

The CRCA is a trans-NIH program that is being administered by the National Heart, Lung and Blood Institute (NHLBI). An NIH Coordinating Committee participates in all phases of the Program. In addition to NHLBI awards, two institutions have received awards separately from the National Center for Complementary and Alternative Medicine (NCCAM). The K30 award is a 5-year renewable award. Further information can be found at: http://grants.nih.gov/training/k30.htm.

Mentored Medical Student Clinical Research Program

In 2001, the NCRR announced a mentored medical student clinical research program to support a small number of medical and dental students at GCRCs. This fellowship program provides supplemental grants to GCRCs to offer 1 year of support for medical and dental students, usually from their third through fourth year of medical school, in the form of salary, supplies, and tuition assistance. See http://www.ncrr.nih.gov/clinical/crguide2001/guidenov2001.pdf for additional information.

NIH Extramural Loan Repayment Program Regarding Clinical Researchers

On December 28, 2001, the NIH implemented the NIH Extramural Loan Repayment Program Regarding Clinical Researchers (LRP-CR). The purpose of the Program is the recruitment and retention of highly qualified health professionals as clinical investigators. It provides for the repayment of up to $35,000 per year of the principal and interest of an individual's educational loans for each year of obligated service. Also, the NIH covers the Federal taxes on the loan repayments, which are considered taxable income to program participants. To be eligible, a clinical investigator must have received an NIH research award, training grant, career development award, or other NIH grant as a first-time PI or a first-time director of a subproject on a grant or cooperative agreement. In addition, participants must be US citizens, nationals, or permanent residents. Awardees contractually agree to engage in clinical research for at least 2 years. The NIH planned to fund 396 loan repayment contracts for a total of $20.2 million by the end of FY 2002. Funding for this program may double in FY 2003. Additional information on the Loan Repayment Program may be found at: http://www.niams.nih.gov/rtac/funding/grants/notice/notod02-024.htm, and the NIAMS IRP at: http://www.lrp.nih.gov.

Summary

This overview of the role of the NIH in extramural clinical studies and trials discusses activities related to: (1) clinical research support and administration, (2) human subject protection, and (3) clinical research training and career development. The URLs identified above are excellent sources for further information. As noted, the NIH Program Officer is an excellent source for additional scientific and technical information or to address questions. Potential applicants are encouraged to look to the NIH for research support and training in the clinical and basic biomedical sciences.

Future Directions

"To The Patient": The Clinical and Marketing Challenges

The introduction of new technology into the clinical arena, including engineered musculoskeletal tissues, represents an exciting opportunity accompanied by considerable associated practical issues. These considerations are important to clarify and address in order to permit new approaches to reach their full potential and provide the increased success and decreased morbidity the public expects and deserves. After all, it is ultimately the public that supports innovation, as well as benefits from these investments.

While considerable overlap exists, the constituent "stakeholders" in the process of bringing new products and approaches to clinical practice include the patient, the physician, the hospital, the payer, industry, the regulator, and the body of science. None is more important than the patient and none is more demanding than the science.

The patient

At the outset, the patient needs a clear understanding of his or her disorder or disease since this determines the nature of the treatment required and the realistic goals of therapy. There is an expectation that the chosen technology will work. More so, the patient anticipates the new approach works better than other available options as judged by the frequency and predictability of satisfactory outcomes, the durability of the treatment response, the time frame required to achieve the desired end point, and the nature of the necessary rehabilitation program. Patients are entitled to feel confident that the technology is safe, or that the risks have been sufficiently well-defined and shared in order to provide the basis for a meaningful risk-benefit decision. Cost, to the patient, is also a factor, including direct medical care as well as lost opportunity time (work, school, leisure).

- Public education programs should be developed that provide patients with timely, comprehensible information concerning the nature of new technologies, indications for its use, expectations concerning outcomes, risks compared with other approaches, as well as expenses (time and money) that may not be covered by third party payers.

The physician

The physician has the obligation to determine the nature of the patient's disorder and the required therapeutic goals. Surgeons also want to know that the approach is technically feasible, and there is reasonable reimbursement available for the time and effort required to provide treatment. Obviously, physicians share with their patients concerns for safety and efficacy. Physicians must play an important role in preclinical and clinical studies during the development and assessment of new approaches, and also have an obligation to participate in the continuing evaluation of new technology after it is introduced to the market.

- Physician education programs should be developed to provide information on the nature of new technologies, their appropriate clinical indications, technical aspects of applying these approaches, as well as evaluating results.
- Physicians should be encouraged to participate in rigorous premarket and postmarket clinical research designed to evaluate the safety and effectiveness of new technologies.

The hospital

The hospital also needs to be able to anticipate reasonable cost recovery for new technology, so the institution can continue to provide other necessary services and invest in additional future advances in care.

- Educational programs should be developed to inform hospitals about the nature of new technologies, including the numbers of patients and types of disorders and diseases that would benefit from these approaches. Physicians should be encouraged to participate in the needs assessments required by hospitals as they consider investing in new technologies.

The payer

The payer is responsible for managing the cost of care. While there are apparent conflicting agendas with other stakeholders, the payer needs the new technology to be cost-effective. This can occur even with expensive approaches, provided these treatment strategies reduce future expenses that fall within the responsibility of the payer. Rigorous outcomes research provides the payers information important in their decisions for coverage. Payers are also unlikely to cover the expenses of treatment not approved by the responsible regulatory authority, such as the Food and Drug Administration in the United States.

- Programs should be developed that provide payers with an understanding of the nature of new technologies and their impact on the health of their covered members.

The regulatory agency

The regulatory agency is responsible for establishing for the public both the safety and effectiveness of the new technology. Since there are multiple approaches to the treatment of most disorders, the question becomes one of effectiveness compared to other accepted therapeutic modalities. Alternatively, a new technology may be associated with less morbidity than other approaches, without a loss of efficacy. There is considerable autonomy amongst regulatory authorities around the world. It would be of benefit, however, in the globalized society if there was more uniformity in the approaches used by regulators around the world to assess the safety and effectiveness of new technologies. Also, it would be a tremendous value if the methods of assessing outcomes (biological, biomechanical, or clinical) were agreed on between regulatory agencies, industry, scientists, and clinicians. This would improve the relevance of study designs and the predictability of interpretation of results.

- Workshops should be developed where outcomes assessment tools and approaches could be established with the consensus of regulatory agencies, industry, scientists, and clinicians. These activities should include broad representation and occur separate from the consideration of individual studies by the regulatory agencies.

Industry

Industry needs to know the technology works reliably and safely, and how this profile compares with currently available alternatives. A clear vision of situations in which the technology is useful and appropriate is paramount. In the end, the manufacturer needs to recover development costs, based on payer acceptance and regulatory approval. Consequently, the technology must be provided at an affordable cost. It is also important that new technologies be evaluated by others, such that industry (or individual investigators) should adequately disclose the nature of the new technologies sufficient to allow for independent review and validation of safety and efficacy.

- Approaches should be developed that enhance the partnership between industry and academia in developing new technologies, designing clinical investigations, and in assessing clinical outcomes.
- Industry should be encouraged to provide training for clinicians as required to select and treat appropriate patients.

Scientific sensitivities

Scientific sensitivities are the heart of understanding the merits of the new technology. Developmental research and clinical evaluation must be accomplished in a manner that clearly defines safety and efficacy. The "product" must be fully and clearly characterized in a manner that can be shared and reproduced (respecting proprietary interests that can be protected by patents rather than secrets). The experimental design must have clear and measurable end points in validated and appropriate models that adequately isolate the contributions of the product. Proper controls and comparisons are invaluable in understanding the results and defining indications for new approaches.

- Development of more reliable and widely used assessment approaches, whether biological, biomechanical, or clinical, is necessary. For clinical outcomes studies, better minimally invasive methods are needed.
- Future studies, especially in the clinical arena, should compare new technology with current best approaches. Biases must be identified and managed.
- More investigators should be trained in rigorous outcomes research methodology.
- Continuing postmarket assessments of safety and efficacy should be encouraged and their results widely shared.

Section Two

Tissue Engineering of Bone

Giordano Bianchi, MD
Scott D. Boden, MD
Ranieri Cancedda, MD
Thomas A. Einhorn, MD
Johnny Huard, PhD
Paul H. Krebsbach, DDS, PhD

Jueren Lou, MD
Maddalena Mastrogiacomo, PhD
Anita Muraglia, PhD
Brian Nussenbaum, MD
Hairong Peng, MD, PhD
Rodolfo Quarto, MD

Chapter 10

Tissue Engineering of Fracture Repair

Thomas A. Einhorn, MD

Abstract

The healing of fractures is one of the most optimized repair processes of the human body. Related in many ways to growth and development, fracture healing involves a series of cellular and molecular signals programmed to achieve tissue regeneration and restoration of mechanical integrity. Although the majority of fractures heal without difficulty, many patients experience problems with healing that result in significant morbidity. Current strategies to enhance or ensure fracture healing include the use of osteoconductive scaffolds (for example, calcium phosphates, calcium sulfates, or composites of these calcium-based substances with demineralized bone matrix or allogeneic bone), human bone marrow or blood concentrates, and osteoinductive proteins of the transformingtissue growth factor-β (TGF-β) superfamily. Recent studies have provided data regarding the use of recombinant bone morphogenetic protein (BMP)-2 or BMP-7 (OP-1) in the clinical setting. Research on the use of recombinant proteins in fracture repair has focused on the osteoinductive BMPs. In addition, angiogenic factors such as vascular endothelial growth factor (VEGF) may play a role in enhancing the healing of fractures either alone or in combination with BMP or other inductive molecules. Future technologies, including gene therapy and the use of systemic compounds, could enhance the overall skeletal response to injury. It is anticipated that many new treatment strategies will be available for orthopaedic surgeons within the next decade.

Introduction

It is estimated that by 2004, more than 1.5 million bone graft operations will be performed in the United States to enhance the healing of spinal fusions, for internal fixation of fractures, for maxillofacial reconstruction, and to restore bone at sites where it has been lost due to trauma or ablative surgery.[1] Autologous bone graft is often effective in the treatment of these conditions. However, for many patients it is insufficiently osteogenic, available in limited supply, and associated with additional surgical morbidity such as blood loss, nerve and arterial injury, and pain at the donor site.[2] Thus, the development of reliable methods to restore bone through tissue engineering is an important goal of orthopaedic research.

In theory, enhancement of fracture healing can be achieved by both physical and biologic means. Physical means include biophysical modalities such as electrical stimulation and ultrasound but may also involve direct application of mechanical loads to the site of fracture healing. One excellent example of how mechanical manipulation can enhance fracture healing is the method of Ilizarov. By applying a tensile force at an optimal rate and with an optimal rhythm, substantial amounts of new bone can be induced. Biologic means involve enhancement of fracture healing by local application of autologous grafts, calcium-based materials, and/or growth factors.

The most simple methods for biologic enhancement of fracture healing involve osteogenic materials such as autologous bone, allogeneic bone, and autologous bone marrow. Of these, the use of autologous bone marrow has substantial potential for research development and could lead to new strategies for tissue engineering of bone. Osteoconductive materials include the calcium phosphate, calcium hydroxyapatite (HA), and calcium sulfate materials, as well as combinations of these materials with demineralized bone matrix (DBM) and freshly harvested human bone marrow. Although a few randomized, controlled trials using calcium HA and calcium phosphate/collagen composites have demonstrated equivalent effectiveness to autologous bone in the healing of fresh fractures, only anecdotal reports show that combinations with other components such as bone marrow or DBM are also effective.

Current research on the use of recombinant proteins in fracture repair has focused on the osteoinductive BMPs. Other growth factors such as fibroblast growth factor (FGF) and platelet derived growth factor (PDGF) that have been shown in preclinical studies to be effective in the enhancement of fracture healing are not being pursued as aggressively by biotechnology companies. Reasons for this are unclear but may relate to economic concerns or allocation of resources to other tissue engineering technologies that may have more immediate application in the clinic. Angiogenic factors such as VEGF may play a role in the direct enhancement of fracture healing or as a co-therapy with a BMP or other inductive molecules in the engineering of fracture healing. Investigations of angiogenic factors are ongoing. The status of development of these technologies for clinical use is the subject of this review.

Role of Autologous Bone Marrow in the Engineering of Fracture Repair

Osteoconduction is defined as "a process that supports the ingrowth of sprouting capillaries, perivascular tissues, and osteoprogenitor cells into the three-dimensional structure of an implant or graft."[3] Current osteoconductive materials based on calcium phosphate or calcium sulfate are effective in certain clinical applications. However, their clinical utility is limited to those skeletal sites where they are surrounded by bone on all sides. The restoration of segmental bone defects, particularly in the diaphyses of the long bones, is difficult to achieve with these osteoconductive materials alone.

One strategy to enhance the clinical performance of osteoconductive materials is to provide osteoprogenitor cells to the site where bone regeneration is needed. However, bone marrow cells with the capacity to be induced to express an osteoblastic phenotype constitute only about 0.01% of the bone marrow stromal cell population.[4] Thus, only 50 to 60 cells per every 1 million nucleated cells in the bone marrow have osteogenic capacity. Bruder and associates[5] demonstrated the ability to heal segmental bone defects in a canine model with the use of a proprietary method for harvesting human bone marrow mesenchymal cells and expanding them in culture. Using a 21-mm long critical-sized defect in canine femora, implantation of HA blocks showed healing at 6 weeks, but the mechanical integrity of the tissue was insufficient to support mechanical loads. However, when these blocks were bathed in culture-expanded mesenchymal stem cells before implantation, not only was healing achieved but the calluses showed remodeling. These results suggest that growth factors produced by the bone marrow stem cells also recruited cells to the microenvironment to participate in healing. Soon after this research was published, a series of three cases in which attempts to use a similar procedure in patients was reported.[6] Both healing and remodeling were noted on radiographs.[6] These preclinical and clinical reports suggest that this is a fruitful area for further investigation and may lead to new technologies that could enhance tissue repair of bone.

The Role of Tissue Repair Factors (Non-BMP) in the Engineering of Fracture Healing

Several growth factors are expressed during different phases of experimental fracture healing. On the basis of these findings, it has been thought that these growth factors represent potential therapeutic agents to enhance the repair of bone. Among these growth factors are TGF-β, BMP, FGF, PDGF, insulin-like growth factor, and VEGF.[7] Although many preclinical studies have tested the effects of these factors, only FGF and BMP have been investigated in clinical trials.

The FGFs are a family of structurally related polypeptides characterized by their affinity for the glycosaminoglycan heparin-binding sites on cells and known to play a critical role in angiogenesis and mesenchymal cell mitogenesis. The most abundant types in normal adult tissue are acidic FGF-1 or α = FGF and basic FGF-2 (or bFGF). Although both FGF-1 and FGF-2 enhance callus repair and have been evaluated in models of fracture healing, only FGF-2 has been developed for clinical application.

The ability of rhFGF-2 to accelerate fracture healing in primates was confirmed in a nonhuman primate fracture model.[8] In that study, rhFGF-2 and hyaluronic acid were combined into a viscous gel formulation that was percutaneously injected into a 1-mm noncritical-sized osteotomy defect in the fibulae of baboons. Intact fibulae from an additional group of baboons was used as a positive control group. The osteotomy sites were treated with three different doses of rhFGF-2 and healing was evaluated by biomechanical testing. Defects treated with FGF-2 showed statistically significant increases in energy to failure and load at failure, suggesting that FGF-2 may have potential in the enhancement of fracture repair.[8] Phase I and II clinical trials of this compound have been initiated but the results are not available. Momentum in the development of this growth factor seems to have diminished and this may be related to nonscientific, economically driven decisions of the sponsor companies. Because this growth factor appears to have potential in the enhancement of human skeletal repair, it is hoped that these investigations are reinitiated in the near future.

Clinical Applications of BMPs: Current Status and Future Development

The ability to truly engineer skeletal tissue, to regenerate entire portions of the skeleton or even limbs, requires tissue induction. Osteoinduction is defined as "a process that supports the mitogenesis of undifferentiated perivascular mesenchymal cells leading to the formation of osteoprogenitor cells with the capacity to form new bone."[3] This property is attributed to a group of proteins known as BMPs, growth factors that are not only involved in bone induction but in the induction of other tissues and the development of the human embryo.

The BMPs are members of the TGF-β superfamily, and 15 individual molecules have been identified in humans. Presently, BMP-2, BMP-4, BMP-7, and BMP-9 are known to play a critical role in bone healing by means of their ability to stimulate differentiation of mesenchymal cells to an osteochondroblastic lineage.[9] The BMPs also play a critical role in cell growth, and mice deficient in BMP-2, BMP-4, and BMP-7 die either during embryonic development or soon after birth. Indeed, the essential role of BMPs in embryologic development is evidenced by the abnormalities seen in the skulls, hind limbs, and kidneys of mice deficient in BMP-2.[10-13] Mice deficient in BMP-5 have short-ear deformities and BMP-7 deficiency has been associated with hind limb polydactyly and renal agenesis.[14]

The concept that there is a substance in bone that can induce new bone formation was recognized by Urist[15] when he observed that a new ossicle of bone had formed after implantation of DBM in a muscle pouch in a rat. More than 20 years later, Wozney and associates[16] identified the genetic sequence of BMP, which led to the identification of its various isoforms. With this genetic information, and with the advent of recombinant gene technology, it is now possible to produce various BMPs. Indeed, two BMPs, rhBMP-2 and rhBMP-7 (OP-1) are currently available for specified clinical uses.

The effectiveness of recombinant human OP-1 (rhOP-1) was assessed in a prospective, randomized, partially blinded clinical trial involving 122 patients with 124 tibial nonunions.[17] Treatment consisted of intramedullary nail fixation and implantation of either recombinant rhOP-1 in a type-1 collagen carrier or autogenous iliac bone graft. Nine months after the surgical procedure, 81% of the 63 nonunions that had been treated with rhOP-1 and 85% of the 61 nonunions that had been treated with autograft were judged to have been treated successfully according to the clinical criteria ($P = 0.524$). In that study, a clinical success was defined as full weight bearing with less than severe pain at the fracture site. At 9 months, radiographic analysis revealed that 75% of the nonunions that had been treated with OP-1 and 84% of those that had been treated with autograft had united ($P = 0.218$). The investigators concluded that there was no significant difference with respect to either clinical or radiographic outcome between the patients who had been treated with rhOP-1 and those who had been treated with autograft. Although the United States Food and Drug Administration did not respond to these data with an approvable letter to the company that markets this device, a Humanitarian Device Exemption for the use of OP-1 to treat recalcitrant nonunions of long bones (nonunions that have failed to respond to prior treatment modalities) was granted.

Over the past year, under the Humanitarian Device Exemption, clinicians have used OP-1 in the treatment of recalcitrant nonunions. In our experience at the Boston University Medical Center (Boston, MA), OP-1 has been used with success in seven patients. There have been no failures of treatment. Evaluations of radiographs of these patients suggest the induction of new bone in conjunction with the use of OP-1 to enhance healing of the nonunion.

Recently, the BMP-2 Evaluation and Surgery for Tibial Trauma (BESTT) study group reported the results of a prospective, randomized controlled trial in 450 open tibial shaft fractures.[18] Patients received either initial irrigation and débridement and a statically locked intramedullary nail (standard care) or standard care plus 0.75 or 1.50 mg/kg rhBMP-2 on an absorbable collagen sponge at the time of wound closure. After 12 months, there was a 44% reduced risk of secondary interventions in the group treated with 1.5 mg/kg rhBMP-2 compared with those treated with standard care (RR = 0.56; pairwise $P = 0.0005$). Among patients treated with rhBMP-2, 58% were healed compared with 38% of patients treated with standard care ($P = 0.0008$). Compared with patients treated with standard care, those treated with rhBMP-2 at the higher dose, had fewer hardware failures, fewer infections, and faster wound healing.[18]

The role of recombinant BMPs has also been evaluated in the engineering of spinal fusions. In a trial of 14 patients who underwent single-level anterior interbody fusion of the fifth lumbar and first sacral vertebrae, 11 patients were treated with a tapered titanium fusion cage filled with 10 mg of rhBMP-2 in a collagen carrier. Three patients in the control group were treated with a cage filled with autogenous bone graft. Six months after the procedure, all 11 patients who had been treated with recombinant protein and two of the three patients who had been treated with autogenous graft had evidence of fusion on plain radiographs and CT scans.[19] These results have led to the approval of rhBMP-2 in a collagen carrier and with a titanium cage in the treatment of single-level interbody fusions. Anecdotal reports from clinicians who have used this material during the 4 months after approval are encouraging.

Future Developments for Tissue Engineering of Fracture Repair

Current use of rhBMPs is by surgical exposure and direct implantation at the surgical site. However, a percutaneous injection of an osteoinductive factor could have obvious clinical advantages. In a study evaluating the ability of a single percutaneous injection of rhBMP-2 to accelerate fracture healing, we demonstrated a 34% acceleration in the acquisition of torsional strength in rats injected with rhBMP-2. Moreover, at 4 weeks after injection, the stiffness and strength of the rhBMP-2 treated fractures were equal to those of the intact unfractured bone in the contralateral femur. The ability of an injection of rhBMP-2 to accelerate fracture repair provides a rationale for its use in fractures that do not require surgical treatment, or that would undergo surgical treatment without direct exposure of the fracture site. In addition, an injectable compound that produces a bone-graft–like effect could have applications in a variety of settings ranging from craniofacial surgery to reconstructive surgery of the spine and extremities. Indeed, the ability to introduce an osteoinductive substance through a percutaneous approach could work in conjunction with advances in surgical treatment to limit the size of surgical exposures and possibly convert major surgical interventions into minimally invasive, short-stay, or outpatient procedures.

Although recombinant proteins are available as therapeutic growth factors for specific clinical indications, there is concern that a single dose of an exogenous protein will not induce an adequate biologic response in patients, particularly in situations where the viability of surrounding soft tissues is compromised. Moreover, to truly regenerate large segments of the skeleton and/or extremities, the introduction of multiple factors in so-called "cocktails" has been proposed. This notion, albeit reasonable in theory, may not be feasible for a variety of reasons. These include the regulatory difficulties in developing such combinations, monetary costs associated with producing multiple recombinant proteins, and the inability to control the effects and expression of multiple molecules. A more logical approach would be gene therapy in which synergistic enhancement of bone, cartilage, tendon, or ligament formation would be possible by the regulated expression of multiple genes.

Using an adenoviral gene transfer protocol to create BMP-2 producing bone marrow cells, Lieberman and associates[20] demonstrated the feasibility of ex vivo gene transfer by comparing critical sized defects implanted with BMP-2-expressing adenovirus-transduced bone marrow cells in comparison to rhBMP-2 and several control implants. Although both rhBMP-2 and BMP-2 engineered bone marrow cells healed these defects, there was greater bone formation and accelerated remodeling in the defects treated by gene therapy. More recently, Peng and associates[21] reported the synergistic enhancement of bone formation in critical sized calvarial defects treated with stem cells engineered to express BMP-4 and VEGF. They noted that the interaction between angiogenic and osteogenic factors was essential to the accelerated enhancement of cartilage resorportion and bone formation. The beneficial effect of VEGF on bone healing elicited by BMP-4 depended on the ratio of VEGF to BMP-4, and an improper ratio of VEGF and BMP-4 was associated with detrimental effects. In addition, the inhibition of VEGF impaired bone formation elicited by BMP-4. Thus, the ability of gene therapy to engineer and coordinate the expression of multiple genes provides an important direction for future tissue engineering research.

Summary

Tissue engineering of fracture repair has made significant progress since the discovery of the bone induction principle in 1965.[15] New technologies are already in our operating rooms and early results with the recombinant BMPs are encouraging. However, future development of this field will require advanced technologies that will lead to better responses in human tissues and the ability to express multiple factors simultaneously. Although gene therapy is currently under review mainly for the treatment of life-threatening diseases, the ability of this powerful technology to enhance the engineering of musculoskeletal tissues will only be possible once the safety issues of gene therapy are adequately resolved. The care of orthopaedic patients will likely be advanced by techniques to enhance the osteogenic potential of human bone marrow preparations, the development of better delivery systems and gene therapy applications for growth factors and osteoinductive substances, and the production of injectable formulations for locally active compounds. These techniques are expected to be available within the next decade.

References

1. Deutche Bank/Alex Brown: 2001.

2. Younger EM, Chapman MW: Morbidity at bone graft donor sites. *J Orthop Trauma* 1989;3:192-195.

3. Urist MR: Bone transplants and implants, in Urist MR (ed): *Fundamental and Clinical Bone Physiology.* Philadelphia, PA, JB Lippincott, 1980, pp 331-368.

4. Muschler GF, Nitto H, Boehm CA, et al: Age- and gender-related changes in the cellularity of human bone marrow and the prevalence of osteoblastic progenitors. *J Orthop Res* 2001;19:117-125.

5. Bruder SP, Kraus KH, Goldberg VM, Kadiyala S: The effect of implants loaded with autologous mesenchymal stem cells on the healing of canine segmental bone defects. *J Bone Joint Surg Am* 1998;80:985-996.

6. Quarto R, Mastrogiancomo M, Cancedda R, et al: Repair of large bone defects with the use of autologous bone marrow stromal cells. *N Engl J Med* 2001;355:385-386.

7. Barnes GL, Kostenuik PJ, Gerstenfeld LC, et al: Growth factor regulation of fracture repair. *J Bone Miner Res* 1999;14:1805-1815.

8. Radomsky ML, Thompson AY, Spiro RC, et al: Potential role of fibroblast growth factor in enhancement of fracture healing. *Clin Orthop* 1998;355:S283-S293.

9. Schmidt JM, Hwang K, Winn SR, et al: Bone morphogenetic proteins: An update on basic biology and clinical relevance. *J Orthop Res* 1999;17:269-278.

10. Dudley AT, Lyons KM, Robertson EJ: A requirement for bone morphogenetic protein-7 during development of the mammalian kidney and eye. *Genes Dev* 1995;9:2795-2807.

11. Luo G, Hofmann C, Bronckers AL, Sohocki M, Bradley A, Karsenty G: BMP-7 is an inducer of nephrogenesis, and is also required for eye development and skeletal patterning. *Genes Dev* 1995;9:2808-2820.

12. Higinbotham KG, Karanova ID, Diwam BA, et al: Deficient expression of mRNA for the putative inductive factor bone morphogenetic protein-7 in chemically initiated rat nephroblastomas. *Mol Carcinog* 1998;23:53-61.

13. Lieberman JR, Daluiski A, Einhorn TA: The role of growth factors in the repair of bone. *J Bone Joint Surg Am* 2002;84:1032-1044.

14. Kingsley DM, Bland AE, Grubber JM, et al: The mouse short ear skeletal morphogenesis locus is associated with defects in a bone morphogenetic member of the TGF beta super-family. *Cell* 1992;71:399-410.

15. Urist MR: Bone: Formation by autoinduction. *Science* 1965;150:893-899.

16. Wozney JM, Rosen V, Celeste AJ, et al: Novel regulators of bone formation: Molecular clones and activities. *Science* 1988;242:1528-1534.

17. Friedlaender GE, Perry CR, Cole JD, et al: Osteogenic protein-1 (bone morphogenetic protein-7) in the treatment of tibial nonunions. *J Bone Joint Surg Am* 2001;83:S151-S158.

18. Govender S, Csimma C, Genant HK, Valentin-Opran A: Recombinant human bone morphogenetic protein-2 for treatment of open tibial fractures. *J Bone Joint Surg Am* 2002;84:2123-2134.

19. Boden SD, Zdeblick TA, Sandhu HS, et al: The use of rhBMP-2 in interbody fusion cages: Definitive evidence of osteoinduction in humans. A preliminary report. *Spine* 2000;25:376-381.

20. Lieberman JR, Daluiski A, Stevenson S, et al: The effect of regional gene therapy with bone morphogenetic protein-2-producing bone-marrow cells on the repair of segmental femoral defects in rats. *J Bone Joint Surg Am* 1999;81:905-917.

21. Peng H, Wright V, Usas A, et al: Synergistic enhancement of bone formation and healing by stem cell-expressed VEGF and bone morphogenetic protein-4. *J Clin Invest* 2002;110:751-759.

Chapter 11

Engineered Cells in Scaffolds Heal Bone

Ranieri Cancedda, MD
Rodolfo Quarto, MD
Giordano Bianchi, MD
Maddalena Mastrogiacomo, PhD
Anita Muraglia, PhD

Abstract

Bone marrow stromal cells (BMSC) are nonhemopoietic cells residing in the marrow cavity. Under appropriate conditions, BMSC differentiate into osteoblasts, chondrocytes, adipocytes, and hematopoiesis-supportive stroma. With in vitro expansion, the multi-differentiation potential of BMSC is gradually lost and their ability to form bone in vivo is dramatically reduced. However, when BMSC are isolated and expanded in the presence of fibroblast growth factor-2 (FGF-2), a greater percentage of cells able to differentiate into the osteogenic, chondrogenic, and adipogenic lineages is maintained. Bone repair has been achieved by local delivery of cells with differentiation potential within a scaffold. Bone formation in response to delivery of engineered cells was first evaluated in small animal models. Large animal models were subsequently developed to prove the feasibility of the tissue engineering approach in a model that more closely mimics the clinical situation. These preclinical studies were followed by pilot clinical studies. Adequate, well-controlled clinical studies have yet to be completed. Available results are very encouraging, but additional preclinical and clinical studies are needed before tissue engineering procedures becomes incorporated into routine orthopaedic clinical practice.

Introduction

Bone damage, either due to pathology or trauma, is a very common occurrence, requires costly medical and/or surgical interventions, and is associated with significant morbidity. Currently available therapeutic approaches include osteotomy followed by distraction osteogenesis (Ilizarov bone transport) and graft transplants (bone grafts or synthetic materials). The Ilizarov technique consisting of external fixators is very inconvenient for the patient, requires a long recovery time, and has a high rate of complications.

Bone grafts are either of autologous source (from the patient himself; autograft) or from other donors (homologous and heterologous grafts). Autologous bone grafts are either nonvascularized or vascularized.

Vascularized grafts are the most widely used. The success rate is high, but complications, including infections and nonunions, frequently occur. Furthermore, the autograft procedure requires a second operation that causes significant donor site morbidity, increases the risk of complications, impairs the patient's recovery, and substantially increases the overall costs of the bone graft procedure.

The use of allografts is associated with the risk of blood-borne diseases. Their use is limited to the countries that are able to correctly store and donor-receiver match fresh frozen grafts. Many allografts are obtained from cadavers by heat and dehydration procedures that are expensive, time-consuming, and strictly regulated by law. The use of synthetic osteoconductive materials (bioceramics) for implants has only been successful for patients with relatively small bone defects.

Tissue engineering represents a feasible and productive approach to repair and reconstitute bone. Tissue engineering procedures have been developed based on the technology of resorbable biomaterials and knowledge of cell biology. The implant of engineered bone graft material in which an osteoconductive scaffold is combined with osteogenic committed cells could represent a valid alternative to the techniques currently used for bone repair.

In Vitro Expanded Bone Marrow Stromal Cells: Stem or Progenitor Cells?

It has been proposed that BMSC represent a population of stem cells, mesenchymal stem cells (MSC), that can generate all different mesenchymal lineages when exposed to different microenvironments.[1-3] In some laboratories, BMSC were shown to behave as "true stem cells" capable of differentiating into cells of ectodermal and endodermal origin astrocytes, oligodendrocytes, neurons, and hepatocytes.[4-6]

The colony-forming units-fibroblastic derived from bone marrow cells clonally undergo osteogenic, chondrogenic, and adipogenic differentiation.[7,8] Analysis of the differentiation potential of single cells indicates that there is heterogeneity of the original BMSC population.[8,9] When clones are isolated and expanded in the presence of FGF-2, there is an increase in the percentage of clones able to differentiate into the osteogenic, chondrogenic, and adipogenic lineages ("tripotential" clones).[8]

The use of BMSC as a source of multipotent mesenchymal cells for tissue engineering is appealing because of the ease with which they can be isolated from the patient's bone marrow and expanded in vitro. However, extensive in vitro proliferation affects both the differentive and the replicative potential of BMSC.[10,11]

Differentiation into the three lineages can be induced by varying the composition of the culture medium and with the use of different culture conditions. We noted that after 18 to 19 doublings, the ability of BMSC to respond to adipogenic stimulation was greatly reduced and was completely lost by the 22nd doubling. The ability for chondrogenesis and osteogenesis were maintained longer but decreased after approximately 22 doublings.[8,11]

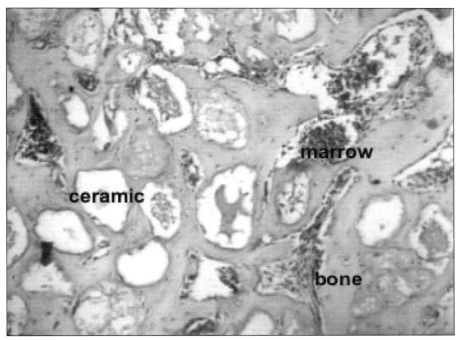

Fig. 1 In vivo bone formation on HA granules. Abundant bone is formed on and around HA granules loaded with BMSC, included in a fibrin glue and subcutaneously injected in an immunocompromised mouse (nude, CD-1 nu/nu). Some bone marrow formation is also evident.

The addition of basic FGF-2 to primary cultures of BMSC helps to maintain their differentiation potential. This effect is associated with a longer telomere size of the cultured cells. Because no telomerase activity is detectable in expanded BMSC, the observed increase in telomere length is due to the selection of a subset of progenitor cells. In low density conditions and in the presence of FGF-2, these cells have an extended in vitro life span of about 70 doublings and they retain their differentiation potential for more than 50 doublings.[12]

The addition of FGF-2 to the culture medium resulted in a sizeable decrease in expression of alkaline phosphatase, an early marker of osteogenic commitment. These finding indicate that FGF-2 delays but does not prevent the expression of a more committed phenotype.[8] In fact, in vitro expansion of BMSC in the presence of FGF-2 and their subsequent induction toward osteogenesis yielded approximately three times more mineral deposition than cultures expanded in the absence of FGF-2.

When implanted into immunocompromised mice (nude, CD-1 nu/nu), BMSC combined with mineralized tridimensional scaffolds form highly vascularized primary bone tissue (Figure 1). With in vitro expansion, the bone forming efficiency of in vivo transplanted BMSC is dramatically reduced.[8,11] When BMSC were cultured in the presence of FGF-2 and subcutaneously transplanted with hydroxyapatite (HA) into immunocompromised mice, in vivo bone formation was maintained. At 4 weeks, bone was detected only in the samples loaded with FGF-2 expanded cells. At 8 weeks, bone formation reached 12.6% of the total tissue observed with FGF-2 expanded BMSC. Bone formation was less than 0.5% in control samples.[13]

Our data support the idea that BMSC are not to be considered as an MSC population. They are better defined as a heterogeneous population of mesenchymal progenitor/precursor cells. The addition of FGF-2 to cell cultures maintains cells in a more immature state that allows in vitro expansion of osteoprogenitors, possibly by selecting a specific cell subset of the expanding population. Although not sufficient by itself to prevent gradual senescence, FGF-2 considerably improves the proportion of multipotential clones and is useful to obtain a large number of cells with differentiation potential for use in mesenchymal tissue repair.

Bone Repair by Bone Marrow Stromal Cells

Bone formation by BMSC was first reported in small animals either othotopically for the repair of small experimentally induced osseous defects in mice and rats[14,15] or ectopically by subcutaneous implantation of small porous ceramic cubes loaded with cells from different animal species, including human, into syngeneic rats[16] and immunodeficient mice.[17-20] In a few cases, osteoprogenitor cells transplanted into ceramic scaffolds were used to create vascularized bone flaps.[21,22]

Large animal models were subsequently developed. Canine autologous BMSC were loaded onto porous ceramic cylinders (65:35 HA:β-tricalcium phosphate) and implanted into critical-sized segmental defects in the femora of adult animals. After 4 months, a strong union formed at the interface between the host bone and the implants. In contrast, an atrophic nonunion occurred in the femora of untreated control animals. Both woven and lamellar bone filled the pores of the implants that had been loaded with MSC. A large collar of bone formed around the BMSC loaded implants.[19-23]

We evaluated the use of porous bioceramics (100% HA) in the repair of extensive bone defects in the tibiae of sheep (Figure 2). Ceramic cylinders were implanted with and without the addition of autologous ex vivo expanded BMSC.[24] Bone formation was observed both in implants with bioceramic alone and in those with added cells. However, the presence of BMSC greatly accelerated callus formation and bone repair and decreased the time of recovery. Cell-loaded HA cylinders were more stiff on the indentation test and most pores were filled with newly formed bone, whereas only fibrous tissue developed inside the implants without cells.

The healing process involves four main steps: (1) bone formation on the outer surface of the implant, (2) bone formation in the inner cylinder canal, (3) formation of fissures and cracks in the implant body, and (4) bone formation in the bioceramics pores. When assessed by radiographs and CT scans, bone formation was most prominent over the external surface and within the inner canal of the implants. This might be due to a higher density of loaded cells and/or to a better survival of cells within the outermost

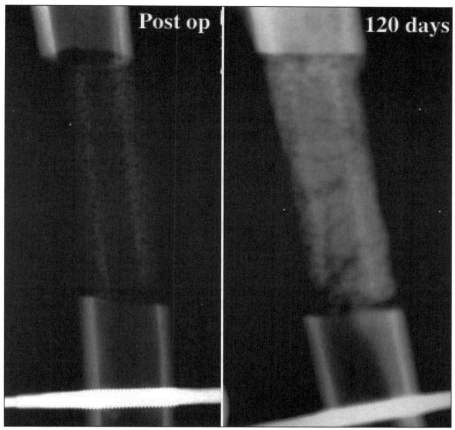

Fig. 2 Second-generation implant of porous bioceramic loaded with cells. In contrast with implants previously used,[23,25] the new scaffolds have a much higher porosity (up to 90%). With the new scaffolds, new bone formation is better followed by radiographs. Tibia sheep model; implant scaffold is made of 100% HA.

portions of the HA bioceramics. Alternatively, the implanted cells could stimulate, by a paracrine loop, resident osteoprogenitor cells that are located within the skeletal tissues at the resection ends.[24] Similar results were obtained by Petite and associates[25] by implanting a combination of a coral scaffold with in vitro-expanded marrow stromal cells in a large segmental defect in sheep tibiae.

We first reported a clinical application of this cell-based tissue engineering approach to treat three patients with long bone segmental (4 to 7 cm) defects.[26] After ex vivo expansion, BMSC were loaded on porous ceramic scaffolds (40% to 50% porosity; 100% HA). Mechanical stability was initially obtained by external fixation. An abundant callus formation along the implant and a good integration at the interface with the host bone was observed on radiographs by the second month after the operation. All three patients recovered limb function 6 and 12 months after the procedure.

In the same year, Vacanti and associates[27] reported the functional restoration of a stable thumb by the replacement of a phalanx with tissue engineered bone. The engineered tissue was obtained by loading autologous periosteal cells expanded in vitro on a scaffold of coral porous HA.

Summary and Future Developments

Our studies showed that BMSC are an easily accessible source of cells able to differentiate toward several mesenchymal lineages. After being in culture, these cells do not behave as true stem cells but lose their proliferative and differentiation potential. The addition of FGF-2 to the culture medium maintains the cells in an undifferentiated state. Cells expanded in the presence of FGF-2 are very efficient in the reconstruction of bone segmental defects in which traditional interventions have proved unsuccessful.

Nevertheless, we consider bone repair achieved with the use of tissue engineering very encouraging but not optimal. Additional preclinical and clinical trials are necessary before tissue engineering can be incorporated into standard orthopaedic clinical practice. There are unanswered questions regarding which cells and scaffolds should be used, how to prepare the implants, how to optimize surgical procedures, and how to implement appropriate patient follow-up.

Questions regarding selection of cells include:

1. Are BMSC the optimal choice? Do we have alternatives? Possible alternatives could be osteogenic cells derived from fat, muscle, or other tissues.

2. Can we use allogeneic cells? How can the immune response to allogeneic material be controlled?

3. Which genetic modifications of cells are needed before implantation? A combination of a cell therapy and a gene therapy approach is a possibility.

The use of scaffolds may be optimized by:

1. Testing scaffolds of various chemical composition, resorption rates, architecture, and other characteristics.

2. Investigation of custom-made implants and evaluation of implants of a variety of predetermined shapes.

The preparation of implants could be optimized by studies of:

1. The ex vivo expansion of the cells cultured on plasticware or within the porous implants.

2. The comparison of the use of autologous serum, calf serum, and serum-free media; there is an increasing demand for serum-free media both for safety reasons and for the development of more physiologic microenvironments for cell growth and differentiation.

3. The development and use of bioreactors; a bioreactor can be extremely useful not only to guarantee safer culture conditions, but also to test the possible influence of applied mechanical forces during the implant preparation.

Issues pertaining to surgical procedures and patient follow-up include the need to:

1. Standardize and validate clinical protocols.

2. Prove efficacy of the tissue engineering approach and to perform multicenter clinical trials to compare this approach with other more traditional procedures.

3. Compare different mechanical stabilizations, such as internal and external fixators.

As additional studies are completed, it is anticipated that tissue engineering techniques will emerge as a viable option to currently available procedures for bone repair.

References

1. Owen M: Marrow stromal stem cells. *J Cell Sci Suppl* 1988;10:63-76.

2. Friedenstein AJ: Osteogenic stem cells in the bone marrow. *J Bone Miner Res* 1990;7:243-272.

3. Caplan AI: Mesenchymal stem cells. *J Orthop Res* 1991;9:641-650.

4. Woodbury D, Schwarz EJ, Prockop DJ, et al: Adult rat and human bone marrow stromal cells differentiate into neurons. *J Neurosci Res* 2000;61:364-370.

5. Sanchez-Ramos J, Song S, Cardozo-Pelaez F, et al: Adult bone marrow stromal cells differentiate into neural cells in vitro. *Exp Neurol* 2000;164:247-256.

6. Dezawa M, Takahashi I, Esaki M, et al: Sciatic nerve regeneration in rats induced by transplantation of in vitro differentiated bone-marrow stromal cells. *Eur J Neurosci* 2001;14:1771-1776.

7. Pittenger MF, Mackay AM, Beck SC, et al: Multilineage potential of adult human mesenchymal stem cells. *Science* 1999;284:143-147.

8. Muraglia A, Cancedda R, Quarto R: Clonal mesenchymal progenitors from human bone marrow differentiate in vitro according to a hierarchical model. *J Cell Sci* 2000;113:1161-1166.

9. Kuznetsov SA, Krebsbach PH, Satomura K, et al: Single-colony derived strains of human marrow stromal fibroblasts form bone after transplantation in vivo. *J Bone Miner Res* 1997;12:1335-1347.

10. Bruder SP, Jaiswal N, Haynesworth SE: Growth kinetics, self-renewal, and the osteogenic potential of purified human mesenchymal stem cells during extensive subcultivation and following cryopreservation. *J Cell Biochem* 1997;64:278-294.

11. Banfi A, Muraglia A, Dozin B, et al: Proliferation kinetics and differentiation potential of ex vivo expanded human bone marrow stromal cells: Implications for their use in cell therapy. *Exp Hematol* 2000;28:707-715.

12. Banfi A, Bianchi G, Notaro R, et al: Replicative aging and gene expression in longterm cultures of human bone marrow stromal cells. *Tissue Eng* 2002;8:901-910.

13. Martin I, Muraglia A, Campanile G, et al: Fibroblast growth factor-2 supports ex vivo expansion and maintenance of osteogenic precursors from human bone marrow. *Endocrinology* 1997;138:4456-4462.

14. Ohgushi H, Goldberg VM, Caplan AI: Repair of bone defects with marrow cells and porous ceramic: Experiments in rats. *Acta Orthop Scand* 1989;60:334-339.

15. Krebsbach PH, Mankani MH, Satomura K, et al: Repair of craniotomy defects using bone marrow stromal cells. *Transplantation* 1998;66:1272-1278.

16. Goshima J, Goldberg VM, Caplan AI: The osteogenic potential of culture-expanded rat marrow mesenchymal cells assayed in vivo in calcium phosphate ceramic blocks. *Clin Orthop* 1991;262:298-311.

17. Muraglia A, Martin I, Cancedda R, et al: A nude mouse model for human bone formation in unloaded conditions. *Bone* 1998;22:131S-134S.

18. Goshima J, Goldberg VM, Caplan AI: The origin of bone formed in composite grafts of porous calcium phosphate ceramic loaded with marrow cells. *Clin Orthop* 1991;262:274-283.

19. Kadiyala S, Young RG, Thiede MA, et al: Culture expanded canine mesenchymal stem cells possess osteochondrogenic potential in vivo and in vitro. *Cell Transplant* 1997;6:125-134.

20. Krebsbach PH, Kuznetsov SA, Satomura K, et al: Bone formation in vivo: Comparison of osteogenesis by transplanted mouse and human marrow stromal fibroblasts. *Transplantation* 1997;63:1059-1069.

21. Casabona F, Martin I, Muraglia A, et al: Prefabricated engineered bone flaps: An experimental model of tissue reconstruction in plastic surgery. *Plast Reconstr Surg* 1998;101:577-581.

22. Mankani MH, Krebsbach PH, Satomura K, et al: Pedicled bone flap formation using transplanted bone marrow stromal cells. *Arch Surg* 2001;136:263-270.

23. Bruder SP, Kraus KH, Goldberg VM, et al: The effect of implants loaded with autologous mesenchymal stem cells on the healing of canine segmental bone defects. *J Bone Joint Surg Am* 1998;80:985-996.

24. Kon E, Muraglia A, Corsi A, et al: Autologous bone marrow stromal cells loaded onto porous hydroxyapatite ceramic accelerate bone repair in critical-size defects of sheep long bones. *J Biomed Mater Res* 2000;49:328-337.

25. Petite H, Viateau V, Bensaid W, et al: Tissue-engineered bone regeneration. *Nat Biotechnol* 2000;18:959-963.

26. Quarto R, Mastrogiacomo M, Cancedda R, et al: Repair of large bone defects with the use of autologous bone marrow stromal cells. *N Engl J Med* 2001;344:385-386.

27. Vacanti CA, Bonassar LJ, Vacanti MP, et al: Replacement of an avulsed phalanx with tissue-engineered bone. *N Engl J Med* 2001;344:1511-1514.

Chapter 12

Bone Engineering With Mesenchymal Stem Cells and Gene Therapy

Jueren Lou, MD

Abstract

Engineered osteogenic material has great potential value for the repair of large, segmental bone loss. Mesenchymal stem cells (MSC) are present in various tissues including bone marrow, muscle, and skin. These progenitor cells have the capability to differentiate into an osseous lineage and to form bone in an appropriate osteogenic microenvironment. However, not all environments induce MSC to differentiate. Gene therapy/gene transfer techniques allow the transfer of osteoinductive genes into MSC. The MSC with transferred genes undergo osteogenic differentiation by autocrine and paracrine induction that leads to bone formation. Animal studies demonstrate that the use of MSC in combination with gene transfer methods can effectively induce bone formation in vivo. Bone formation and bone repair have been successfully achieved in several different animal models including mice, rats, rabbits, and pigs, either in ectopic or orthopaedic sites. The results demonstrate that this combined approach is effective, independent of the local osteogenic environment and the availability of local osteoprogenitor cells. The use of MSC with transferred genes has the potential for future clinical applications.

Introduction

More than one million orthopaedic operations involving bone repair are performed annually in the United States.[1] Autologous bone grafts are often necessary to achieve bone healing and repair. However, harvesting bone for autografts is invasive and is associated with a risk of serious morbidity. In addition, the use of this procedure in clinical practice is often limited by the failure of bone graft to completely resorb, difficulty in shaping the bone graft to fit the osseous defect in the healing bone, or an insufficient amount of bone graft material to completely fill the defect. The combination of the use of MSC and gene therapy may provide an alternative to conventional methods of bone grafting for bone repair.

The Potential of MSC

Several types of MSC have been isolated from different tissues. Bone marrow-derived MSC are considered part of the nonhematopoietic portion of bone marrow. These MSC have been isolated from the whole marrow of chicks, mice, rats, rabbits, goats, and humans.[2] Unlike hematopoietic stem cells, marrow-derived MSC are self-renewing and can be conditioned to grow in vitro. Human bone marrow-derived MSC have been subcultured for as many as 15 passages without losing their potential to differentiate. Neither cryopreservation nor thawing has any effect on the growth or differentiation of these MSC.[3] Marrow-derived cells have the potential to differentiate into bone, cartilage, tendon, muscle, and fat both in vitro and in vivo.[4,5] Injection of a single MSC into an early blastocyst of mouse demonstrated that the MSC contributed to and differentiated into almost all somatic cell types, including cells with visceral mesoderm, neuroectoderm, and endoderm characteristics.[6] The osteogenic potential of marrow-derived MSC has been well defined as evidenced by bone formation following transplantation of MSC in vivo.[4,7,8] In addition, in vitro experiments demonstrate the formation of mineralized nodules, elevated production of alkaline phosphatase (AP), osteopontin, bone sialoprotein, osteocalcin, types I and III collagen, and generation of parathyroid hormone (PTH) and estrogen receptors during the differentiation of MSC into osteoblasts.[4,9-11] Marrow-derived MSC were influenced by a variety of factors during osteogenic differentiation. For example, AP activity is decreased when marrow-derived MSC are treated with transforming growth factor-β1 (TGF-β1) or platelet derived growth factor-BB in culture, and TGF-β1 further blocks the osteochondrogenic potential of the cells.[12] In contrast, dexamethasone[4,11] and bone morphogenetic protein-2 (BMP-2)[10] facilitate the osteogenic potential of these cells.

Multipotential MSC have also been identified in skeletal muscle. Several cell phenotypes, such as satellite cells, side population cells, and muscle-derived stem cells, were isolated from muscles of various animals and humans. These cells all demonstrated the ability to differentiate into multiple mesodermal phenotypes including bone, cartilage, adipocyte, and myotubes.[13,14] Similar MSC were found in the skin and adipose tissue.[15,16]

Although it is well established that MSC possess the potential to differentiate into osteoblastic cells, an adequate inducement to direct MSC differentiation into osteolineage is essential. Adding chemicals and recombinant growth factors to cell culture medium was effective in inducing MSC to differentiate into osteoblastic cells in vitro. However, continuous induction was lost when the cells were harvested from the culture and implanted in vivo. An in vivo environment can provide a sufficient osteoinduction to produce bone. In many situations, however, the in vivo environment may not be adequate to induce endogenous MSC to repair large musculoskeletal defects, such as segmental loss from trauma, tumors, or necrosis. In contrast, gene transfer/gene therapy techniques are able to transduce genes into MSC and enable the cells to continuously express the genes, which influence cell differentiation, either in vitro or in vivo by both autocrine and paracrine mechanisms, on an ongoing basis. By providing both functional molecules and target cells, the combination of MSC and gene therapy techniques is much more effective than either the functional molecule or the target cells alone.

Gene Transfer

Gene transfer can be accomplished with the use of either nonviral or viral vectors. Nonviral vector methods include naked DNA plasmid, liposomes, and electroporation-mediated gene transfer. Viral vector methods include adenovirus, adeno-associated virus, and retroviral vector-mediated gene transfer.

Adenovirus is the one of the most popular viral vectors used for bone repair. Adenovirus is a double-stranded DNA virus. It is effective in a wide range of hosts and is capable of infecting either dividing or nondividing cells. The adenoviral genome is positioned epichromosomely in the host cells, which reduces the risk of insertional mutagenesis. The adenovirus-mediated gene expression is transient. This transient expression, although problematic for many systemic gene therapy applications, has been sufficient to induce bone formation in both immunodeficient and immunocompetent animals.[17-21] The transient gene expression also avoids potential side effects from long-term expression of the transgenes.

Several genes have been tested as transgenes to enhance bone repair in animal studies. The genes of BMP-2,[17-21] BMP-4,[22] BMP-6,[23] BMP-7,[24] and BMP-9,[25] transcription factor LIM mineralization protein-1,[26] and PTH[27] were proven effective for bone repair and bone formation in different animal models. In most studies, single gene transfer was effective in enhancing bone repair and bone healing.[17-27] The delivery of two genes simultaneously was also tested and the synergistic enhancement of bone formation from two gene transfers was observed.[28]

Either in vivo or ex vivo gene transfer can be used in animal studies. The in vivo strategy of gene transfer involves directly delivering the gene into cells at specific sites. The ex vivo strategy involves collecting cells from animals, transducing the collected cells in vitro, and then delivering the transduced cells back to the animals. In general, the ex vivo strategy has been effective in enhancing bone formation because it provides both functional molecules and target cells in a closed system whereas direct gene delivery may not target enough MSC.

The transfer of the BMP-2 gene using an ex vivo strategy has been well studied for bone repair and bone formation in different animal models. The BMP-2 transduced MSC underwent osteoblastic differentiation in in vitro culture.[17-21] The transduced cells showed osteoblastic marker elevation, such as increased AP activity, overexpression of bone matrix proteins, and formation of mineralized nodules. Stem cells also showed a dose-dependent proliferation in response to transfer of the BMP-2 gene. These changes were attributed to the transgene product BMP-2 molecule because these changes were not detected in control stem cells. Osteoblastic differentiation induced by the transfer of the BMP-2 gene in vitro was observed in MSC from different species, including mice,[17] rats,[18,29] rabbits,[19,20] pigs,[21] and humans.[30] In

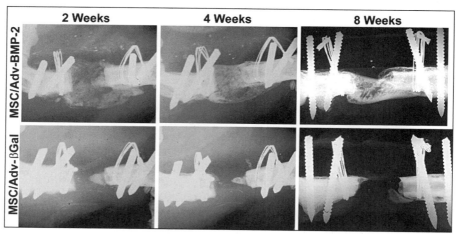

Figure 1 Radiographic view following Isograft (Syngeneic group) in rat femoral defect model. MSC/Adv-BMP-2 induced bone formation and repaired bone defect. No significant bone formation was observed by MSC/Adv-βGal.

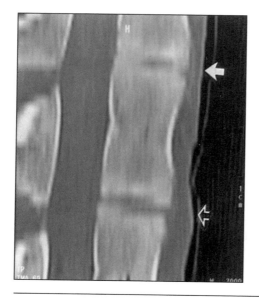

Figure 2 MRI analysis of pig anterior spinal fusion. Solid arrow indicates: Adv-BMP-2 transduced MSC; outline arrow indicates: control. Adv-BMP-2 transduced MSC induced spinal fusion.

vivo studies from independent laboratories further demonstrated that a combination strategy of MSC and BMP-2 gene transfer effectively repaired critical bone defects in the rat femur[18,29] (Figure 1) and mouse skull,[30] and induced anterior spinal fusion in pigs[21] (Figure 2). Significant amounts of bone formation in ectopic sites[17,21] or orthopaedic sites[18,20,29] was observed when BMP-2 gene transferred MSC were injected or implanted into ani-

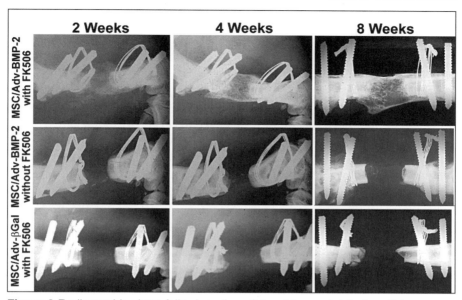

Figure 3 Radiographic views following allograft in rat femoral defect model. MSC/Adv-BMP-2 with FK506 induced bone formation and repaired bone defect. No significant bone formation was observed by MSC/Adv-BMP-2 without FK506 or MSC/Adv-βGal with FK506 treatment.

mals. In these studies, various types of MSC were used, including bone-derived or muscle-derived MSC.

Implantation of Allogeneic MSC

Although implantation of autologous MSC is a promising technique, culturing individual MSC for autologous therapy is difficult and often ineffective. In the clinical setting, in vitro cultures of MSC can be limited because the patient's marrow is damaged or the yield of cells from healthy marrow is reduced due to a variety of factors such as chemotherapy, myelofibrosis, and lipid storage disease.

As an alternative to autologous MSC, BMP-2 gene transferred allogeneic marrow-derived MSC were studied in a rat femoral defect model.[29] With short-term administration of the immunosuppressant FK506 (one injection daily for 2 weeks and one injection every other day for an additional week), 6-mm segmental defects of rat femur were completely repaired 4 to 8 weeks after implantation of BMP-2 gene engineered allogeneic marrow-derived MSC (Figure 3). Up to 24 weeks postimplantation, radiographs showed that the density of the newly formed bone in the defect sites continued to increase and a complete cortical bone structure was formed. No decrease in bone formation, bone remodeling, or resorption of newly formed bone was observed after discontinuation of the administration of immunosuppressant

FK506 in the group receiving BMP-2 gene engineered allogeneic marrow-derived MSC. No significant repair of bone defect in the control group occurred without FK506 administration.

Summary

Animal studies demonstrate that the use of MSC in combination with gene transfer methods can induce bone formation in vivo. Bone formation and bone repair have been successfully achieved in several different animal models. Based on these studies, we conclude that MSC and gene transfer/gene therapy is effective for enhancing bone repair in vivo. With additional testing to confirm the safety of these techniques, it may become an established option for patients who would otherwise require bone grafting.

References

1. Chaput C, Selmani A, Rivard CH: Artificial scaffolding materials for tissue extracellular matrix repair. *Curr Opin Orthop* 1996;7:62-68.

2. Caplan AI: The mesengenic process. *Clin Plast Surg* 1994;21:429-435.

3. Bruder SP, Neelam J, Haynesworth SE: Growth kinetics, self-renewal, and the osteogenic potential of purified human mesenchymal stem cells during extensive subcultivation and following cryopreservation. *J Cell Biol* 1997;64:278-294.

4. Cheng SL, Yang JW, Rifa L, et al: Differentiation of human bone marrow osteogenic stromal cells in vitro: Induction of the osteoblast phenotype by dexamethasone. *Endocrinology* 1994;134:277-286.

5. Caplan AI: Mesenchymal stem cells. *J Orthop Res* 1991;9:641-650.

6. Jiang Y, Jahagirdar BN, Reinhardt RL, et al: Pluripotency of mesenchymal stem cells derived from adult marrow. *Nature* 2002;418:41-49.

7. Goshima J, Goldberg VM, Caplan AI: The osteogenic potential of culture-expanded rat marrow mesenchymal cells assayed in vivo in calcium phosphate ceramic blocks. *Clin Orthop* 1991;262:298-311.

8. Goshima J, Goldberg VM, Caplan AI: The origin of bone formed in composite grafts of porous calcium phosphate ceramic loaded with marrow cells. *Clin Orthop* 1991;269:274-283.

9. Haynesworth SE, Goshima J, Goldberg VM, et al: Characterization of cells with osteogenic potential from human marrow. *Bone* 1992;13:81-88.

10. Lecanda F, Avioli LV, Cheng SL: Regulation of bone matrix protein expression and induction of differentiation of human osteoblasts and human bone marrow stromal cells by bone morphogenetic protein-2. *J Cell Biochem* 1997;67:386-398.

11. Jaiswal N, Haynesworth SE, Caplan AI, et al: Osteogenic differentiation of purified, culture-expanded human mesenchymal stem cells in vitro. *J Cell Biochem* 1997;64:295-312.

12. Cassiede P, Dennis JE, Ma F, Caplan AI: Osteochondrogenic potential of marrow mesenchymal progenitor cells exposed to TGF-B1 or PDGF-BB as assayed in vivo and in vitro. *J Bone Miner Res* 1996;11:1264-1273.

13. Lee JY, Qu-Petersen Z, Cao B, et al: Clonal isolation of muscle-derived cells capable of enhancing muscle regeneration and bone healing. *J Cell Biol* 2000;150:1085-1100.

14. Williams JT, Southerland SS, Souza J, et al: Cells isolated from adult human skeletal muscle capable of differentiating into multiple mesodermal phenotypes. *Am Surg* 1999;65:22-26.

15. Toma JG, Akhavan M, Fernandes KJ, et al: Isolation of multipotent adult stem cells from the dermis of mammalian skin. *Nat Cell Biol* 2001;3:778-784.

16. Zuk PA, Zhu M, Mizuno H, et al: Multilineage cells from human adipose tissue: Implications for cell-based therapies. *Tissue Eng* 2001;7:211-228.

17. Lou J, Xu F, Merkel K, et al: Gene therapy: Adenovirus-mediated human bone morphogenetic protein-2 gene transfer induces mesenchymal progenitor cell proliferation and differentiation in vitro and bone formation in vivo. *J Orthop Res* 1999;17:43-50.

18. Lieberman JR, Daluiski A, Stevenson S, et al: The effect of regional gene therapy with bone morphogenetic protein-2-producing bone-marrow cells on the repair of segmental femoral defects in rats. *J Bone Joint Surg Am* 1999;81:905-917.

19. Riew KD, Wright NM, Cheng S-L, et al: Induction of bone formation using a recombinant adenoviral vector carrying the human BMP-2 gene in a rabbit spinal fusion model. *Calcif Tissue Int* 1998;63:357-360.

20. Cheng SL, Lou J, Wright NM, et al: In vitro and in vivo induction of bone formation using a recombinant adenoviral vector carrying the human BMP-2 gene. *Calcif Tissue Int* 2001;68:87-94.

21. Lou J, Wright NM, Cheng SL, et al: Pig anterior spine fusion induced by Adv-BMP2 transduced autologous bone marrow derived mesenchymal stem cells. *Trans Orthop Res Soc,* 2000, p 266.

22. Gysin R, Wergedal JE, Sheng MH, et al: Ex vivo gene therapy with stromal cells transduced with a retroviral vector containing the BMP4 gene completely heals critical size calvarial defect in rats. *Gene Ther* 2002;9:991-999.

23. Jane J, Dunford B, Kron A, et al: Ectopic osteogenesis using adenoviral bone morphogenetic protein (BMP)-4 and BMP-6 gene transfer. *Mol Ther* 2002;6:464-470.

24. Franceschi RT, Wang D, Krebsbach PH, et al: Gene therapy for bone formation: in vitro and in vivo osteogenic activity of an adenovirus expressing BMP7. *J Cell Biochem* 2000;78:476-486.

25. Helm GA, Alden TD, Beres EJ, et al: Use of bone morphogenetic protein-9 gene therapy to induce spinal arthrodesis in the rodent. *J Neurosurg* 2000;92:191-196.

26. Boden SD, Titus L, Hair G, et al: Lumbar spine fusion by local gene therapy with a cDNA encoding a novel osteoinductive protein (LMP-1). *Spine* 1998;23:2486-2492.

27. Fang J, Zhu YY, Smiley E, et al: Stimulation of new bone formation by direct transfer of osteogenic plasmid genes. *Proc Natl Acad Sci USA* 1996;93:5753-5758.

28. Peng H, Wright V, Usas A, et al: Synergistic enhancement of bone formation and healing by stem cell-expressed VEGF and bone morphogenetic protein-4. *J Clin Invest* 2002;110:751-759.

29. Tsuchida H, Hashimoto J, Crawford E, et al: Engineered allogeneic mesenchymal stem cells repair femoral segmental defect in rats. *J Orthop Res* 2003;21:44-53.

30. Lee JY, Peng H, Usas A, et al: Enhancement of bone healing based on ex vivo gene therapy using human muscle-derived cells expressing bone morphogenetic protein 2. *Hum Gene Ther* 2002;13:1201-1211.

Chapter 13

Induction of Bone Formation by Stem Cells

Johnny Huard, PhD
Hairong Peng, MD, PhD

Abstract

We isolated a population of muscle-derived stem cells that maintain long-term proliferative capability and express hematopoietic stem cell markers. These muscle-derived stem cells retain their phenotype for more than 30 passages and can differentiate into muscle, neural, and endothelial lineages both in vitro and in vivo. The transplantation of these muscle-derived stem cells improved the efficiency of muscle regeneration and delivery of dystrophin in dystrophic muscle. The ability of the muscle-derived stem cells to proliferate in vivo for an extended period of time combined with their strong capacity for self-renewal, multipotent differentiation, and immune-privileged behavior contributes to the results observed with muscle-derived stem cell transplantation. Our results suggest that muscle-derived stem cells may significantly improve the efficacy of muscle cell–mediated therapies. Indeed, we demonstrated that muscle-derived stem cells retrovirally transduced to express the osteogenic factors bone morphogenetic protein-2 (BMP-2) and BMP-4 are capable of differentiating into the osteogenic lineage and accelerating the healing of skull defects in mice. The interaction between angiogenic and osteogenic factors in bone formation and bone healing was investigated with the use of ex vivo gene therapy approaches based on muscle-derived stem cells genetically engineered to express human BMP-4, vascular endothelial growth factor (VEGF), or VEGF-specific antagonist (soluble Flt1). Although VEGF alone did not improve bone regeneration, it acted synergistically with BMP-4 to increase recruitment of mesenchymal stem cells, enhance cell survival, and augment cartilage formation during the early stages of endochondral bone formation. These early effects, coupled with accelerated cartilage resorption, eventually led to a significant enhancement of bone formation and healing. The beneficial effect of VEGF on bone healing elicited by BMP-4 depends on the ratio of VEGF to BMP-4, with an improper ratio leading to detrimental effects on bone healing. Finally, we demonstrated that soluble Flt1 inhibits bone formation elicited by BMP-4. Our studies have important implications for the identification of new strategies to improve bone healing with the use of osteogenic and angiogenic factors in combination with muscle-derived stem cells-based gene therapy.

Introduction and Background

The healing of fractures continues to pose significant challenges in the field of orthopaedic surgery. Although the majority of fractures heal well with treatment, complications associated with delayed union or nonunion of fractures can be devastating. Among the 5.6 million fractures that occur annually in the United States, 5% to 10% display delayed or impaired healing.[1] Orthopaedic surgeons may treat complex fractures, including fractures resulting in segmental bone loss, with bone autograft, vascularized bone graft, allograft supplemented with osteogenic proteins, bone transport, or amputation.[2,3] Unfortunately patients frequently endure a lengthy recovery period characterized by numerous medical procedures, donor site morbidity, and, often, an unsatisfactory outcome.[4,5] Thus there is intense interest in the development of new mechanical and biologic approaches to improve the treatment of fractures.

Research has recently focused on the role of biologic factors in fracture healing, particularly during the osteoconductive and osteoinductive phases of healing. To facilitate bone conduction, a graft must serve as a scaffold for the ingrowth of native bone, which incrementally replaces the graft structure. Osteoinduction occurs when bioactive proteins, such as the family of BMPs, recruit host osteogenitor cells into the osseous defect through induction of pluripotent stem cells and proliferation of already committed pre-osteoblasts. The relative efficacy of current treatment techniques, such as autogenous bone grafting, is already predicated on these principles. Further refinement of current treatment modalities will likely occur through improvement of the ability to modulate these biologic principles.

In 1965, Urist[6] demonstrated the ability of cancellous bone to induce ectopic bone formation, a seminal finding in orthopaedics. Many of the proteins responsible for inducing bone formation, including BMP-2, have since been characterized, cloned, and made available as recombinant human proteins.[7] The use of these recombinant proteins to improve bone healing is already challenging the gold standard, autologous bone grafting, based on results generated by numerous animal studies[8-14] and limited human trials.[15-17] The potential role of BMPs in spine fusion,[18,19] the healing of osteochondral defects,[13] implant stabilization,[13] and the treatment of osteoporosis[20] continues to be evaluated.

Both the findings of Urist[6] and the standard in vivo assay used to assess bioactivity of the various BMP preparations are based on the process of ectopic ossification in rodent skeletal muscle. More recently, numerous studies have demonstrated orthotopic bone formation. In 1986, the ability of purified bovine BMP delivered in gelatin capsules to enhance fracture healing and bone defect filling was confirmed in a canine ulnar model[10] and in a sheep trephine skull defect model.[12] The use of rhBMP to fill segmental defects has since been studied in canine,[21] rat femur,[14] and rabbit forearm[8,13,22] models. The utilization of BMP in clinical trials has provided results comparable to those in animal models. Human BMP has been delivered in the clinical setting using polylactic-polyglycolic strips or gelatin, either alone or with allogenic bone grafts, for the treatment of defects of the tibia,[17] femoral nonunions,[15] and spinal fusion.[16] A dose-dependent response to BMP has been observed, with higher doses resulting in more rapid bone formation.[8,14]

Like BMP-2, BMP-4 can induce bone and cartilage formation when implanted into extraskeletal sites.[7,9,23-27] These BMPs are thought to be functionally interchangeable.[28,29] Knockout mice for BMP-2 and BMP-4 die between 6.5 and 10.5 days postcoitum, before the initiation of chondrogenesis/osteogenesis.[30,31] As endochondral ossification proceeds, BMP-2 and BMP-4 are expressed at high levels in primitive mesenchymal and chondrocytic cells, indicating that these osteogenic proteins are important regulators of cell differentiation during fracture repair.[32] More relevant to our topic, these osteogenic proteins are expressed in cultures of fetal rat calvarial osteoblasts before these cells form mineralized bone nodules and as they express alkaline phosphatase (AP), osteocalcin, and osteopontin.[33] Although BMP-2 and BMP-4 seem to play very similar roles during bone fracture repair, they display some differential activity in the injured site. Unlike BMP-4, BMP-2 plays a role in the chemotactic recruitment of undifferentiated osteoblasts during bone remodeling and bone healing.[34] In addition, during the early phases of fracture repair, BMP-4 is expressed by osteoprogenitor cells residing in rather specific sites, such as the cambium layer of the periosteum, the marrow cavity, and skeletal muscles proximate to the fracture site.[35] The BMP-4 gene is expressed by less differentiated osteoprogenitor cells rather than by mature osteoblasts. The expression of BMP-4 is enhanced in more severe fractures (ie, the greater the physical impact, the greater the expression of BMP-4) and is localized in callus-forming tissue before callus formation.[35] The overexpression of BMP-4 has been observed in fibroproliferative lesions (skeletal muscle) of patients with fibrodysplasia ossificans progressiva,[36,37] a heritable disorder of connective tissue characterized by postnatal formation of ectopic bone, primarily in skeletal muscle.[36-38] These findings suggest that: (1) BMP-4 plays a major role during the early phases of bone healing during which the osteoprogenitor cells are activated, and (2) this particular osteogenic protein (BMP-4) very efficiently induces ectopic bone formation within skeletal muscle.

In addition to osteogenic factors, angiogenesis is believed to affect bone formation, fracture healing, and endochondral ossification in the growth plate.[39,40] Indeed, the use of antiangiogenic factors such as TNP-470 (a synthetic analog of fumagillin, an antiangiogenic agent that blocks neovascular formation) inhibits ectopic bone formation mediated by BMP-2.[39] Although the exact phase of bone formation that requires angiogenesis remains unclear, these findings suggest that combining an angiogenic factor with osteogenic proteins may further improve bone healing. VEGF is the most important essential mediator of angiogenesis.[41,42] Although various types of cells secrete VEGF, osteoblasts and chondrocytes are the cells that produce this cytokine in the skeleton.[43,44] In addition to stimulating angiogenesis, VEGF is involved in early hematopoietic development and chemotaxis of monocytes.[45,46]

The ability to deliver recombinant osteogenic and angiogenic proteins by injection is hindered by the short half-lives and rapid clearance of these products. Therefore, a carrier matrix is often required to achieve satisfactory results. Gene therapy based on ex vivo gene transfer of these proteins represents an alternative means to promote persistent delivery of osseous proteins. Osteoblasts and bone marrow stromal cells are logical choices to serve as gene delivery vehicles for ex vivo gene transfer to improve bone healing, but the requirement for harvesting, isolation, and cultivation of these cells may limit their clinical utility. The use of muscle cells for delivery of osteogenic proteins offers some advantages over methods based on osteoblasts or bone marrow stromal cells. Muscle tissue contains inducible osteoprogenitor cells. Muscle cells can be harvested using noninvasive techniques, are relatively easy to cultivate, and may display high levels of expression of active osteogenic proteins.

The major goal of our research is to investigate the use of muscle-derived cells as delivery vehicles for osteogenic and angiogenic proteins and as a source of osteoprogenitor cells to improve bone healing. The remainder of this chapter summarizes our accomplishments, which justify the use of muscle-derived cells in gene therapy and tissue engineering applications to improve bone healing.

Isolation of a Population of Muscle-Derived Stem Cells from Adult Skeletal Muscle

We used the preplate technique to isolate various populations of myogenic cells from normal mouse skeletal muscle based on their adhesion characteristics, proliferation behavior, and myogenic and stem cell marker expression profiles.[47] Although many of these cell populations display characteristics of satellite cells, a unique population of long-term proliferating (LTP) cells expressing hematopoietic stem cell markers was also isolated. These LTP cells retain their phenotype (Sca-1+, CD34 low/-, cKIT-, and CD45-) for more than 30 passages with apparently normal karyotype, and can differentiate into muscle, neural, endothelial, and osteogenic lineages both in vitro and in vivo. The transplantation of LTP cells, in contrast to that of other myogenic cell populations, significantly improves the efficiency of muscle regeneration and delivery of dystrophin in dystrophic muscle. The ability of these cells to proliferate in vivo for an extended period of time—combined with their strong capacity for self-renewal, their multipotency, and their immune-privileged behavior—reveals, at least in part, a basis for the benefits observed following their transplantation into skeletal muscle.[47] These findings suggest that LTP cells represent a population of muscle stem cells that will significantly improve muscle cell–mediated therapies.[47]

The Use of Muscle-Derived Stem Cells to Improve Bone Healing

We demonstrated that a population of muscle-derived stem cells isolated from mouse skeletal muscle produced AP in response to rhBMP-2 and rhBMP-4 in a dose-dependent manner.[48-51] These mouse muscle-derived

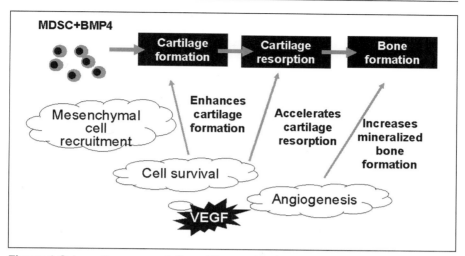

Figure 1 Schematic representation of the mechanism by which VEGF accelerates bone formation and healing mediated by BMP-4–expressing muscle-derived stem cells.

cells induced and participated in ectopic bone formation after being genetically engineered to express the osteogenic proteins BMP-2 and BMP-4 and injected into skeletal muscle. More importantly, these muscle-derived stem cells were able to differentiate toward the osteogenic lineage and consequently improved bone healing in calvarial defects in immunodeficient and immunocompetent mice.[48-51]

Although it is established that angiogenesis is important for normal bone development, it is not known whether bone regeneration can be improved by the use of angiogenic factors, either alone or in combination with BMPs, which are the most promising growth factors for inducing bone formation. To address this question, we transduced muscle-derived stem cells to express either human BMP-4 or VEGF. By enhancing angiogenesis, VEGF significantly improved the efficacy of bone formation and regeneration elicited by BMP-4. However, VEGF alone did not improve bone regeneration. We further characterized the mechanism by which VEGF accelerates bone formation mediated by genetically engineered muscled-derived stem cells expressing BMP-4 (Figure 1). Compared with BMP-4 alone, the combination of BMP-4 and VEGF resulted in increased mesenchymal cell infiltration and cartilage formation, decreased cell apoptosis at the injured site, earlier cartilage resorption, and a subsequent increase in mineralized bone formation within regenerated bone.[52] Thus VEGF enhances BMP-4–induced endochondral bone formation by affecting steps before and after cartilage formation. The beneficial effect of VEGF on bone healing elicited by BMP-4 appears to require proper dosing of VEGF, as high doses of VEGF lead to detrimental effects on bone healing.[52] The synergistic effects between osteogenic and angiogenic factors delivered via muscle-derived stem cells offer a new method to improve bone repair.

We are currently focusing on the isolation of a population of muscle-derived stem cells from human skeletal muscle to test the use of gene therapy and tissue engineering to support bone regeneration and repair in the clinical setting. We explored the use of primary human muscle-derived cells isolated from adult subjects (approximately age 50 years) using the preplate technique. Certain cells isolated in a primary cell culture taken from human skeletal muscle are responsive to BMP-2, suggesting that they are osteocompetent. These cells can also be transduced via ex vivo gene transfer to secrete BMP-2 at levels sufficient to induce ectopic bone formation within skeletal muscle that is detectable on radiographs. Histologic assessment of the injected, transduced, human muscle-derived cells revealed that the cells may respond to the secreted BMP-2 in an autocrine manner by becoming osteoblasts, thereby contributing to the induction of bone formation.[53]

In addition, we genetically engineered freshly isolated human skeletal muscle cells with adenovirus and retrovirus constructs to express human BMP-2. These genetically engineered cells were then implanted into non-healing bone defects (skull defects) in severe combined immunodeficient mice. The closure of the defect was monitored by gross examination and by histologic methods. Mice that received BMP-2–producing human muscle-derived cells displayed full closure of the defect by 4 to 8 weeks posttransplantation.[53] Remodeling of the newly formed bone was evident on histologic examination during this 4- to 8-week period.[54] When analyzed by fluorescent in situ hybridization, a small fraction of the injected human muscle-derived cells was found within the newly formed bone where osteocytes normally reside.[54] These results indicate that genetically engineered human muscle-derived cells enhance bone healing primarily by delivering BMP-2. In addition, a small fraction of the cells appears to differentiate into osteogenic cells.

Summary

Using a modified preplate technique, we isolated a population of muscle-derived stem cells that maintain long-term proliferative capability and the ability to express hematopoietic stem cell markers. These cells may be useful as vehicles to deliver osteogenic and angiogenic proteins, and as a source of osteoprogenitor cells to improve bone healing. Our work focused on osteogenic proteins, such as BMPs, and the essential mediator of angiogenesis, VEGF. By transducing MDSC to express either human BMP-4 or VEGF, we showed that VEGF significantly improved the efficacy of BMP-4–elicited bone formation and regeneration by enhancing angiogenesis. However, the proper ratio of VEGF to BMP-4 is of critical importance, as an improper ratio leads to detrimental efects on bone healing. These results have important implications for the development of new strategies to improve bone healing through the use of osteogenic and angiogenic factors in combination with muscle-derived stem cell–based gene therapy.

Acknowledgments

We wish to thank Ryan Sauder for excellent editorial assistance and Dr. Zhuqing Qu-Petersen for helpful suggestions. This work was supported in part by grants to Johnny Huard (1 R01 DE13420-01; 1 R01 AR49684-02; 1 P01 AR45925-04) and by a grant to Hairong Peng (1 R03 AR050201-01) from the National Institutes of Health, as well as by grants to Drs. Peng and Huard from the Pittsburgh Tissue Engineering Initiative and the Albert B. Ferguson, Jr, MD Orthopaedic Fund of The Pittsburgh Foundation. The Growth and Development Laboratory also receives financial support from the William F. and Jean W. Donaldson Chair at the Children's Hospital of Pittsburgh and the Henry J. Mankin Endowed Chair for Orthopaedic Research at the University of Pittsburgh.

References

1. Einhorn TA: Enhancement of fracture-healing. *J Bone Joint Surg Am* 1995;77:940-956.

2. Feibel RJ, Oliva A, Jackson RL, Louie K, Buncke HJ: Simultaneous free-tissue transfer and Ilizarov distraction osteosynthesis in lower extremity salvage: Case report and review of the literature. *J Trauma* 1994;37:322-327.

3. Prokuski LJ, Marsh JL: Segmental bone deficiency after acute trauma: The role of bone transport. *J Orthop Trauma* 1994;25:753-763.

4. Banwart JC, Asher MA, Hassanein RS: Iliac crest bone graft harvest donor site morbidity: A statistical evaluation. *Spine* 1995;20:1055-1060.

5. Summers BN, Eisenstein SM: Donor site pain from the ilium: A complication of lumbar spine fusion. *J Bone Joint Surg Br* 1989;71:677-680.

6. Urist MR: Bone: Formation by autoinduction. *Science* 1965;150:895-899.

7. Wozney JM, Rosen V, Celeste AJ, et al: Novel regulators of bone formation: molecular clones and activities. *Science* 1988;242:1528-1534.

8. Bostrom M, Lane JM, Tomin E, et al: Use of bone morphogenetic protein-2 in the rabbit ulnar nonunion model. *Clin Orthop* 1996;327:272-282.

9. Wang EA, Rosen V, D'Assandro JS, et al: Recombinant human bone morphogenetic protein induces bone formation. *Proc Natl Acad Sci USA* 1990;87:2220-2224.

10. Nilsson OS, Urist MR, Dawson EG, et al: Bone repair induced by bone morphogenetic protein in ulnar defects in dogs. *J Bone Joint Surg Am* 1986;68:635-642.

11. Heckman JD, Sarasohn-Kahn J: The economics of treating tibia fractures: The cost of delayed unions. *Bull Hosp Joint Dis* 1997;56:63-72.

12. Lindholm TC, Lindholm TS, Alitalo I, et al: Bovine bone morphogenetic protein (bBMP) induced repair of skull trephine defects in sheep. *Clin Orthop* 1988;227:265-268.

13. Cook SD, Baffes GC, Wolfe MW, et al: The effects of recombinant human osteogenic protein-1 on healing of large segmental bone defects. *J Bone Joint Surg Am* 1994;76:827-838.

14. Yasko AW, Lane JM, Fellinger EJ, et al: The healing of segmental bone defects, induced by recombinant human bone morphogenetic protein (rhBMP-2). *J Bone Joint Surg Am* 1992;74:659-670.

15. Johnson EE, Urist MR, Finerman GAM: Bone morphogenetic protein augmentation grafting of resistant femoral nonunions. *Clin Orthop* 1988;230:257-265.

16. Boden SD, Kang J, Sandhu H, et al: Use of recombinant human bone morphogenetic protein-2 to achieve posterolateral lumbar spine fusion in humans: A prospective, randomized clinical pilot trial: 2002 Volvo award in clinical studies. *Spine* 2002;27:2662-2673.

17. Johnson CEE, Urist MR, Finerman GAM: Repair of segmental defects of the tibia with cancellous bone grafts augmented with human bone morphogenetic protein: A preliminary report. *Clin Orthop* 1988;236:249-257.

18. Sandhu HS, Kanim LE: Kabo JM, Toth JM, Zeegen EN, Liu D, Delamarter RB, Dawson EG: Effective doses of recombinant human bone morphogenetic protein-2 in experimental spinal fusion. *Spine* 1996;21:2115-2122.

19. Boden SD, Moskovitz PA, Morone MA, et al: Video-assisted lateral intertransverse process arthrodesis: Validation of a new minimally invasive lumbar spinal fusion technique in the rabbit and non-human primate (rhesus) models. *Spine* 1996;21:2689-2697.

20. Mundy GR: Regulation of bone formation by morphogenetic proteins and other growth factors. *Clin Orthop* 1996;323:24-28.

21. Heckman JD, Boyan BD, Aufdemorte TB, et al: The use of bone morphogenetic protein in the treatment of non-union in a canine model. *J Bone Joint Surg Am* 1991;73:750-764.

22. Zellin G, Linde A: Treatment of segmental defects in long bone using osteopromotive membranes and recombinant human bone morphogenetic protein-2. *Scand J Plast Reconstr Hand Surg* 1997;31:97-104.

23. Hammonds RG Jr, Schwall R, Dudley A, et al: Bone-inducing activity of mature BMP-2b produced from a hybrid BMP-2a/2b precursor. *Mol Endocrinol* 1991;5:149-155.

24. Sampath TK, Maliakai JC, Hauschka PV, et al: Recombinant human osteogenic protein-1 (hOP-1) induces new bone formation in vivo with a specific activity comparable with natural bovine osteogenic protein and stimulates osteoblast proliferation and differentiation in vitro. *J Biol Chem* 1992;267:20352-20362.

25. Sato M, Ochi T, Nakase T, et al: Mechanical tension-stress induces expression of bone morphogenetic protein (BMP-2) and BMP-4, but not BMP-6, BMP-7, and GDF-f mRNA, during distraction osteogenesis. *J Bone Miner Res* 1999;14:1084-1095.

26. Fang J, Zhu YY, Smiley E, et al: Stimulation of new bone formation by direct transfer of osteogenic plasmid genes. *Proc Natl Acad Sci USA* 1996;93:5753-5758.

27. Leong LM, Brickell PM: Molecules in focus: Bone morphogenetic protein-4. *Int J Biochem Cell Biol* 1996;28:1293-1296.

28. Sampath TK, Rashka KE, Doctor JS, et al: Drosophila transforming growth factor beta superfamily proteins induce endochondral bone formation in mammals. *Proc Natl Acad Sci USA* 1993;90:6004-6008.

29. Vainio S, Karavanova I, Jowett A, et al: Identification of BMP-4 as a signal mediating secondary induction between epithelial and mesenchymal tissues during early tooth development. *Cell* 1993;75:45-58.

30. Winnier G, Blessing M, Labosky PA, et al: Bone morphogenetic protein-4 is required for mesoderm formation and patterning in the mouse. *Genes Dev* 1995;9:2105-2116.

31. Zhang H, Bradley A: Mice deficient for BMP2 are nonviable and have defects in amnion/chorion and cardiac development. *Development* 1996;122:2977-2986.

32. Bostrom MPG, Lane JMK, Berberian WS, et al: Immunolocalization and expression of bone morphogenetic proteins 2 and 4 in fracture healing. *J Orthop Res* 1995;13:357-367.

33. Harris SE, Sabatini M, Harris MA, et al: Expression of bone morphogenetic protein messenger RNA in prolonged cultures of fetal rat calvarial cells. *J Bone Miner Res* 1994;9:389-394.

34. Lind M, Eriksen EF, Bünger C: Bone morphogenetic protein-2 but not bone morphogenetic protein-4 and -6 stimulates chemotactic migration of human osteoblasts, human marrow osteoblasts, and U2-OS cells. *Bone* 1996;18:53-57.

35. Nakase T, Nomura S, Yoshikawa H, et al: Transient and localized expression of bone morphogenetic protein 4 messenger RNA during fracture healing. *J Bone Miner Res* 1994;9:651-659.

36. Shafritz AB, Shore EM, Gannon FH, et al: Overexpression of an osteogenic morphogen in fibrodysplasia ossificans progressiva. *N Engl J Med* 1996;335:555-561.

37. Lanchoney TF, Olmstead EA, Shore EM, et al: Characterization of bone morphogenetic protein 4 receptor in fibrodysplasia ossificans progressiva. *Clin Orthop* 1998;346:38-45.

38. Gannon FH, Valentine BA, Shore EM, et al: Acute lymphocytic infiltration in an extremely early lesion of fibrodysplasia ossificans progressiva. *Clin Orthop* 1998;346:19-25.

39. Mori S, Yoshikawa H, Hashimoto J, et al: Antiangiogenic agent (TNP-470) inhibition of ectopic bone formation induced by bone morphogenetic protein-2. *Bone* 1998;22:99-105.

40. Ferguson C, Alpern E, Miclau T, et al: Does adult fracture recapitulate embryonic skeletal formation? *Mech Dev* 1999;87:57-66.

41. Nakagawa M, Kaneda T, Arakawa T, et al: Vascular endothelial growth factor (VEGF) directly enhances osteoclastic bone resorption and survival of mature osteoclasts. *FEBS Lett* 2000;473:161-164.

42. Leung DW, Cachianes G, Kuang WJ, et al: Vascular endothelial growth factor is a secreted angiogenic mitogen. *Science* 1989;246:1306-1309.

43. Goad DL, Rubin J, Wang H, et al: Enhanced expression of vascular endothelial growth factor in human SaOS-2 osteoblast-like cells and murine osteoblasts induced by insulin-like growth factor I. *Endocrinology* 1996;137:2262-2268.

44. Gerber HP, Vu TH, Ryan AM, et al: VEGF couples hypertrophic cartilage remodeling, ossification and angiogenesis during endochondral bone formation. *Nat Med* 1999;5:623-628.

45. Hidaka M, Stanford WL, Bernstein A: Conditional requirement for the Flk-1 receptor in the in vitro generation of early hematopoietic cells. *Proc Natl Acad Sci USA* 1999;96:7370-7375.

46. Mitola S, Sozzani S, Luini W, Primo L, Borsatti A, Welch H, Bussolino F: Tat-human immunodeficiency virus-1 induces human monocyte chemotaxis by activation of vascular endothelial growth factor receptor-1. *Blood* 1997;90:1365-1372.

47. Qu-Petersen Z, Deasy B, Jankowski R, et al: Identification of a novel population of muscle stem cells in mice: potential for muscle regeneration. *J Cell Biol* 2002;157:851-864.

48. Bosch P, Musgrave DS, Lee JY, et al: Osteoprogenitor cells within skeletal muscle. *J Orthop Res* 2002;18:933-944.

49. Wright V, Peng H, Usas A, Young B, Gearhart B, Cummins J, Huard J: BMP4-expressing muscle-derived stem cells differentiate into osteogenic lineage and improve bone healing in immunocompetent mice. *Mol Ther* 2002;6:169-178.

50. Lee JY, Musgrave DS, Pelinkovich D, et al: Effect of bone morphogenetic protein-2 expressing muscle derived cells on healing of critical sized bone defects in mice. *J Bone Joint Surg Am* 2001;83:1032-1039.

51. Peng H, Wright V, Usas A, Gearhart B, Shen HC, Cummins J, Huard J: Synergistic enhancement of bone formation and healing by stem cell-expressed VEGF and bone morphogenetic protein-4. *J Clin Invest* 2002;110:751-759.

52. Musgrave DS, Pruchnic R, Bosch P, et al: Human skeletal muscle cells in ex vivo gene therapy to deliver bone morphogenetic protein-2. *J Bone Joint Surg Br* 2002;84:120-127.

53. Lee JY, Peng H, Usas A, et al: Enhancement of bone healing based on ex vivo gene therapy using muscle derived cells expressing bone morphogenetic protein 2. *Hum Gene Ther* 2002;13:1201-1211.

Chapter 14

Clinical Trials in Bone Tissue Engineering: Spine Applications Update

Scott D. Boden, MD

Abstract

A variety of tissue engineering products to biologically control bone induction are currently under development. Many of the early clinical trials of bone tissue engineering involved various types of spinal fusions as these represent approximately 60% of all current bone grafting procedures. Bone tissue engineering is particularly important for spinal fusion procedures because even the most successful graft material, autogenous iliac crest bone, is associated with an unacceptably high failure rate and donor site morbidity. At present, there are two examples of bone tissue engineered morphogenetic proteins (TEMPs) for spinal fusion that have consistently achieved success in preclinical and clinical studies, and one example of a TEMP that has been less consistently successful. Healing of spinal fusion is one of the most difficult challenges for a bone TEMP and thus serves as an excellent proving ground for this technique. This review provides an update of results from clinical trials of bone tissue engineering for spinal fusion procedures.

Introduction and Background

Most of the first-generation bone tissue engineering products for spinal fusion will likely include one or more members of the family of osteoinductive proteins called bone morphogenetic proteins (BMPs).[1,2] It has been more than 35 years since the concept of a BMP was described by Urist and associates[1,3-5] and more than 15 years since the first BMPs were isolated, cloned, and sequenced.[6-9] Early studies with recombinant BMPs and purified BMP extracts were successful in a variety of rodent models, but initial clinical trials resulted in disappointing outcomes.[10-12] Recombinant BMP-7 has demonstrated variable success in rodents,[13-16] but recombinant BMP-2 has been more consistently successful in animal models.[17-20] Ne-Osteo (Sulzer Biologics, Austin, TX), a bovine-derived mixture of BMPs, has also demonstrated consistent bone induction in a rodent model of spinal fusion.[21-25]

In retrospect, early clinical failures were likely the result of suboptimal carrier matrices to deliver the BMP and a failure to recognize the requirement for substantially increased doses of BMP to induce bone formation in nonhuman primates as compared with rodents.[21-26] To date, only two clinical pilot studies that show the ability of a BMP to consistently induce bone in the anterior and posterolateral lumbar spine have been published.[27,28] Thus, the update on clinical trials will be largely based on information presented at clinical meetings and provided by investigators and companies rather than the articles published in peer-reviewed journals.

Studies of Ne-Osteo, a Bovine-derived Extract With Multiple BMPs

The first bone tissue engineering product to be studied in clinical trials was Ne-Osteo, a bovine-derived bone extract containing multiple BMPs delivered on a scaffold of collagen.[29] The pilot trial initiated in 1996 was designed to determine the dose and carrier required for consistent bone induction in humans as assessed by radiographs. This pilot study was divided into three phases over 4 years and included 22 patients with lumbar spinal stenosis and/or spondylolisthesis requiring spine arthrodesis. To minimize the risk of nonunion, patients received autogenous iliac bone graft on one side and Ne-Osteo growth factor on the other. The doses used were 12.5 mg, 25 or 50 mg, or 25 mg Ne-Osteo per side in each of the three phases, respectively. In the first phase, two of six patients showed bone induction by radiographs and/or CT scans read in a blinded manner on the Ne-Osteo side (12.5 mg dose). In the second phase, both sides were graded as fused in five of six patients. Although graded as fused, CT scans obtained at 6 months demonstrated a ring of new bone, with the center filling in more slowly (12 to 24 months) than predicted by studies of nonhuman primate (4 to 6 months).

Based on these results, the carrier for the third phase of the study was designed to have a more porous or open early fusion mass than the dense demineralized bone matrix paste. This was accomplished by mixing local bone or cancellous allograft chips with the demineralized bone matrix paste. Results using the 25-mg and 50-mg doses were the same, so the 25-mg dosage was used in the third phase. In this last phase, nine of ten autografts were fused by 12 months. Five of five patients treated with Ne-Osteo plus local bone and four of five patients with allograft chips were fused by 6 months. The one patient in the group who did not heal on either the autograft or Ne-Osteo side was a smoker.

These pilot studies showed that the use of Ne-Osteo achieved a continuous spinal fusion mass in 15 of 16 patients at a dose of at least 25 mg/side without iliac crest bone graft. This result was at least as good as those for autografts (94%) in this side-by-side model. The next step would be to confirm these results on both sides of the spine in pivotal trials with Ne-Osteo. The design of these trials would eliminate the need to harvest bone graft from the iliac crest. However, the initiation of additional trials is uncertain due to financial issues of the sponsor company. In any case, this work serves as a proof of concept that an extracted mixture of bone-derived proteins can serve as the bioactive factor in a bone tissue engineering device to promote spinal fusion.

Studies of Recombinant Human Osteogenic Protein-1 (rhOP-1)

There are two recombinant BMPs currently being evaluated as components of bone tissue engineering strategies for spinal fusion. The first, rhOP-1, is marketed by Stryker Biotech Inc. (Hopkinton, MA). In the commercial form, 3.5 mg of lyophilized rhOP-1 is combined with a carrier consisting of 1 g of type 1 bovine bone collagen for a final concentration of rhOP-1 of 0.875 mg/mL. The dry powdered mixture is reconstituted by the addition of saline to form a paste just before implantation. The addition of 230 mg of carboxymethylcellulose (CMC) to the OP-1 device forms a putty that is currently being used in clinical trials of spinal fusion.[30]

A study of the safety and effectiveness in humans compared autograft alone to autograft augmented with OP-1 for posterolateral spinal arthrodesis.[30] Sixteen patients with degenerative lumbar spondylolisthesis and spinal stenosis were randomized to uninstrumented posterolateral fusion with autograft plus OP-1 (3.5 mg rhOP-1/1 g bovine bone collagen/230 mg CMC) or autograft alone. Clinical outcome was assessed using the Oswestry score, and fusion status was graded by two radiologists reading dynamic and static radiographs in a blinded manner. At 6 months follow-up, nine of 12 patients (75%) in the group who received the autograft plus OP-1 were graded as fused. Of four patients in the autograft alone group, 2 (50%) achieved fusion ($P = 0.547$). Clinical success, defined by at least a 20% improvement in the Oswestry score, was achieved in 83% of the autograft plus OP-1 group but only 50% of the group who received autografts alone ($P = 0.245$). No adverse effects were observed in the patients treated with OP-1.

A similar study of patients with degenerative spondylolisthesis is being conducted in Australia. Following decompression, patients were treated with an uninstrumented posterolateral arthrodesis with autograft on one side and OP-1 putty (3.5 mg rhOP-1/1 g bovine bone collagen/230 mg CMC) on the contralateral side. Preliminary results from 6 months of study have been reported by Vaccaro and associates.[30] Using CT to measure bone formation between the transverse processes, the fusion masses of five patients have been evaluated. In all patients, bone formation was noted to be equal or greater on the OP-1 side compared with the autograft side.

An additional pilot trial of OP-1 for posterolateral spinal fusion is ongoing. Patients with lumbar spinal stenosis and degenerative spondylolisthesis (L3 through L5 segments) are treated with decompression, followed by uninstrumented posterolateral fusion using either iliac crest autograft alone or OP-1 putty (3.5 mg rhOP-1/1 g bovine bone collagen/230 mg CMC) alone. Clinical outcome is assessed by the Oswestry scale, and fusion status is graded by radiologists reading radiographs in a blinded manner. To be considered a clinical success, patients must achieve at least a 20% improvement in their Oswestry score and demonstrate a solid arthrodesis on radiographs by demonstration of bony bridging and no motion on flexion/extension radiographs. The results from 36 enrolled patients evaluated for 6 months have been reported.[30] The overall clinical success rate of patients treated with OP-1 was 32% greater than in patients treated with autografts. Although the difference in clinical success between the groups treated with autografts or OP-1 was not statistically significant in the preliminary analysis, enrollment of additional patients may show that OP-1 treatment provides a statistically significant benefit compared with autografts. To date, no adverse events attributed to OP-1 treatment have been reported.

Studies of recombinant human (rh)BMP-2

The other bone TEMP being evaluated for spinal fusion uses rhBMP-2 (Medtronic Sofamor Danek, Minneapolis MN) as the bioactive factor. Consistent success in rodents and in nonhuman primates has been documented.[17,20] Early work with rhBMP-2 demonstrated that the carrier had a profound effect on the effectiveness of the active molecule for specific applications. It was determined that the carrier or scaffold component serves three very important functions: (1) providing a three-dimensional space occupier in which de novo bone formation can occur, (2) maintaining a critical threshold concentration of BMP at the site, and (3) binding to the BMP to prevent heterotopic bone formation.

To date, more than 480 patients have received rhBMP-2 in clinical trials of spinal fusion.[31] In addition, more than 280 patients have been included in these studies as part of control groups. The composite device consisting of rhBMP-2 carried by the absorbable collagen sponge is known as InFUSE Bone Graft (Medtronic Sofamor Danek, Minneapolis, MN).

Pilot Trial of rhBMP-2 in Patients With Symptomatic Degenerative Disk Disease

In January 1997, an Investigational Device Exemption (IDE) was initiated following US Food & Drug Administration approval to study the use of rhBMP-2 in patients with symptomatic degenerative disk disease. The rhBMP-2 was combined with an absorbable collagen sponge as the carrier. This composite was then placed into a tapered lordotic titanium interbody fusion device (LT Cage). Four investigational centers received regulatory approval to enroll 14 patients in a prospective nonblinded randomized controlled trial. Eleven of these patients were implanted with rhBMP-2 and three were enrolled in the control cohort. Patients in the control group were implanted with a lordotic tapered cage filled with autogenous bone graft from iliac crest. Based on the results of the nonhuman primate study in which the rhBMP-2/collagen sponge composite was also implanted in a tapered titanium cage for anterior lumbar interbody fusion (ALIF), a dose of 1.5 mg/mL of rhBMP-2 on same collagen sponge was used. All 11 patients implanted with rhBMP-2 went on to successful fusion as determined by thin-cut CT scans read in a blinded manner by radiologists. Two patients (67%) of three in the control group also achieved successful fusion.[27]

Pivotal Trial for ALIF

The pilot clinical study of InFUSE placed within a lordotic tapered cage for ALIF provided safety and efficacy data that were used to support final regulatory approval of a pivotal trial that began enrolling patients in 1998.[31] This pivotal clinical study of rhBMP-2 for ALIF was a prospective, multi-center, open-label, randomized trial. A total of 145 patients were enrolled in the investigational treatment group and 136 patients were enrolled in the control group. Sixteen investigational sites were involved in the study. All patients in the study underwent open retroperitoneal approach to the spinal column. The experimental group received the lordotic tapered cage with InFUSE for single level ALIF. The control group received the same cage filled with autogenous bone graft, which was harvested from the iliac crest. More than 90% of the patients completed the 2-year follow-up. The surgical time and blood loss were significantly less in the experimental (InFUSE) group than in the control group ($P < 0.05$). Of patients who underwent harvest of iliac crest bone, 31% experienced some degree of donor site pain 2 years after the procedure. The two groups did not differ with regard to Oswestry scores at each of the follow-up time points.

The criteria for fusion in this trial were based on analysis of plain radiographs and thin-cut CT scans. Successful fusion required the presence of bridging trabecular bone, less than or equal to 3 mm of sagittal translation, and less than 5° of angulation. Furthermore, the presence of lucent lines around more than 50% of either one of the two metallic implants was considered an indication of pseudarthrosis. Radiographic fusion success was achieved in more than 99% of patients in both groups. Bayesian statistical analysis of these data showed that the posterior probability of equivalence of the experimental treatment to the control treatment was 99.9%.

Concurrent with this study, a laparoscopic surgical approach study cohort was added under the IDE. A total of 136 patients from 14 investigational sites were involved in the laparoscopic study. Results from the laparoscopic arm were similar to those of the experimental group of rhBMP-2 for ALIF with regard to fusion success and clinical outcome. The mean surgical time was 2.0 hours and the average hospital stay was 1.3 days. This is in contrast to a hospital stay of 3.0 days previously reported in a clinical study of laparoscopic ALIFs with LT Cages filled with autograft. If success was defined as at least a 15-point improvement, 81.8% of the patients implanted with InFUSE/LT Cage through a laparoscopic approach had a successful pain and disability outcome. A similar number of patients in both the experimental and historical control groups returned to work. However, a significantly greater percentage of patients in the laparoscopic experimental group returned to work compared with the control group in the open-label trial ($P < 0.05$).

Clinical Trial of Single-Level Anterior Diskectomy and Fusion

A nonhuman primate study demonstrated the effectiveness of the rhBMP-2/absorbable collagen sponge within a machined allograft dowel.[31] Based on this study, a prospective, multicenter, randomized trial was initiated in 1998. A total of 24 and 23 patients were enrolled in the experimental treatment and control groups, respectively. All patients underwent a single-level anterior diskectomy and fusion through an open approach. The blood loss was significantly less among patients implanted with InFUSE compared with those undergoing iliac crest bone harvest ($P < 0.05$). Furthermore, the incidence of persistent donor site pain 2 years following surgery was 38.9% in the autograft group. Patients implanted with InFUSE had significantly greater improvement in Oswestry scores at 3 months ($P = 0.03$) and 6 months ($P = 0.04$) after surgery, but the difference was not significant after 12 months ($P = 0.17$) compared with patients in the control group. Eighty-three percent of patients implanted with InFUSE had more than a 15-point improvement in Oswestry score by the reexamination at month 12. Ninety percent of patients in the experimental group achieved fusion by 6 months, and all patients in this group achieved fusion by 1 year following surgery. In contrast, 65% of patients in the control group achieved fusion by 6 months and 90% of these patients achieved fusion after 1 year.

Clinical Trial of Posterior Lumbar Laminectomy

There has been one prospective, controlled, and randomized trial of intervertebral fusion using a posterior lumbar laminectomy approach.[31] The fixation devices were cylindrical, hollow, and threaded cages designed to house graft material and to fix the intervertebral segment through a laminectomy approach. The cages were placed as "stand-alone" devices and were done only for single level fusion. Seventy-one patients were enrolled at 14 sites. Thirty-six patients underwent posterior lumbar interbody infusion (PLIF) procedures with implantation of the Interfix device (Medtronic Sofamor Danek, Minneapolis, MN) filled with autogenous bone graft, and 35 patients underwent the procedure implanted with the Interfix device filled with InFUSE. More than half of the patients in both groups were tobacco users. The perioperative data indicated a shorter hospital stay in the experimental group. However, Oswestry pain and function, back pain, and leg pain scores were similar in both groups. The fusion rate was lower in the InFUSE group at the 12-month postoperative time point ($P < 0.05$). In several patients, heterotopic bone was observed in the spinal canal posterior to the fixation devices and in the tract of their insertion. It is not clear whether the formation of this bone was related to the technique used. Because of this radiologic finding and the uncertainty of the clinical relevance, this study was stopped. Although the presence of this intracanal bone did not appear to be associated with negative clinical sequelae, this study was not resumed because stand-alone PLIFs without supplemental pedicle screw fixation are no longer commonly performed.

Clinical Trial in Posterolateral Lumbar Fusion

One of the most challenging clinical problems in bone healing is posterolateral lumbar fusion. A prospective randomized pilot trial was performed to determine if the dosages and carrier of rhBMP-2 that were successful in rhesus monkeys could induce consistent spinal fusion in humans.[28] Twenty-five patients undergoing lumbar arthrodesis were randomized to three groups based on arthrodesis technique using a 1:2:2 ratio: (1) autograft/Texas Scottish Rite Hospital (TSRH) pedicle screw instrumentation (n = 5), (2) rhBMP-2/TSRH (n = 11), and (3) rhBMP-2 only without internal fixation (n = 9). A dosage of 20 mg rhBMP-2/side was delivered on a carrier consisting of

60% hydroxyapatite and 40% tricalcium phosphate granules (10 cm³/side). Patients had single level disk degeneration, grade I or less spondylolisthesis, mechanical low back pain with or without leg pain, and failure of at least 6 months of nonsurgical treatment. All 25 patients were available for follow-up (mean 17 months, range 12 to 27). The rate of fusion observed on radiographs was 40% (2 of 5) in the autograft/TSRH group and 100% (20 of 20) with rhBMP-2 with or without TSRH internal fixation ($P = 0.004$). A statistically significant improvement in Oswestry score was seen at 6 weeks in the rhBMP-2 only group (-17.6, $P = 0.009$) and at 3 months in the rhBMP-2/TSRH group (-17.0, $P = 0.003$), but not until 6 months in the Autograft/TSRH group (-17.3, $P = 0.041$). At final follow-up, Oswestry improvement was greatest in the rhBMP-2 only group (-28.7, $P < 0.001$). The results on the SF-36 Pain Index and Physical Component Subscale subscales showed similar changes. This is the first trial with at least 1-year follow-up to demonstrate successful posterolateral spinal fusion using a BMP-based bone tissue engineering product as determined by radiographs and CT scans. The rhBMP-2 consistently induced formation of bone in the posterolateral lumbar spine when delivered at a dose of 20 mg/side. There was a statistically greater and quicker improvement in patient-derived clinical outcome measures in the groups receiving rhBMP-2.

Clinical Trial of Spinal Fusion of the Cervical Spine

The only clinical study examining the efficacy of InFUSE for spinal fusion of the cervical spine involves the implantation of machined fibular ring grafts (Cornerstone, Medtronic Sofamor Danek, Inc) filled with either autograft or InFUSE.[31] This prospective, randomized, and controlled study involved 33 patients enrolled at four study sites. The patients underwent anterior diskectomy and fusion procedures and were evaluated at the same time points as in the other studies. Eighteen patients enrolled in the InFUSE treatment arm. A total of 15 patients enrolled in the control cohort were implanted with the machined fibular ring filled with autograft from the iliac crest. In the control groups, donor site pain was negligible 3 months postoperatively as noted by a 20-point analog scale. All patients achieved fusion by 6 months after surgery as assessed by evaluation of radiographs.

Summary

In summary, spinal fusion procedures represent an excellent proving ground for bone TEMPs because of the high rate of spontaneous failure with the current standard of autograft. Clinical trials of TEMP must be designed to account for difficulty with the noninvasive assessment of the success or failure of the spinal fusion. The possible dissociation between technically successful bone formation and positive clinical outcome also represents a challenge in the design of clinical trials. Studies of each bioactive factor/carrier combination at each anatomic site in the spine may be required because results at one site may not necessarily predict results at another site. Furthermore, there may be confounding issues based on the specific clinical diagnosis such as degenerative spondylolisthesis or isthmic spondylolisthesis with different success rates in each of those groups. Finally, the effects of steroids, other pharmacologic agents, smoking, or diabetes mellitus on bone healing with TEMPs are unknown. Patients who have conditions thought to impair healing may require higher doses of growth factor or different carriers to achieve positive clinical outcomes. In any case, if these devices are as successful in clinical trials as in the preclinical studies, the frequency of nonunions may be greatly reduced in orthopaedic practice.

References

1. Urist MR, Strates BS: Bone morphogenetic protein. *J Dent Res* 1971;50:1392-1406.

2. Wozney JM: Bone morphogenetic proteins and their gene expression, in Noda M (ed): *Cellular and Molecular Biology of Bone*. Boston, MA, Academic Press, 1993, 131-167.

3. Urist MR: Bone: Formation by autoinduction. *Science* 1965;150:893-899.

4. Urist MR, Jurist JM Jr, Dubuc FL, et al: Quantitation of new bone formation in intramuscular implants of bone matrix in rabbits. *Clin Orthop* 1970;68:279-293.

5. Urist MR, Strates BS: Bone formation in implants of partially and wholly demineralized bone matrix: Including observations on acetone-fixed intra and extracellular proteins. *Clin Orthop* 1970;71:271-278.

6. Wozney JM, Rosen V, Celeste AJ, et al: Novel regulators of bone formation: Molecular clones and activities. *Science* 1988;242:1528-1534.

7. Wang EA, Rosen V, Cordes P, et al: Purification and characterization of other distinct bone-inducing factors. *Proc Natl Acad Sci USA* 1988;85:9484-9488.

8. Rosen V, Wozney JM, Wang EA, et al: Purification and molecular cloning of a novel group of BMPs and localization of BMP mRNA in developing bone. *Connect Tissue Res* 1989;20:313-319.

9. Sampath TK, Coughlin JE, Whetstone RM, et al: Bovine osteogenic protein is composed of dimers of OP-1 and BMP-2A, two members of the transforming growth factor-b superfamily. *J Biol Chem* 1990;265:13198-13205.

10. Boden SD: Clinical application of the BMPs. *J Bone Joint Surg Am* 2001;83:S161.

11. Jeppsson C, Saveland H, Rydholm U, et al: OP-1 for cervical spine fusion: bridging bone in only 1 of 4 rheumatoid patients but prednisolone did not inhibit bone induction in rats. *Acta Orthop Scand* 1999;70:559-563.

12. Laursen M, Hoy K, Hansen ES, et al: Recombinant bone morphogenetic protein-7 as an intracorporal bone growth stimulator in unstable thoracolumbar burst fractures in humans: Preliminary results. *Eur Spine J* 1999;8:485-490.

13. Cook S: Preclinical and clinical evaluation of osteogenic protein-1 (BMP-7) in bony sites. *Orthopaedics* 1999;22:669-671.

14. Cunningham B, Kanayama M, Parker L, et al: Osteogenic protein versus autologous interbody arthrodesis in the sheep thoracic spine: A comparative endoscopic study using the Bagby and Kuslich interbody fusion device. *Spine* 1999;24:509-518.

15. Magin MN, Delling G: Improved lumbar vertebral interbody fusion using rhOP-1: A comparison of autogenous bone graft, bovine hydroxylapatite (Bio-Oss), and BMP-7 (rhOP-1) in sheep. *Spine* 2001;26:469-478.

16. Grauer JN, Patel TC, Erulkar JS, Troiano NW, Panjabi MM, Friedlaender GE: 2000 Young Investigator Research Award winner: Evaluation of OP-1 as a graft substitute for intertransverse process lumbar fusion. *Spine* 2001;26:127-133.

17. Boden SD, Moskovitz PA, Morone MA, et al: Video-assisted lateral intertransverse process arthrodesis: Validation of a new minimally invasive lumbar spinal fusion technique in the rabbit and nonhuman primate (rhesus) models. *Spine* 1996;21:2689-2697.

18. Holliger EH, Trawick RH, Boden SD, et al: Morphology of the lumbar intertransverse process fusion mass in the rabbit model: A comparison between two bone graft materials - rhBMP-2 and autograft. *J Spinal Disord* 1996;9:125-128.

19. Martin GJ, Boden SD, Titus L: Recombinant human bone morphogenetic protein-2 reverses the inhibitory effect of ketorolac, a non-steroidal anti-inflammatory drug (NSAID) on posterolateral lumbar intertransverse process spine fusion. *Spine* 1999;24:2188-2194.

20. Schimandle JH, Boden SD, Hutton WC: Experimental spinal fusion with recombinant human bone morphogenetic protein-2 (rhBMP-2). *Spine* 1995;20:1326-1337.

21. Boden SD, Schimandle JH, Hutton WC: 1995 Volvo Award in Basic Sciences: The use of an osteoinductive growth factor for lumbar spinal fusion. Part II: Study of dose, carrier, and species. *Spine* 1995;20:2633-2644.

22. Boden SD, Schimandle JH, Hutton WC, et al: In vivo evaluation of a resorbable osteoinductive composite as a graft substitute for lumbar spinal fusion. *J Spinal Disord* 1997;10:1-11.

23. Boden SD, Schimandle JH, Hutton WC: Lumbar intertransverse process spine arthrodesis using a bovine-derived osteoinductive bone protein. *J Bone Joint Surg Am* 1995;77:1404-1417.

24. Boden SD, Martin GJ, Morone MA, et al: The use of coralline hydroxyapatite with bone marrow, autogenous bone graft, or osteoinductive bone protein extract for posterolateral lumbar spine fusion. *Spine* 1999;24:320-327.

25. Silcox DH, Boden SD, Schimandle JH, et al: Reversing the inhibitory effect of nicotine on spinal fusion using an osteoinductive protein extract. *Spine* 1998;23:291-297.

26. Martin GJ, Boden SD, Morone MA, et al: Posterolateral intertransverse process spinal fusion arthrodesis with rhBMP-2 in a non-human primate: Important lessons learned regarding dose, carrier, and safety. *J Spinal Disord* 1999;12:179-186.

27. Boden SD, Zdeblick TA, Sandhu HS, et al: The use of rhBMP-2 in interbody fusion cages. Definitive evidence of osteoinduction in humans: A preliminary report. *Spine* 2000;25:376-381.

28. Boden SD, Kang JD, Sandhu HS, Heller JG: 2002 Volvo Award for Low Back Pain Research: Use of rhBMP-2 to achieve posterolateral lumbar spine fusion in humans: A prospective and randomized clinical pilot trial. *Spine* 2002;27:2662-2673.

29. Damien CJ, Grob D, Boden SD, et al: Purified bovine BMP extract and collagen for spine arthrodesis: Preclinical safety and efficacy. *Spine* 2002;27:S50-S58.

30. Vaccaro AR, Anderson DG, Toth CA: Recombinant human osteogenic protein-1 (bone morphogenetic protein-7) as an osteoinductive agent in spinal fusion. *Spine* 2002;27:S59-S65.

31. McKay B, Sandhu HS: Use of recombinant human bone morphogenetic protein-2 in spinal fusion applications. *Spine* 2002;27:S66-S85.

Chapter 15

Practical Matters in the Application of Tissue Engineered Products for Skeletal Regeneration in the Head and Neck Region

Brian Nussenbaum, MD
Paul H. Krebsbach, DDS, PhD

Abstract

Patients with large craniomaxillofacial or mandible defects would certainly benefit if tissue engineering approaches could be used successfully and safely. The incorporation of tissue regenerative techniques into the clinical practice of head and neck reconstructive surgeons has been problematic because the defects are often associated with a local wound environment that is suboptimal for supporting tissue regeneration. Several factors, as discussed in this review, will need to be considered before considering the use of bone regenerative approaches in the head and neck region. It is hoped that the barriers to using tissue engineering techniques in this region will be overcome, and thus provide a new therapeutic option for treating head and neck bone defects.

Introduction

There have been many advances in skeletal regeneration, beginning with the discovery of bone morphogenetic proteins (BMPs) by Urist in 1965[1] and continuing with the isolation, cloning, and purification of these molecules in the late 1980s.[2,3] During the past 10 years, there has been great effort to incorporate the use of these osteoinductive proteins into clinical practice. Based on the success of multicenter, prospective, randomized clinical trials, therapy using BMP-2 and BMP-7 is now approved by the US Food and Drug Administration for marketing (as class III device) and is commercially available as an alternative to free bone grafts for limited orthopaedic indications.[4,5] Compared with standard therapies, the use of BMPs has the potential advantages of avoiding donor site morbidity from bone graft harvest, decreasing the incidence of delayed union and nonunion, and accelerating the rate of fracture healing.

The incorporation of osteoinductive protein therapy into clinical practice for repair of craniomaxillofacial and mandibular defects has lagged behind other advances in orthopaedic treatments. The application of osteoinductive protein therapy for head and neck defects is still in the experimental stages because the characteristics typical of these defects result in a local environment that is suboptimal for tissue regeneration. Although these characteristics are of little concern in the treatment of many orthopaedic defects, they will need to be considered before expanding the use of BMPs for more complex skeletal wounds, such as those resulting from tumor surgery or occurring in compromised tissue beds.

This review explores the possibility of using bone regenerative approaches to treat defects in the head and neck region. Although the mandible discontinuity defect is discussed, the issues described could readily apply to other defects involving the craniomaxillofacial region.

Background and Current Therapy

Significant mandible defects that require reconstructive surgery most commonly arise from resection of advanced malignancies. A much smaller percentage of these defects are caused by resection of benign neoplasms, osteoradionecrosis, or trauma. These defects are anatomically complex, usually having both osseous and soft-tissue components. The goals of reconstruction of significant mandible defects are to restore both form and function, and to facilitate wound healing to allow timely treatment with adjuvant therapies.

The need to reconstruct oromandibular defects was recognized as early as 1962 when Conley described the concept of "the crippled oral cavity."[6] Implications of not performing adequate reconstruction include cosmetic deformities, swallowing and mastication problems, loss of oral competence, and tracheostomy dependence. Studies using both objective testing[7] and outcomes measures[8] have shown that restoring mandibular continuity is beneficial for patients. As many as 30% of cancer survivors request a secondary reconstruction of the mandible if it is not performed at the time of the initial surgery.[9] The success of secondary reconstruction is limited by scarring, fibrosis, and distorted anatomic relationships.

Prior to 1976, options for mandible reconstruction were limited to metal prosthetic implants and free nonvascularized autologous bone grafts. Skeletal defects resulting from the resection of head and neck cancers are likely to be contaminated with oral bacteria and treated with pre- or postoperative radiation therapy. These wound characteristics, along with the few available options for reconstruction of adjacent soft-tissue defects, limited the success of primary bone reconstruction using nonvascularized free bone grafts. Complications such as resorption, extrusion, wound dehiscence, orocutaneous fistula, and osteoradionecrosis frequently required hospital admission and further surgery.[10] These morbidities curbed the enthusiasm for reestablishing mandibular continuity with the use of bone grafts until microsurgical techniques were introduced in the 1980s.

The detrimental effects of bacterial contamination, inflammation, and radiation therapy on wound healing are frequently overcome by microvascular free tissue transfer because the bone is immediately revascularized and the graft includes a vascularized soft-tissue component for reliable closure of accompanying mucosal or skin defects. Although free tissue transfer has gained acceptance for primary reconstruction of oromandibular defects because of the high success rate,[11] this approach has some limitations that restrict its universal use. Microvascular reconstructive surgery requires expert skills obtained from specific training in this technique and requires several additional hours of surgical time. Some patients are not medically capable of tolerating this additional surgery. Despite the surgeon's best efforts, there is a 5% to 10% failure rate due to clot formation in the reconnected blood vessels. The mandible is three-dimensionally complex making precise restoration of the shape and height of the defect challenging. Multiple defect- and patient-related factors are important for surgical planning, but the need for bone restricts the flap choices to the scapula, iliac crest, fibula, or radial sites. Finally, harvest of bone from these donor sites can cause significant donor site morbidity such as chronic pain, paresthesias, extremity movement problems, gait disturbances, abdominal wall hernia, and limb ischemia. Given these limitations of current therapy, patients requiring mandible reconstruction would benefit greatly from the successful application of tissue engineering approaches.

Factors Limiting Tissue Engineering Approaches

General Considerations

Preclinical studies have demonstrated successful healing of critical-sized mandible defects in canines[12] and monkeys[13] using BMP delivered on biodegradable carriers. There is also a case report of a patient who had a 6-cm lateral mandibular discontinuity defect from an ameloblastoma resection reconstructed using a BMP bioimplant rather than a microvascular free tissue transfer.[14] Although these findings are promising, the wound environment in these animal models and the case report were optimized to provide an ideal milieu for bone regeneration. This is not typical of patients with ablative head and neck surgical defects in whom the local vascular and mesenchymal cell environment is likely to be adversely affected by adjuvant radiation therapy, microbiologic contamination, advanced patient age, tobacco exposure, comorbidities, and past cancer treatments. The defects also tend to be "super" critical-sized and include the overlying periosteum and soft tissues. Therapy with osteoinductive proteins has successfully regenerated bone in patients requiring alveolar ridge[15] or sinus floor[16] augmentation, but the magnitude of these defects is not comparable to those resulting from ablative cancer surgery.

Given the typical wound characteristics, there is concern that a single exposure to an exogenous morphogen may not provide an adequate osteoinductive stimulus to the cells within the wound microenvironment to heal these defects. Gene therapy and mesenchymal stem cell therapy are alternative tissue regenerative approaches that might be advantageous for healing compromised wounds because the gene and/or cells that directly participate in the reparative process are delivered to the wound. The use of mesenchy-

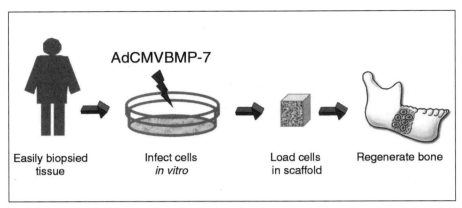

AdCMVBMP-7

Easily biopsied tissue   Infect cells *in vitro*   Load cells in scaffold   Regenerate bone

Figure 1 Ex vivo gene therapy directed osteogenesis.

mal stem cells to regenerate craniofacial defects has been successful in animal models.[17] Gene therapy strategies that deliver BMPs using viral vectors have also been successful in regenerating skeletal defects in animal models.[18]

Gene therapy can result in a period of sustained BMP production, thus enabling the host's wound to respond to the osteoinductive stimulus in a more robust fashion than after delivery of a single dose of BMP. Although in vivo gene therapy approaches have been successful,[19] an ex vivo gene therapy approach (Figure 1) has several advantages including direct delivery of the osteoinductive gene to the desired site, targeting cells for gene delivery, supplying cells that directly participate in the osteoinductive process, and the potential for controlling the rate or extent of gene expression by using inducible delivery vectors. Different types of differentiated mesenchymal cells and adult stem cells can be genetically modified to express BMP.[20] Dermal fibroblasts are an example of cells that can be easily harvested with minimal morbidity and readily expanded in cell culture. Although fibroblasts lack an intrinsic responsiveness to BMP-2,[20] these cells are capable of converting into other tissue lineages, such as osteoblasts, when genetically modified to express BMP-7.[21] Thus, in addition to providing the osteoinductive signal, these cells may also directly participate in the reparative process.

Prefabrication is another tissue engineering approach that can be used for mandibular reconstruction. This technique allows for bone regeneration in an ectopic, easily accessible soft-tissue site, which is subsequently transferred to the defect as a microvascular tissue transfer once the newly formed bone has inosculated to the surrounding tissue. This approach has already been shown to be successful in a minipig model using the latissimus dorsi muscle as the soft-tissue pouch and BMP-7 as the osteoinductive agent.[22] The prefabrication approach allows for custom design of the regenerated bone into the three-dimensional shape of the defect and avoids trying to induce new bone growth in a compromised tissue bed. The time required to induce new bone at the ectopic site (6 weeks in the minipig model), however, would likely make this approach applicable for secondary rather than primary reconstruction. Studies also show that the bone regenerated at this ectopic, nonbiomechanically loaded site is woven rather than lamellar in nature (Wurzler and associates, Sacramento, CA, unpublished data, 2002). It is not known if this bone later undergoes functional adaptation after tissue transfer to the defect site.

Radiation Therapy

The negative effects of radiation therapy on the wound will need to be considered when devising strategies to regenerate skeletal elements lost because of head and neck cancer. To date, studies have been limited. In vitro irradiation of up to 8 Gy did not affect BMP-2 induced osteoblast differentiation of C2C12 cells,[23] and bone growth still occurred on a BMP-2 treated hydroxyapatite disk implanted into a rabbit snout subperiosteal pocket radiated preoperatively with a fractionated dose of 20 Gy.[24] Despite these encouraging findings, 3-mm rat calvarial defects preoperatively treated with a 12 Gy radiation dose and subsequently with BMP-2 had successful bone regeneration but incomplete healing of the defect.[25] In a rat 4-mm mandible defect treated with a fractionated 45 Gy radiation dose and 2 weeks later with demineralized bone powder, only 39% of defects had more than 50% bone fill.[26] Using a 7-mm rat calvarial defect model, Khouri and associates[27] showed that microvascular transfer of a muscle flap, in addition to treatment with BMP-3, allowed for complete healing of the defect after receiving a single preoperative 15 Gy radiation dose. Treatment with no implant, a microvascular muscle flap alone, or BMP-3 alone did not heal the defect. These results suggest that it is necessary to restore the population of responsive mesenchymal cells in addition to delivering an osteoinductive protein to heal bone defects compromised by preoperative radiation. Studies to replicate these findings in larger animal models and to evaluate postoperative radiation therapy in the defect model need to be performed.

Bacterial Contamination

Invariably, the delivery of tissue engineered products for mandible reconstruction will involve exposure to the oral bacterial flora, which contains 10^6-10^7 colony forming units per milliliter (CFU/mL) of saliva. In the healthy oral cavity, the mucosal surfaces are colonized by approximately 90% anaerobic (mostly *Bacteroides* species, *Peptostreptococcus*, and *Fusobacterium*) and 10% aerobic (*Streptococcus*, *Staphlococcus*, and *Corynebacterium*) bacteria. Whether BMPs are capable of inducing new bone formation in the presence of bacterial contamination or infection is unknown. Using an ectopic trunk site in rats, Parmer and associates showed that local inoculation with 10^6-10^9 CFU/mL of *Staphlococcus aureus* inhibited new bone formation induced by BMP-7 as measured by a 60% to 80% inhibition of alkaline phosphatase activity (V Parmar and associates, Dallas, TX, unpublished data, 2002). Both live and dead bacteria had the same effect. In a rat segmental femur defect inoculated with 105 CFU/mL of *Staphlococcus aureus*, Chen and associates[28] showed significant inhibition of bone formation induced by BMP-7 based on radiographs and histologic evaluations. This inhibition was somewhat, but not completely, overcome with a fivefold dose of BMP-7. Additional investigations are needed to determine whether a more robust osteoinductive stimulus can successfully overcome the inhibitory effect of bacterial contamination on new bone growth. The effects of antibiotics also need to be determined.

Oncogenesis

A contraindication to the use of commercially available BMPs is a patient history of malignancy or application near a resected tumor. Various cancers such as osteosarcoma, breast, prostate, pancreas, and esophagus have been found to express BMPs or BMP-receptors. In an evaluation of 29 specimens from head and neck squamous cell carcinomas, Jin and associates[29] observed that 26 were positive for BMP-2 and/or BMP-4 and 28 were positive for BMP-5 as determined by immunostaining. Another study used reverse transcriptase-polymerase chain reaction to detect expression of BMP-2 mRNA from tongue and gingival squamous cell carcinoma cell lines.[30] Expression of BMPs may also have value for predicting poor prognosis, as shown for both pancreatic[31] and esophageal[32] cancer. Given these findings, there is concern that BMPs may have a role in carcinogenesis.

However, there is no direct evidence that BMPs are carcinogenic. In vitro studies have shown the BMP-2 and BMP-7 have an antiproliferative effect on breast and anaplastic thyroid cancer cell lines, respectively.[33,34] At doses of 10, 100, and 1,000 ng/mL, BMP-2 did not stimulate tumor cell proliferation in cell lines from breast, ovarian, lung, and prostate cancer.[35] Significant inhibition of tumor cell proliferation was observed in 25% of the cell lines at a dose of 1,000 ng/mL. No mutations for BMP receptors or smads specific for the signal transduction of BMPs have been identified in human cancers, except for the C-terminal truncation of smad 5 associated with some leukemias. Evidence of carcinogenesis has not been revealed in extensive preclinical experience and the incidence of cancer in patients treated with BMP-7 for orthopaedic indications does not appear to be greater than that of the general population.[36]

Depending on the concentration and microenvironment, BMPs have pleiotrophic activity. In addition to promoting mesenchymal cell differentiation, other known actions of BMPs include mediation of angiogenesis, chemotaxis, and apoptosis. Although evidence suggests that BMPs do not cause cancer, their expression may be part of the pathogenic processes conferring a survival advantage or metastatic potential to cancer cells. Additional studies need to be performed to fully evaluate the safety of using BMPs in cancer patients.

Summary

Patients with large craniomaxillofacial or mandible defects would certainly benefit if tissue engineering approaches can be used successfully and safely in the clinical setting. To successfully apply a tissue engineering approach, the recipient bed needs to be well vascularized and contain responsive mesenchymal cells, the defect needs to have stable fixation, and local bacterial contamination or infection needs to be addressed. Although head and neck reconstruction will likely include tissue engineered products in the future, many obstacles to successful implementation require additional investigation.

Acknowledgments

This work was supported in part by grant DE 13835 (PHK). PHK is a recipient of National Institutes of Health Independent Scientist Award DE00426 sponsored by NIDCR.

References

1. Urist MR: Bone formation by autoinduction. *Science* 1965;150:893-899.

2. Wang EA, Rosen V, Cordes P, et al: Purification and characterization of other distinct bone-inducing factors. *Proc Natl Acad Sci USA* 1988;85:9484-9488.

3. Celeste AJ, Lannazzi JA, Taylor RC, et al: Identification of transforming growth factor-beta superfamily members present in bone-inductive protein purified from bovine bone. *Proc Natl Acad Sci USA* 1990;87:9843-9847.

4. Friedlaender GE, Perry CR, Cole JD, et al: Osteogenic protein-1 (bone morphogenetic protein-7) in the treatment of tibial nonunions. *J Bone Joint Surg Am* 2001;83:151-158.

5. Burkus JK, Transfeldt EE, Kitchel SH, et al: Clinical and radiographic outcomes of anterior lumbar interbody fusion using recombinant human bone morphogenetic protein-2. *Spine* 2002;21:2396-2408.

6. Conley JJ: The crippled oral cavity. *Plast Reconstr Surg* 1962;30:469-478.

7. Urken ML, Buchbinder D, Weinberg H, et al: Functional evaluation following microvascular oromandibular reconstruction of the oral cancer patient: A comparative study of reconstructed and nonreconstructed patients. *Laryngoscope* 1991;101:935-950.

8. Wilson KM, Rizk NM, Armstrong SL, et al: Effects of hemimandibulectomy on quality of life. *Laryngoscope* 1998;108:1574-1577.

9. Tucker HM: Nonrigid reconstruction of the mandible. *Arch Otolaryngol Head Neck Surg* 1989;115:1190-1192.

10. Adamo AK, Szal RL: Timing, results, and complications of mandibular reconstructive surgery: Report of 32 cases. *J Oral Surg* 1979;37:755-763.

11. Urken ML, Buchbinder D, Costantino PD, et al: Oromandibular reconstruction using microvascular composite flaps. *Arch Otolaryngol Head Neck Surg* 1998;124:46-55.

12. Toriumi DM, Kotler HS, Luxemberg DP, et al: Mandibular reconstruction with a recombinant bone-inducing factor. *Arch Otolaryngol Head Neck Surg* 1991;117:1101-1112.

13. Boyne PJ: Animal studies of application of rhBMP-2 in maxillofacial reconstruction. *Bone* 1996;19:83-92.

14. Moghadam HG, Urist MR, Sandor GKB, et al: Successful mandibular reconstruction using a BMP bioimplant. *J Craniofac Surg* 2001;12:119-127.

15. Howell TH, Fiorellini J, Jones A, et al: A feasibility study evaluating rhBMP-2/absorbable collagen sponge device for local alveolar ridge preservation or augmentation. *Int J Periodontics Restorative Dent* 1997;17:124-139.

16. Boyne PJ, Marx RE, Nevins M, et al: A feasibility study evaluating rhBMP-2/absorbable collagen sponge device for maxillary sinus floor augmentation. *Int J Periodontics Restorative Dent* 1997;17:11-25.

17. Krebsbach PH, Mankani MH, Satomura K, et al: Repair of craniotomy defects using bone marrow stromal cells. *Transplantation* 1998;66:1272-1278.

18. Krebsbach PH, Gu K: Franceschi, et al: Gene therapy-directed osteogenesis: BMP-7 transduced human fibroblasts form bone in vivo. *Hum Gene Ther* 2000;11:1201-1210.

19. Lindsey WH: Osseous tissue engineering with gene therapy for facial bone reconstruction. *Laryngoscope* 2001;111:1128-1136.

20. Musgrave DS, Bosch P, Lee JY, et al: Ex vivo gene therapy to produce bone using different cell types. *Clin Orthop* 2000;378:290-305.

21. Rutherford RB, Moalli M, Franceschi RT, et al: Bone morphogenetic protein-transduced human fibroblasts convert to osteoblasts and form bone in vivo. *Tissue Eng* 2002;8:441-452.

22. Terheyden H, Warnke P, Dunsche A, et al: Mandibular reconstruction with prefabricated vascularized bone grafts using recombinant human osteogenic protein-1: An experimental study in miniature pigs. Part II: Transplantation. *Int J Oral Maxillofac Surg* 2001;30:469-478.

23. Ikeda S, Hachisu R, Yamaguchi A, et al: Radiation retards muscle differentiation but does not affect osteoblastic differentiation induced by bone morphogenetic protein-2 in C2C12 myoblasts. *Int J Radiat Biol* 2000;76:403-411.

24. Howard BK, Brown KR, Leach JL, et al: Osteoinduction using bone morphogenetic protein in irradiated tissue. *Arch Otolaryngol Head Neck Surg* 1998;124:985-988.

25. Wurzler KK, DeWeese TL, Sebald W, et al: Radiation-induced impairment of bone healing can be overcome by recombinant human bone morphogenetic protein-2. *J Craniofac Surg* 1998;9:131-137.

26. Lorente CA, Song BZ, Donoff RB: Healing of bony defects in the irradiated and unirradiated rat mandible. *J Oral Maxillofac Surg* 1992;50:1305-1309.

27. Khouri RK, Brown DM, Koudsi B, et al: Repair of calvarial defects with flap tissue: Role of bone morphogenetic proteins and competent responding tissues. *Plast Reconstr Surg* 1996;98:103-109.

28. Chen X, Kidder LS, Lew WD: Osteogenic protein-1 induced bone formation in an infected segmental defect in the rat femur. *J Orthop Res* 2002;20:142-150.

29. Jin Y, Tipoe GL, Liong EC, et al: Overexpression of BMP-2/4, -5 and BMPR-1A associated with malignancy of oral epithelium. *Oral Oncol* 2001;37:225-233.

30. Hatakeyama S, Gao Y, Ohara-Nemoto Y, et al: Expression of bone morphogenetic proteins of human neoplastic epithelial cells. *Biochem Mol Biol Int* 1997;42:497-505.

31. Kleeff J, Maruyama H, Ishiwata T, et al: Bone morphogenetic protein 2 exerts diverse effects on cell growth in vitro and is expressed in human pancreatic cancer in vivo. *Gastroenterology* 1999;116:1202-1216.

32. Raida M, Sarbia M, Clement JH, et al: Expression, regulation and clinical significance of bone morphogenetic protein 6 in esophageal squamous cell carcinoma. *Int J Cancer* 1999;83:38-44.

33. Ghosh-Choudhury N, Ghosh-Choudhury G, Celeste A, et al: Bone morphogenetic protein-2 induces cyclin kinase inhibitor p21 and hypophosphorylation of retinoblastoma protein in estradiol-treated MCF-7 human breast cancer cells. *Biochim Biophys Acta* 2000;1497:186-196.

34. Franzen A, Heldin N: BMP-7 induced cell cycle arrest of anaplastic thyroid carcinoma cells via p21 and p27. *Biochem Biophys Res Commun* 2001;285:773-781.

35. Soda H, Raymond E, Sharma S, et al: Antiproliferative effects of recombinant human bone morphogenetic protein-2 on human tumor colony-forming units. *Anticancer Drugs* 1998;9:327-331.

36. Poynton AR, Lane JM: Safety profile for the clinical use of bone morphogenetic proteins in the spine. *Spine* 2002;16S:S40-S48.

Future Directions

Tissue Engineering of Bone

Bone regeneration is crucially important for a large number of skeletal conditions, including fracture repair and spinal fusion. There is a demonstrated need for developing clinically applicable procedures of bone tissue engineering, capable of either inducing regenerate bone at the desired tissue site or the ex vivo production of bone constructs suitable for implantation. A fundamental guiding principle for future research directions in bone tissue engineering is the development of products that require minimal ex vivo manipulations and restore function efficiently to the patient.

Biologics

The identification and recombinant production of a significant number of osteoactive factors, for example, members of the bone morphogenetic protein (BMP) family, have shown promising results in the induction of new bone formation for fracture repair and spinal fusion. Future directions for research on osteoinductive biologics for bone tissue engineering are:

- Optimize the composition, dose, and time of administration.
- Prime the tissue site for osteoinductive response via systemic/local preadministration of osteoanabolic agents, such as parathyroid hormone.
- Identify the responsive cells in the implant site and elucidate the mechanisms of the osteoinductive response.
- Use the mechanistic information to identify earlier acting and enabling factors involved in osteoinduction.
- Develop valid criteria for evaluating the long-term success of tissue engineered bone by correlating properties of early regenerate bone to its long-term remodeling and performance.

Bone progenitor cells

Mesenchymal progenitor cells isolated from multiple adult tissues, including bone marrow stroma, exhibit multilineage differentiation potential and represent a candidate cell type for ex vivo tissue engineering. Future directions on developing the application of adult tissue-derived mesenchymal progenitor cells for bone tissue engineering include:

- Optimize culture conditions for large-scale cell expansion with retention of multilineage differentiation potential.
- Optimize culture conditions, including both biological and physical stimulation for efficient, homogeneous osteogenic differentiation.
- Establish end point markers for osteogenesis.
- Evaluate additional application of such cells for the delivery of genes and/or gene products.
- Examine the host immunoresponse to allogeneic cells, for the possible application of immunoprivileged allogeneic cells to develop off-the-shelf, prefabricated, tissue engineered bone grafts.

Scaffolds

Scaffolds provide appropriate three-dimensional shape and form for proper bone formation in vivo and ex vivo. Future research directions for the use of scaffolds for bone tissue engineering include:

- Determine the optimal structural and compositional requirements for bone formation in various environments (eg, cancellous non–weight-bearing, submuscular paraspinal, weight-bearing interbody, etc).
- Assess how cell loading of scaffolds may enhance bone formation and determine which cell type, differentiated or progenitors, is more effective.
- Use computer-aided design methods to fabricate scaffolds for the engineering of bone tissues of specific shape and dimension (custom-designed bone).
- Design and functionalize scaffolds and carrier matrices to activate and condition target cells for optimal new bone formation.

Animal and clinical models

- Compare tissue engineered bone in a standardized animal model of clinical relevance (eg, rhesus monkey for posterolateral spinal fusion).
- Standardize outcome measures for clinical trials that correlate with successful bone formation.
- Develop and establish natural and mutation-based human and animal models to elucidate the mechanism of bone formation, eg, in humans, fibrodysplasia ossificans progressive and Ilizarov distraction osteogenesis, and various naturally occurring and transgenically produced mutations in mice that affect skeletal development and bone formation.
- Design valid randomized clinical studies to evaluate efficacy of bone formation.

Application of tissue engineered bone

Successful production of functionally adequate tissue engineered bone generates a potentially highly efficient model system for rapid assessment of the activity of candidate osteoactive and therapeutic agents.

- Tissue engineered bone could be used as in vitro tissue analogs for drug testing (pharmaceutical/nutriceuticals) and to optimize conditions for effective responsiveness to osteoactive factors.

It is recommended that the successful translation of tissue engineering technologies to the clinical setting is crucially dependent on the close collaboration and communication among clinicians, scientists, bioengineers, industry, and government regulatory agencies.

Section Three
Tissue Engineering of Cartilage

H. Davis Adkisson, PhD
Kyriacos A. Athanasiou PhD, PE
Frank Barry, PhD
John Bogdanske, BS
Regina Cheung, BS
Joseph Feder, PhD
Alan J. Grodzinsky, ScD
Keith A. Hruska, PhD
Jerry C. Hu, BS
Ernst B. Hunziker, MD

Neil Kizer, PhD
Yan Lu, MD
Mark Markel, DVM, PhD
Koichi Masuda, MD
Tom Minas, MD, MS
Robert L. Sah, MD, ScD
Michael Sittinger, PhD
Eugene J. Thonar, PhD
Rocky S. Tuan, PhD

Chapter 16

Experimental Principles and Future Perspectives of Skeletal Tissue Engineering

Rocky S. Tuan, PhD

Abstract

The treatment of skeletal diseases is challenging as these conditions often involve tissue degeneration or failure to heal. Given the drawbacks of methods currently available to repair articular cartilage injury and degeneration, the use of functional tissue substitutes developed by tissue engineering techniques has emerged as a potential therapeutic option. There are three fundamental components of successful tissue engineering: cells, scaffold, and environment. Both differentiated cells and tissue progenitor cells are candidates for tissue engineering application. The use of a scaffold allows engineered tissues to be produced with the appropriate shape and form. And, finally, an appropriate local environment is required for the engineering tissue to develop or maintain its appropriate structure and function. Tissue engineering methods based on the use of mesenchymal stem cells (MSC) are being evaluated for use in cartilage repair. This review summarizes recent advances in tissue engineering for repair of cartilage.

Introduction

Current tissue engineering strategies combine experimental approaches of biology and engineering to develop functional tissue substitutes. A recent increase in the number of scientific publications on tissue engineering indicates that this is one of the most exciting areas of biomedical research. Tissue engineering techniques are particularly attractive for the treatment of skeletal diseases, most of which involve tissue degeneration or failure to heal. Chondral defects resulting from injury or degenerative joint diseases, such as osteoarthritis, are especially challenging in terms of natural tissue healing and repair. Due to the acellularity of the tissue, damage to the articular cartilage often results in failure of adequate tissue repair. Clonal proliferation of articular chondrocytes often occurs in damaged cartilage, resulting in the production of mechanically inferior fibrocartilage. Extensive degeneration of the articular surface eventually necessitates total joint arthroplasty; therefore, treatment of these conditions with tissue engineering is of particular interest.

Current Reparative Approaches for Articular Cartilage Injury and Degeneration

Damage to the articular cartilage may be caused by physical trauma, compression of a joint under heavy load, application of angular or shear forces to the surface, or degenerative joint diseases, such as osteoarthritis. The consequences of degenerative joint diseases may include pain, swelling, joint stiffness, and, ultimately, loss of mobility. Osteoarthritis commonly occurs in the hips, knees, and spine. The finger joints, the joint at the base of the thumb, and the joint at the base of the big toe may also be affected. Osteoarthritis affects approximately 20 million men and women in the United States. The chance of developing osteoarthritis increases with age, and most individuals older than age 60 years have some degree of the disease with at least one joint affected. Because the adult articular cartilage does not repair itself, lesions or degenerations are more or less permanent. Therefore, there is a need for engineering cartilage for articular cartilage resurfacing.

The most commonly used method to approximate articular cartilage resurfacing, for focal chondral defects, is microfracture of the subchondral bone to elicit the production of a fibrocartilaginous tissue which refills the surface. However, because fibrocartilage is structurally inferior to the hyaline articular cartilage, lesions usually reappear, resulting in further damage of the articular surface.

Osteochondral autograft, or mosaicplasty, is a surgical method to repair cartilage.[1] This technique is analogous to a hair-plug transfer. The surgeon removes a small section of the patient's cartilage along with the underlying bone plug. This tissue is obtained from an area that does not participate in high loading, such as parts of the femoral trochlea. The bone and cartilage (hence osteochondral) local graft is then transferred to the defect where a receiving hole has been prepared. Obviously, there is a limit to the amount of tissue available for "harvesting." Furthermore, if the site of harvest is also damaged, this technique may not be possible.

Autologous chondrocyte implantation (ACI) is a method of articular cartilage resurfacing.[2] This technique originated in Sweden more than a decade ago and recently has gained acceptance at orthopaedic centers in the United States. A small amount of the patient's articular cartilage is harvested. Cell culture techniques are then used to increase the cell number from a few hundred thousand to more than 10 million cells. These prepared cells are then reimplanted in the knee to repair and resurface areas of cartilage loss.

Although mosaicplasty and ACI have been effective in treating articular defects of a limited size, donor site morbidity continues to be problematic. Therefore, there is a need for tissue-engineered constructs that use cells derived from a site of nonarticular cartilage. Furthermore, it would be preferable to generate implants that would allow repair of large, nonfocal defects.

Requirements of Tissue Engineering

Tissue engineering techniques are being developed and applied to enhance the repair of cartilage. Before discussing specific techniques, the general requirements of tissue engineering will be discussed. In general, there are three fundamental components in successful tissue engineering: cells, scaffold, and environment.

Cells

Both differentiated cells and tissue progenitor cells are candidates for tissue engineering applications. Differentiated cells have the advantage of being fully committed to and engaged in the specific cellular activities of the desired tissue. However, differentiated cells can only be obtained by harvesting from a healthy donor site (with inherent donor site morbidity), followed by subsequent expansion in culture, which often results in loss of cellular differentiation. This loss of differentiation results from either the inability of the differentiated cells to maintain their phenotype in vitro, or their low rate of proliferation that allows fibroblast-like cells to replace them in culture. A potential solution is to use progenitor cells, such as MSC, that can proliferate to yield clinically significant cell numbers. These cells could then be induced to differentiate into the desired phenotype under controlled conditions.

Scaffold

The three-dimensional nature of tissues, particularly those of the skeleton, require that target engineered tissues be produced with the appropriate shape and form. A biocompatible scaffold is generally used for this purpose. Tissue engineering is essentially "reverse tissue development," in that an artificial scaffold is first assembled and then the cells are seeded. In contrast, normal histogenesis results from differentiated cells producing a tissue-specific extracellular matrix in which the cells reside. Additional modification of the scaffold may involve bioactivation, in that the cells are not merely lodged in the matrix but are exposed to an environment that is conducive to tissue generation. The desired characteristics of a tissue engineering scaffold would include biocompatibility, bioresorbability, and biodegradability upon tissue healing. In addition, the scaffold should be highly porous to permit cell penetration and tissue impregnation, permeable enough to facilitate nutrient delivery and gas exchange, and adaptable to the mechanical environment. Moreover, the scaffold should have a surface that is conducive to cell attachment and migration to permit appropriate extracellular assembly and the transmittal of signaling molecules.

Environment

Tissue engineering is dependent on the local environment to initiate and/or maintain the functional cell/tissue type. Regardless of whether differentiated or progenitor cells are used, the environment plays a key role in the successful outcome of tissue engineering. In general, the environment consists of bioactive factors, including signaling molecules, the appropriate extracellular matrix, and, for skeletal tissues, a mechanobiologically active component, such as mechanical loading. Members of the transforming growth factor-β (TGF-β) superfamily are the most commonly used bioactive factors for cartilage tissue engineering. Members of the TGF-β family are secreted growth factors that interact with specific membrane-bound receptors to activate signaling pathways mediated by the Smad family of signaling molecules, resulting in the regulation of specific gene expression events. These intracellular activities subsequently modulate cellular interactions with TGF-β and other growth factors, as well as with extracellular matrix components. Dynamic loading is often used to simulate the compressive environment of cartilage and to activate chondrocytes.

Tissue Engineering Approaches to Cartilage Repair

Mesenchymal Stem Cells

Derived from adult tissues, MSC exhibit the potential to differentiate into various mesenchymal lineage cell types, including chondrocytes, osteoblasts, adipocytes, fibroblasts, marrow stroma, and other tissues of mesenchymal origin. These MSC reside in diverse tissue types and are able to "regenerate" cell types specific for these tissues (Figure 1). Examples of these tissues include adipose,[3] periosteum,[4,5] synovial membrane,[6] muscle,[7] dermis,[8] pericytes,[9-11] blood,[12] bone marrow,[13] and trabecular bone.[14,15] Currently, bone marrow aspirate is the most accessible and enriched source of MSC. We showed that adult human trabecular bone may be an alternative source of clinically significant numbers of MSC.[16] Given the wide distribution of MSC, the bone marrow stroma is likely to be the source of multipotent cells that circulate to various tissues where they subsequently adopt characteristics to maintain and repair the specific tissue type. In fact, the presence of MSC in tissues other than the marrow stroma strongly suggests the existence of cell populations with more limited capacity for differentiation. These MSC outside of the marrow stroma may have differentiation potentials developmentally adapted to, and perhaps restricted to, the tissues in which they reside.

Bone marrow contains three main cell types: endothelial cells, hematopoietic stem cells, and stromal cells. In a groundbreaking study, Friedenstein and associates[15] isolated colony forming unit-fibroblasts from whole bone marrow and demonstrated that these cells were capable of forming bone and cartilage-like colonies. Subsequent studies have confirmed the multipotent nature of cells isolated with the use of the method developed by Friedenstein and associates.[17] These studies have prompted interest not only in the differentiation potential of MSC, but in the mechanisms of lineage specific differentiation, particularly to bone and cartilage. For example, Pittenger and associates[13] demonstrated that cells isolated from human marrow aspirates were capable of remaining in a stable undifferentiated state even upon long-term maintenance in culture, up to 30 to 40 passages. Colonies derived from single isolated cells from these cultures could be induced to differentiate along osteogenic, adipogenic, and chondrogenic lineages under appropriate conditions. Growth factors that have been used in

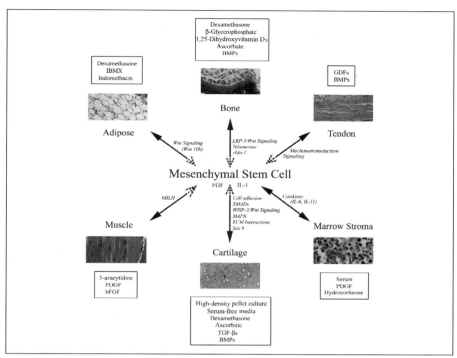

Figure 1 Multilineage differentiation of adult-tissue derived MSC. This diagram demonstrates the in vitro differentiation potential of MSC in vitro, and some of the culture conditions necessary for directing the respective differentiation pathways. *(Reproduced with permission from Tuan RS, Boland G, Tuli R: Mesenchymal stem cells and cell-based tissue engineering. Arthritis Res Ther 2003;5:32-45.)*

cultures of MSC include platelet-derived growth factor, TGF-β, basic fibroblast growth factor, and epidermal growth factor.[18] Techniques for the isolation and in vitro expansion of bone marrow-derived MSC include aspiration, density gradient centrifugation, simple direct plating methods, and size sieving.[19,20]

MSC are characterized as multipotent cells based on their potential for differentiation into a variety of different cell/tissue lineages, even as clonally isolated cells (Figure 1). However, in most studies, it has not been determined if true stem cells are present, or whether the population is instead a diverse mixture of lineage-specific progenitors. Inconsistency in the reported growth characteristics and differentiation potential of MSC underscores the need for a functional definition of MSC. At present, there is lack of a uniform definition of MSC and little data on specific markers that define the cell types characterized as MSC. Thus, MSC are defined by their ability to: (1) differentiate along specific mesenchymal lineages when induced, (2) remain in a quiescent undifferentiated state until provided the signal to

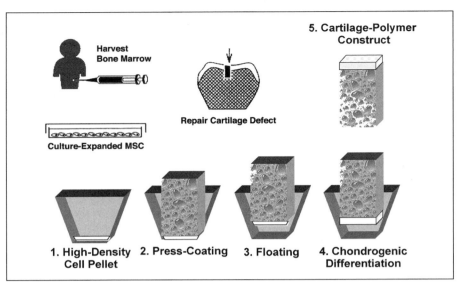

Figure 2 Schematic of the fabrication of cartilage construct by press-coating chondrogenic MSC onto a poly-L-lactide plug. Possible application of such tissue engineered cartilage construct for articular cartilage repair is indicated.

divide asymmetrically, and (3) undergo many more replicative cycles than normal, fully differentiated cells.

Ex vivo Skeletal Tissue Engineering Using Mesenchymal Stem Cells

Cartilage tissue engineering has become one of the most active areas of tissue engineering, given the potential market needs and the scientific challenges of generating a three-dimensional, pressure-bearing tissue structure. Some of the findings from our laboratory are provided below.

In vitro Engineered Cartilage Constructs Produced by Coating Biodegradable Polymer With Human Mesenchymal Stem Cells

We developed an in vitro engineered cartilage construct consisting of biodegradable polymer coated with bone marrow-derived human MSC for the repair of cartilage defects.[21] The construct was fabricated by press-coating a D,D-L,L-polylactic acid polymer block of $1 \times 0.5 \times 0.5$ cm onto a high-density pellet MSC culture consisting of 1.5×10^6 cells (Figure 2). Following attachment of the cells to the polymer surface, chondrogenesis was induced by culturing the construct for 3 weeks in a serum-free, chemically defined, chondrogenic differentiation medium supplemented with ITS-plus (insulin, transferin, and selenium supplement), dexamethasone, ascorbate, sodium pyruvate, praline, and TGF-β1. The coated MSC formed a homogeneous cell layer composed of morphologically distinct, chondrocyte-like cells, surrounded by a fibrous, sulfated proteoglycan-rich extracellular matrix. Type II collagen and cartilage proteoglycan link protein were detected with immunohistochemical analyses. Expression of the cartilage marker genes, collagen types II, IX, X, and XI, and aggrecan, was detected by reverse transcriptase-polymerase chain reaction. Scanning electron microscopy and histology revealed organized and spatially distinct zones of cells within the cell-polymer construct, with the superficial layer resembling compact hyaline cartilage. The fabrication method of coating polymer surfaces with MSC allows the in vitro production of cartilage-polymer constructs of different sizes and shapes, which may be applicable as a prototype for the reconstruction of partial- or full-thickness cartilage defects.

Three-Dimensional Cartilage Formation by MSC Seeded in Polylactide/Alginate Amalgam

We examined the potential of MSC for cartilage tissue engineering by examining their chondrogenic properties within a three-dimensional amalgam scaffold consisting of the biodegradable polymer, poly-L-lactic acid (PLA), alone and with the polysaccharide gel, alginate. Cells were suspended either in alginate or medium and loaded into porous PLA blocks. Alginate was used to improve cell loading and retention within the construct, whereas the PLA polymeric scaffold provided appropriate mechanical support and stability to the composite culture. Cells seeded in the PLA/alginate amalgams and in the plain PLA constructs were treated with different concentrations of TGF-β1 either continuously (10 ng/mL), or during the first 3 days of culture only (50 ng/mL). Chondrogenesis was assessed at weekly intervals with cultures maintained for up to 3 weeks. Histologic and immunohistochemical analyses of the TGF-β1-treated PLA/alginate amalgam and PLA constructs showed development of a cartilaginous phenotype from Day 7 to Day 21, as demonstrated by localization of Alcian blue staining with both collagen type II and cartilage proteoglycan link protein. Expression of cartilage-specific genes, including collagen types II and IX, and aggrecan, was detected in TGF-β1-treated cultures. The initiation and progression of chondrogenic differentiation within the polymeric macrostructure occurred with both the continuous and the 3-day treatment with TGF-β1, suggesting that key regulatory events of chondrogenesis take place during the early period of cell growth and proliferation. Scanning electron microscopy revealed abundant cells with a rounded morphology in the PLA/alginate amalgam. These findings suggest that the three-dimensional PLA/alginate amalgam is a potential bioactive scaffold for cartilage tissue engineering applications.

Development of Nanofibrous Scaffold for Cartilage Tissue Engineering

Using an electrospinning technology, we successfully produced nanofibrous scaffolds using biodegradable polymers, including PLA and poly (ε-caprolactone)[22] (Figure 3). These nanofibrous scaffolds have fiber ranging from 300 to 700 nm in diameter and a 90% porosity. When MSC are seeded onto these scaffolds, they adhere readily to the nanofibers and display prominent expression of several cartilage-associated matrix genes, including aggrecan, collagen types II, IX, and XI. We observed enhanced incorporation of

Figure 3 Scanning electron microscopic view of nanofibrous scaffold fabricated by electrospinning of poly-(e-caprolactone). Bar = 1 mm.

metabolic sulfate as compared with MSC cultured on tissue culture substrateor seeded as high density cell pellets. This finding is indicative of the synthesis of matrix sulfated proteoglycan. Efforts are now being directed to test the ability of these three-dimensional constructs to repair experimentally produced full-thickness articular cartilage defects in animal models.

Future Prospects of Tissue Engineering

Although there is a great deal of interest in tissue engineering as a viable, effective approach in regenerative medicine, there are several challenges to successful development of tissue engineering products. The most important are: (1) safety, (2) satisfactory biologic and mechanical properties, including outcomes, and (3) cost-effectiveness. Starting in the early 1990s, there has been considerable investment in tissue engineering ventures. However, the market value of tissue engineering companies has decreased in relation to the initial investments. Thus, there is a great deal of uncertainty regarding the development of tissue engineering products. Added to this concern is the recent bankruptcy announcement of at least two tissue engineering-based companies in the United States. Therefore, there are challenges for tissue engineering from both the intellectual and market perspectives. The intellectual challenges include identifying and optimizing usage of MSC, fabricating ideal scaffolds, developing functionality postimplantation, using engineered tissues for the delivery of genes and gene products, establishing relevant experimental models, and developing off-the-shelf products. To meet these challenges, active and productive collaborations must exist between scientists, engineers, clinicians, governmental research and regulatory agencies, and device and biologics industries. From a marketing perspective, tissue-engineering products must be scaleable, reproducible, safe, cost-effective, and, at a minimum, represent a reasonable alternative to current treatments. In parallel to the scientific and product development issues, a great deal of attention must also be paid to develop effective education programs for all involved in the tissue engineering field. Patient education is particularly important, given that they must be convinced that tissue engineering products truly represent an improved treatment regimen that will restore function in an effective manner. Tissue engineering is an exciting technique that will alter the field of regenerative medicine.

Summary

Healing of damaged cartilage is a challenging clinical problem because of tissue degeneration or failure to heal. Given the drawbacks of currently available methods to repair articular cartilage injury and degeneration, the feasibility of tissue engineering techniques is being evaluated. It is essential that the requirements of tissue engineering (cells, scaffold, and environment) be addressed to achieve optimal clinical outcomes. Significant progress has been made to date, including studies of the use of an in vitro engineered cartilage construct and methods for the seeding of MSC. Although techniques have yet to be optimized, tissue engineering strategies offer hope for new therapies to manage several skeletal diseases.

Acknowledgment

The author thanks members of his laboratory who have contributed toward the understanding of mesenchymal stem cells and their application in tissue engineering, particularly EJ Caterson, Ulrich Noth, Richard Tuli, Wan-ju Li, David Hall, Genevieve Boland, and Keith Danielson.

References

1. Hangody L, Feczko P, Bartha L, et al: Mosaicplasty for the treatment of articular defects of the knee and ankle. *Clin Orthop* 2001;391:S328-S336.

2. Brittberg M, Lindhal A, Nilsson A, et al: Treatment of deep cartilage defects in the knee with autologous chondrocyte transplantation. *N Engl J Med* 1994;331:889-895.

3. Zuk PA, Zhu M, Mizuno H, et al: Multilineage cells from human adipose tissue: Implications for cell-based therapies. *Tissue Eng* 2001;7:211-228.

4. Nakahara H, Goldberg VM, Caplan AI: Culture-expanded human periosteal-derived cells exhibit osteochondral potential in vivo. *J Orthop Res* 1991;9:465-476.

5. De Bari C, Dell'Accio F, Luyten FP: Human periosteum-derived cells maintain phenotypic stability and chondrogenic potential throughout expansion regardless of donor age. *Arthritis Rheum* 2001;44:85-95.

6. De Bari C, Dell'Accio F, Tylzanowski P, et al: Multipotent mesenchymal stem cells from adult human synovial membrane. *Arthritis Rheum* 2001;44:1928-1942.

7. Bosch P, Musgrave DS, Lee JY, et al: Osteoprogenitor cells within skeletal muscle. *J Orthop Res* 2000;18:933-944.

8. Young HE, Steele TA, Bray RA, et al: Human reserve pluripotent mesenchymal stem cells are present in the connective tissues of skeletal muscle and dermis derived from fetal, adult, and geriatric donors. *Anat Rec* 2001;264:51-62.

9. Diefenderfer DL, Brighton CT: Microvascular pericytes express aggrecan message which is regulated by BMP-2. *Biochem Biophys Res Commun* 2000;269:172-178.

10. Brighton CT, Lorich DG, Kupcha R, et al: The pericyte as a possible osteoblast progenitor cell. *Clin Orthop* 1992;275:287-299.

11. Reilly TM, Seldes R, Luchetti W, et al: Similarities in the phenotypic expression of pericytes and bone cells. *Clin Orthop* 1998;346:95-103.

12. Zvaifler NJ, Marinova-Mutafchieva L, Adams G, et al: Mesenchymal precursor cells in the blood of normal individuals. *Arthritis Res* 2000;2:477-488.

13. Pittenger MF, Mackay AM, Beck SC, et al: Multilineage potential of adult human mesenchymal stem cells. *Science* 1999;284:143-147.

14. Noth U, Osyczka AM, Tuli R, et al: Multilineage mesenchymal differentiation potential of human trabecular bone-derived cells. *J Orthop Res* 2002;20:1060-1069.

15. Osyczka AM, Noth U, Danielson KG, et al: Different osteochondral potential of clonal cell lines derived from adult human trabecular bone. *Ann N Y Acad Sci* 2002;961:73-77.

16. Tuli R, Seghatoleslami MR, Tuli S, et al: A simple, high-yield method for obtaining multipotential mesenchymal progenitor cells from trabecular bone. *Mol Biotechnol* 2002;23:37-49.

17. Friedenstein AJ, Chailakhyan RK, Gerasimov UV: Bone marrow osteogenic stem cells: in vitro cultivation and transplantation in diffusion chambers. *Cell Tissue Kinet* 1987;20:263-272.

18. Keating A, Horsfall W, Hawley RG, et al: Effect of different promoters on expression of genes introduced into hematopoietic and marrow stromal cells by electroporation. *Exp Hematol* 1990;18:99-102.

19. Wakitani S, Saito T, Caplan AI: Myogenic cells derived from rat bone marrow mesenchymal stem cells exposed to 5-azacytidine. *Muscle Nerve* 1995;18:1417-1426.

20. Kuznetsov SA, Friedenstein AJ, Robey PG: Factors required for bone marrow stromal fibroblast colony formation in vitro. *Br J Haematol* 1997;97:561-570.

21. Noth U, Tuli R, Osyczka AM, et al: In vitro engineered cartilage constructs produced by press-coating biodegradable polymer with human mesenchymal stem cells. *Tissue Eng* 2002;8:131-144.

22. Li WJ, Laurencin CT, Caterson EJ, et al: Electrospun nanofibrous structure: A novel scaffold for tissue engineering. *J Biomed Mater Res* 2002;60:613-621.

Chapter 17

Engineering Cartilage Structures

Michael Sittinger, PhD

Abstract

The basic principle of tissue engineering for repair of musculoskeletal tissue is the delivery of functionally active cells within an appropriate carrier system to the damaged site. Ideally, the implanted construct will restore the original architecture and function of the pathologically altered tissue. This approach includes the interactive triad of responsive cells, a supportive matrix, and bioactive molecules promoting differentiation and regeneration. New developments in cell culture techniques, delivery systems and materials, and regenerative concepts for the engineering of cartilage tissue are discussed.

Introduction

After more than a decade of research of tissue engineering, several cell-based therapies to achieve physiologic regeneration of damaged tissues or organs are at various stages of development. The current technical approaches in cartilage tissue engineering are typically based on artificial tissue constructs of autologous cells and biomaterials that function as a cell-embedding component and/or as a scaffold for the formation of three-dimensional tissues (Figure 1).

Experimental approaches to engineer skeletal tissue transplants vary considerably for different intended clinical indications. In vitro formation of external ear cartilage demands particular attention to tissue preshaping and in vivo shape stabilization.[1] Joint cartilage regeneration involves special techniques to handle and anchor cells and tissues in the joint defect. Successful preliminary clinical studies have shown that in vitro engineering and transplantation of autologous cartilage tissues is possible.

However, new strategies for the regeneration of severely degenerated cartilage surfaces in osteoarthritis and in chronic inflammatory joint diseases are still experimental. The next generation of tissue engineering therapies is expected to rely on the regenerative potential of stem cells and morphogenic growth factors to induce healing in vivo.

Culture Techniques and Cells

Although frequently used in tissue engineering approaches, conventional monolayer cultures of autologous cells have limitations in the generation of highly differentiated structures. The reasons for this are as follows: (1) metabolic conditions within the culture medium are unstable and not comparable

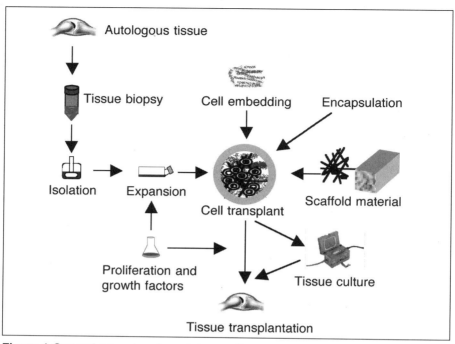

Figure 1 General procedure for autologous tissue replacement. The isolation and selection of autologous cells from tissue biopsies are followed by ex vivo expansion. Expanded cells are embedded within a suitable biomatrix. Cell/matrix composites are either cultured in perfusion chambers or directly implanted into a lesion.

to the in vivo situation, (2) no extracellular matrix (ECM) is formed and consequently, (3) important cell-cell and cell-ECM interactions cannot take place, and (4) mesenchymal cells tend to dedifferentiate in monolayer cultures, including the transformation of some chondrocytes into fibroblasts.[2,3]

To optimize the medium composition, artificial tissue constructs can be grown in culture in bioreactors. This approach also addresses the specific challenges related to the culture of organoid arrangements, including high nutrient consumption due to high densities and the presence of supportive delivery materials. Perfusion of these reactors permits the stabilization of culture medium components, the maintenance of secreted autocrine factors (such as morphogenic signals) at a desired level, and the avoidance of an excess in synthesized paracrine factors. Control of these factors in a bioreactor results in a culture that more closely mimics the in vivo situation than standard culture methods.[4] Additional progress might be achieved by the use of gradient chambers. The use of a concentration gradient of differentiating morphogenic factors established across the artificial tissues would resemble the conditions during embryonic development.

Three-dimensional cell culture systems have been successfully used to support ECM formation and normal cell-cell interactions.[5,6] Such systems, including scaffolds or gels, mimic the in vivo situation quite well. Cells grown in these settings do not dedifferentiate or even redifferentiate to their original phenotype.[3] New developments in this area will be discussed below.

Most tissue repair approaches currently used involve autologous cells obtained by biopsy of healthy sites of the appropriate tissue type. Although the use of these differentiated cells has proved to be successful in many approaches, there are challenges to the use of autologous cells. The availability of cells might be restricted, especially if the desired site of biopsy is severely damaged, and there is morbidity associated with the biopsy procedures. With respect to their functional termination, not all cell types are able to proliferate in vitro. In view of these problems, research has focused on tissue regeneration involving precursor or multipotent stem cells. Autologous mesenchymal stem cells from the bone marrow appear to be good candidates for tissue repair therapies. These cells are relatively easy to obtain, are autologous, have the capacity for unlimited but controlled expansion, and have the potential to differentiate into various mesenchymal tissues depending on the microenvironment. Indeed, under defined culture conditions, uncommitted mesenchymal progenitor cells from bone marrow have been shown to differentiate into chondrocytes.[7]

Delivery Materials

Pressure resistance and fixation of the transplant onto the bone is both important and difficult in joint cartilage replacement. In theory, the artificially grown cartilage layers could be attached directly to the defective joint surface using fibrin glue, resorbable sutures, pins, or staples. Other promising approaches focus on osteochondral replacements for subchondral bone repair, such as artificial cartilage attached directly on porous calcium carbonate.[8]

Until recently, focal articular cartilage defects were primarily treated by injecting suspensions of cartilage cells into a lesion covered by a small periosteal flap.[9] Today, there is progress in the use of fully developed or preshaped cartilage transplants. The formation of such three-dimensional tissues is achieved using different scaffolds. These scaffolds provide mechanical stability, particularly during the initial posttransplantation phase. Scaffolds also help to ensure a homogeneous and efficient distribution of the cells within the constructs and thus within the cartilage lesion to be repaired. The matrix should allow the homogeneous distribution of a sufficient number of cells within its structure as close contact between newly synthesized ECM molecules and cells is required for spatial maturation of the tissue.

The materials used for these matrices should be biocompatible. When implanted in vivo, it is important that scaffold materials do not evoke immunologic reactions or alter the normal metabolism of cells. The matrix should also be flexible but sufficiently stable to permit the firm fixation or arthroscopic implantation of cultured cartilage material. After prolonged cultivation, the tissue becomes stabilized by the newly formed ECM and the supportive matrix is no longer required. Therefore, use of biodegradable scaf-

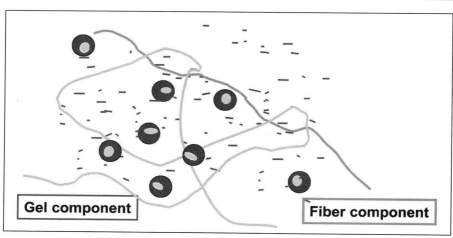

Figure 2 Schematic depiction of a construct for the engineering of cartilage tissue based on fibers and an embedding substance. The embedding substance ensures the three-dimensional immobilization and uniform distribution of cells within the fibrous meshwork.

fold material that can be resorbed after a certain period of time without interfering with the metabolic activity of the maturing tissue is recommended.[10]

Fibers

Polymer fiber constructs consisting of poly(α-hydroxyesters) meet the aforementioned requirements most closely. These resorbable fibers can form a three-dimensional fleece that is robust, but flexible so that it can be shaped to meet anatomic requirements. The low material-to-volume ratio of these fibers effectively promotes the three-dimensional growth of cells.

When loading the fleeces with cells, the latter tend to leach out of the fiber structure. To avoid this, chondrocytes can either be directly attached to the fibers using adhesion factors or be embedded in gel-like substances[10] such as agarose, fibrin, or hyaluronan, which permit a homogeneous distribution of cells within the carrier. Gel-like substances increase the average distance between fibers, which further reduces the total amount of polymer per tissue volume. In this combination, fibers serve as a scaffold for the tissue and the gels form the three-dimensional distribution of the cells (Figure 2). An alternative approach to accumulate cells and matrix molecules within the fiber structure involves encapsulating the whole structure in a semipermeable membrane.[11]

Collagen matrices exhibit a high cell-binding capacity. However, collagen is known to rapidly loose its mechanical stability when maintained in culture media. Indeed, matrices formed with the use of collagen type I often have inconsistent characteristics. Furthermore, weak interconnections between cavities, as well as low porosity, tend to impede the formation of the optimal three-dimensional distribution of cells. Implantation of collagen

Figure 3 In vitro engineered cartilage construct for the treatment of joint defects. Such an engineered construct has been applied arthroscopically to cartilage lesions.

into articular cartilage lesions without the addition of cells leads to wound closure. However, the newly formed tissue is rather low in proteoglycans, which are decisive for the elasticity of the transplant.[13]

Hydrogels

In contrast to polymer fibers, gels do not provide the stability that is required for the in vitro formation of cartilage transplants. However, the direct injection of cell-carrying gels into a lesion may be used to treat smaller defects.[12] Hydrogels mimic several distinct structural and physicochemical characteristics of the ECM, which is a gel consisting of hyaluronan, proteoglycans, and supportive collagen fibers.

Agarose gels are frequently used for studies of chondrogenesis. However, they are rarely used to generate transplants because the immunologic aspects of agarose and its degradation products have yet to be clarified. Within agarose gels, the distance between cells is relatively large. Although this gives ECM components sufficient space to aggregate,[14] large molecules such as aggrecan remain in the pericellular domain and impede the formation of a connecting intercellular matrix.[15]

Fibrin gels are commonly used in surgical medicine and they provide the biocompatibility required of tissue engineered products. Because the mechanical stability of fibrin gels is rather limited, a combination of these gels with more stable components is recommended. Promising results have been achieved using a chondrocyte-carrying fibrin gel that was mechanically stabilized with a resorbable polymer fleece (Figure 3). This combination facilitated optimal handling during surgery, exhibited satisfactory mechanical stability, and permitted a homogeneous three-dimensional distribution of cells within the transplant.[16]

Alginate is an immobilization matrix, within which chondrocytes are able to generate an ECM and maintain their phenotype in vitro. The subcutaneous implantation of alginate beads covered with chondrocytes has yielded tissue (30% to 40% cartilage) that became wrapped by a fibrous tissue capsule within 6 weeks after implantation.[17] When alginate beads are combined with hyaluronan or fibrin gels, the number of cells that can be trapped within the beads increases substantially.[18] However, the use of alginate as an implant carrier is still under careful consideration because of doubts about the purity of the material and its immunologic behavior.

Hyaluronan chains form networks that are useful carrier matrices.[19] Depending on the type of linking reaction, either fibers or gels can be generated. Low levels of linkage are associated with advanced water uptake into the gels. Although this weakens the mechanical stability of a hyaluronan matrix, it increases its biodegradability and has a positive impact on the generation of a physiologically relevant tissue pressure within the transplant.

Evolution of Regenerative Concepts

Despite the promising approaches described above, the treatment of more extensive joint lesions is still problematic. In particular, the restricted potential of cultured chondrocytes to proliferate and subsequently differentiate impedes the generation of larger portions of replacement tissue of sufficient stability and adequate structure. Moreover, musculoskeletal structures such as the joint comprise a morphologic and functional unit, and pathologic changes are never restricted to a single tissue. Thus, an ideal intervention must address the full extent of the pathophysiology.

Novel strategies to circumvent these shortcomings have focused on the use of osteoinductive proteins and other mediators to stimulate the regeneratation of cartilage and other joint tissues in a concerted manner in vivo. Preliminary investigations using bone morphogenetic protein-2 (BMP-2) or transforming growth factor-β-1 have been promising. Injection of these factors into the joint induced the formation of type II collagen and the synthesis of proteoglycans by chondrocytes, thereby stimulating cartilage growth and maturation.[20] In another approach, osteochondral defects of rabbit knees were treated with collagen sponges coated with human BMP-2. Histologic staining for type II collagen and proteoglycans revealed the formation of normal cartilage tissue after 24 weeks in vivo.[21] Thus, BMP-2 may be a promising candidate for use in the engineering of cartilage tissue in joint defects. For destructive joint diseases such as osteoarthritis and rheumatoid arthritis, BMP-7 has successfully promoted the differentiation and stabilization of cartilage tissue.[22] It is likely that inductive regenerative treatments will involve a sequential cascade of more than one factor to first stimulate cell migration and proliferation, and, subsequently, enhance the differentiation and maturation of regenerating tissues. The combination of such approaches with the cell/scaffold composites described above would represent an important advance in engineering of cartilage tissue.

Summary

There continues to be much interest in the use of tissue engineering to replace damaged cartilage. Approaches in cartilage tissue engineering are usually based on artificial tissue constructs of autologous cells and biomaterials that function as a cell-embedding component and/or as a scaffold for the formation of three-dimensional tissues. Much work has been done to optimize methods of producing cartilage structures in vitro. Tissue constructs can be grown in culture in bioreactors, and three-dimensional cell culture systems have been successfully used to support ECM formation and normal cell-cell interactions. Various materials, including fiber and hydrogels, have been evaluated for use in the delivery of tissue engineered cartilage. It is anticipated that tissue engineered products for clinical use will also involve the use of osteoinductive proteins and/or other growth factors to stimulate repair of tissue in vivo.

References

1. Haisch A, Klaring S, Groger A, Gebert C, Sittinger M: A tissue-engineering model for the manufacture of auricular-shaped cartilage implant. *Eur Arch Otorhinolaryngol* 2002;259:316-321.

2. Minuth WW, Sittinger M, Kloth S: Tissue engineering: Generation of differential artificial tissues for biomedical applications. *Cell Tissue Res* 1998;291:1-11.

3. Benya PD, Shaffer JD: Dedifferentiated chondrocytes re-express the differentiated collagen phenotype when cultured in agarose gels. *Cell* 1982;30:215-224.

4. Sittinger M, Schultz O, Keyszer G, Minuth WW, Burmester GR: Artificial tissues in perfusion culture. *Int J Artif Organs* 1997;20:57-62.

5. Schultz O, Sittinger M, Haeupl T, Burmester GR: Emerging strategies of bone and joint repair. *Arthritis Res* 2000;2:433-436.

6. Risbud MV, Sittinger M: Tissue engineering: Advances in in vitro cartilage generation. *Trends Biotechnol* 2002;20:351-356.

7. Kadiyala S, Young RG, Thiede MA, Bruder SP: Culture expanded canine mesenchymal stem cells possess osteochondrogenic potential in vivo and in vitro. *Cell Transplant* 1997;6:125-134.

8. Kreklau B, Sittinger M, Mensing MB, et al: Tissue engineering of biphasic joint cartilage transplants. *Biomaterials* 1999;20:1743-1749.

9. Brittberg M, Lindahl A, Nilsson A, Ohlsson C, Isaksson O, Peterson L: Treatment of deep cartilage defects in the knee with autologous chondrocyte transplantation. *N Engl J Med* 1994;331:889-895.

10. Sittinger M, Bujia J, Rotter N, Reitzel D, Minuth WW, Burmester GR: Tissue engineering and autologous transplant formation: Practical approaches with resorbable biomaterials and new cell culture techniques. *Biomaterials* 1996;17:237-242.

11. Sittinger M, Lukanoff B, Burmester GR, Dautzenberg H: Encapsulation of artificial tissues in polyelectrolyte complexes: Preliminary studies. *Biomaterials* 1996;17:1049-1051.

12. Risbud M, Ringe J, Bhonde R, Sittinger M: In vitro expression of cartilage-specific markers by chondrocytes on a biocompatible hydrogel: Implications for engineering cartilage tissue. *Cell Transplant* 2001;10:755-763.

13. Speer DP, Chvapil M, Volz RG, Holmes MD: Enhancement of healing in osteochondral defects by collagen sponge implants. *Clin Orthop* 1979;144:326-335.

14. von Schroeder HP, Kwan M, Amiel D, Coutts RD: The use of polylactic acid matrix and periostal grafts for the reconstruction of rabbit knee articular defects. *J Biomed Mater Res* 1991;25:329-339.

15. Verbruggen G, Veys EM, Wieme N, et al: The synthesis and immobilization of carti-
    lage-specific proteoglycan by human chondrocytes in different concentrations of
    agarose. *Clin Exp Rheumatol* 1990;8:371-378.

16. Perka C, Sittinger M, Schultz O, Spitzer RS, Schlenzka D, Burmester GR: Tissue
    engineered cartilage repair using cryopreserved and noncryopreserved chondrocytes.
    Clin Orthop 2000;378:245-254.

17. Cao Y, Rodriguez A, Vacanti M, Ibarra C, Arevalo C, Vacanti CA: Comparative study
    of use of poly (glycolic acid), calcium alginate and pluronics in the engineering of
    autologous porcine cartilage. *J Biomater Sci Polym Ed* 1998;9:475-487.

18. Lindenhayn K, Perka C, Spitzer R, et al: Retention of hyaluronic acid in alginate
    beads: Aspects for in vitro cartilage engineering. *J Biomed Mater Res* 1998;44:149-
    155.

19. Tomihata K, Ikada Y: Preparation of cross-linked hyaluronic acid films of low water
    content. *Biomaterials* 1997;18:189-195.

20. van Beuningen HM, Glansbeek HL, Kraan van der PM, Berg van den WB:
    Osteoarthritis-like changes in the murine knee joint resulting from intra-articular
    transforming growth factor-(beta) injection. *Osteoarthritis Cartilage* 2000;8:25-33.

21. van Beuningen HM, Glansbeek HL, Kraan van der PM, Berg van den WB:
    Differential effects of local application of BMP-2 or TGF-(beta)1 on both articular
    cartilage compositions and osteophyte formation. *Osteoarthritis Cartilage*
    1998;6:306-317.

22. Kaps C, Bramlage C, Smolian H, et al: Bone morphogenetic proteins promote carti-
    lage differentiation and protect engineered artificial cartilage from fibroblast invasion
    and destruction. *Arthritis Rheum* 2002;46:149-162.

Chapter 18

The Role of the Microenvironment in Cartilage Tissue Engineering

Ernst B. Hunziker, MD

Abstract

Lesions of articular cartilage do not heal spontaneously. Although defects that span this layer and penetrate the subchondral bone plate and trabeculae can undergo spontaneous repair, the tissue formed is fibrous in nature, of poor mechanical competence, and short-lived.[1,2] Furthermore, only defects that fall within a narrow range of critical size elicit even this unsatisfactory healing response. These observations and the need for methods to heal cartilage in the clinical setting are the driving force behind attempts to engineer articular cartilage. The ultimate goal is to use tissue engineering to produce articular cartilage with structural, compositional, mechanical, and endurance qualities similar to those of native tissue. However, success in these endeavors requires the consideration and creation of many aspects of the in vivo microenvironment. Before an engineered cartilage construct is implanted within a lesion, the defect bed is first prepared by the removal not only of diseased but also of some adjacent, healthy tissue. This procedure is associated with some morbidity. Even routine surgical suturing of articular cartilage can have deleterious effects, and alternatives to this mode of tissue fixation are suggested. This article addresses the problem of eliminating undesired and unpredictable spontaneous repair activities within full-thickness defects. For example, successful healing can be adversely affected by reactivity of the host to the implanted material. Measures to ensure a compartment-specific repair result within cartilaginous and bony carrels will also be discussed.

Introduction

The ultimate goal of tissue engineering is to reestablish the structure, composition, physical properties, and functionality of tissues and organs. Various approaches have been used, but all basically involve the use of an appropriate cell pool, a suitable scaffolding material, and signaling substance(s). However, attempts to engineer articular cartilage have been suboptimal. Given the numerous variables that are now known to play important roles within the in vivo microenvironment, an empirical approach to the problem must now be adopted. Improvements in the quality of engineered articular cartilage can be achieved only if more rational, systematic, and step-by-step methods are applied.

There are several issues regarding the microenvironment that need to be addressed to provide optimal clinical results for replacement or repair of cartilage with tissue-engineered products. For example, the defect type and scaling must be considered. For full-thickness defects, means of overcoming adverse and unpredictable reactions stemming from the bone marrow and vascular spaces by compartment-specific tissue engineering are still being studied. Micromechanical environments must be established to test the use of various constructs in appropriate lesion models and suitable species. The local mechanical and deformational forces are of great importance because these physical parameters have a considerable bearing on the time course of healing and on the physiotherapeutic recommendations made after an intervention. Local factors influence the integration of an engineered construct with native tissue, and several osteoinductive and growth factors are being evaluated. In addition, healing is affected by the reactivity of the host to the implanted construct. These issues are discussed in this review chapter.

Defect Type and Scaling

Before embarking on any tissue-engineering approach, the investigator must have a clear idea of the size and type of defect to be treated in the patient. In humans, the knee and hip joints are the most common sites of structural lesions. In these locations, the articular cartilage layer may be 2 to 4 mm in height and thus exceeds by approximately tenfold that in most large experimental animals, such as goats, sheep, or dogs. In smaller animals, such as rats, the difference in height is obviously even greater. Hence, when planning experiments, it is necessary to consider not only the size of the construct that must be engineered in vitro for human use, but the scale of animal models (Figure 1) and the effect of scale on the in vivo microenvironment.[3]

A partial-thickness defect is confined to the layer of articular cartilage tissue and may just touch the zone of calcified cartilage.[4] A construct implanted within such a defect will be exposed only to cartilage tissue, and very little cell migration or signaling from this compartment is to be expected. In humans, such a defect is typically 2 to 4 mm in height.[5] This height can be reproduced in animal models only by penetrating the subchondral bone tissue. To preserve the microenvironment of the partial-thickness defect, the portions of the lesion walls that penetrate the bony compartment must be sealed off from the cells, signaling substances, and vascular ingrowth emanating from the bone marrow and vascular spaces. This can be achieved, for example, by applying a fibrin glue[6] (Figure 2, A). By this means, the scale and microenvironment of a human partial-thickness defect can be simulated in an animal model. These animal models permit the physiologic behavior of a large human-scale construct to be tested in vivo under realistic conditions, at least during the early postimplantation phase.

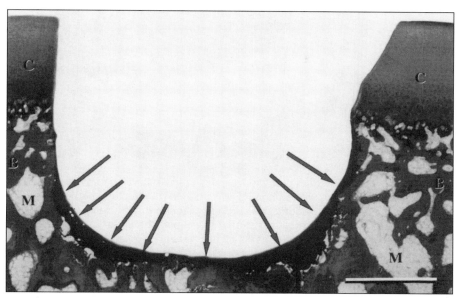

Figure 1 Light micrograph of a full-thickness defect in mature bovine articular cartilage. A virtual partial-thickness defect has been created by lining its floor and walls with fibrin glue (arrows), which blocks cell migration and vascular invasion from osseous tissue and the bone-marrow spaces into the defect void for a considerable time (up to 1 week). B = bone tissue; C = articular cartilage; M = bone-marrow spaces. Thick section, surface-stained with McNeil's Tetrachrome, Toluidine Blue 0, and basic Fuchsine. Bar = 1 mm. *(Reproduced with permission from Hunziker EB: From the preclinical model to the patient, in* Tissue Engineering of Cartilage and Bone. *John Wiley & Sons, in press.)*

Special Considerations Relating to Full-Thickness Articular Cartilage Defects

Unlike partial-thickness defects, full-thickness defects involve the entire depth of the articular cartilage layer and open into the underlying bone tissue, bone marrow, vascular spaces, connective tissue, and adipose tissue. Clearly, the biologic microenvironment of an implanted construct for a full-thickness defect differs dramatically from that of a partial-thickness defect.[7] Each of the aforementioned tissue compartments can be the source of one or several populations of cells that may contribute to the repair response in an uncontrolled and unpredictable manner. The construct will also be invaded by blood vessels and infiltrated by a multitude of ill-defined signaling substances that influence healing. The repair cartilage formed under these influences has been shown to be structurally, compositionally, and functionally inferior to native tissue and to be short-lived.[1,2] Moreover, these factors compromise the ability to analyze the true effect of a construct on repair. Thus, measures must be taken to eliminate these factors and to simplify the situation.

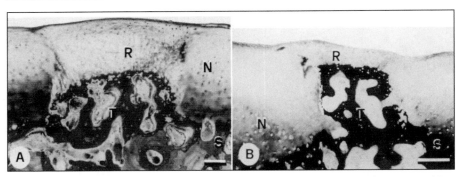

Figure 2 A, Light micrograph of a full-thickness articular cartilage defect created in a miniature pig and filled with a chondrogenic matrix. No structural barrier was inserted at the presumptive border between cartilage and bone compartments. Eight weeks after surgery, repair cartilage (R) occupies the upper half of the cartilaginous defect space. This tissue has a higher density of cells and is more fibrous than native cartilage (N). The lower half of the cartilaginous defect space is occupied by repair bone tissue (T), which has grown upward from and completely fills the underlying bone compartment. Repair bone (T, dark red) is principally of the woven type, whereas native subchondral bone tissue (S, light red) is lamellar. Bar = 100 μm. **B,** Light micrograph of a full-thickness articular cartilage defect created in a miniature pig and filled with a chondrogenic matrix. No structural barrier was inserted at the presumptive border between cartilage and bone compartments. Eight weeks after surgery, repair cartilage (R) occupies no more than approximately one fifth of the depth of the cartilaginous defect space. As in Figure 1, this tissue has a higher density of cells and is more fibrous than native cartilage (N). The bulk of the cartilaginous defect space is occupied by repair bone tissue (T), which has grown upward from and completely fills the underlying bone compartment. Repair bone (T) is of the woven type, whereas subchondral bone tissue (S) is lamellar. Bar = 200 μm. *(Reproduced with permission from Hunziker EB, Driesang IM, Saager C: Structural barrier principle for growth factor-based articular cartilage repair.* Clin Orthop *2001;391:S182-S189.)*

Furthermore, if the tissue engineering approach involves the use of chondrogenic differentiation factors, such as transforming growth factor β or a bone morphogenetic protein, then these substances, which are also osteogenic, may likewise stimulate the upgrowth of blood vessels and consequently the formation of bone within the cartilaginous compartment[8] (Figure 2, *B*). To ensure differential, compartment-specific repair, measures must be taken to thwart vascular upgrowth even in the presence of chondrogenic differentiation factors. One approach is to insert a structural barrier at the prospective interface between the cartilaginous and bony compartments of the defect. This barrier would thus be filled in two stages with the engineered construct.[8] A more elegant solution to the problem would be to adopt the "functional" barrier principle, which involves incorporating an antiangiogenic factor exclusively within the construct destined for the bony compartment of the defect. This measure would prevent the upgrowth of vessels into the cartilaginous compartment and the formation of bone therein without interfering with signaling or the flow of nutrients, which are important for the long-term maintenance of repair cartilage tissue. The functional barrier principle has indeed been applied with success in a miniature pig model of a full-thickness defect.[9]

Mechanical Microenvironment

During normal joint usage, the articular cartilage layer undergoes considerable deformation and compression. If the physical properties of the engineered construct or repair tissue differ significantly from those of the surrounding native cartilage, then shear forces will be generated along the interface between the two compartments during normal usage. Cracks will develop at this junction and the two compartments will consequently fail to act in concert. Recent finite-element modeling computations indicate that unless the stiffness of the repair tissue is at least 80% of that of native articular cartilage, it will fail to survive.[10] Although more data are required, it is clear that the stiffness of an engineered construct should approximate that of native articular cartilage at the time of implantation if it is fully-differentiated, or very soon thereafter (ie, within a few weeks), if it is rudimentary and composed of cells and a matrix requiring differentiation and maturation in vivo.

Tissue Integration

If the mechanical stiffness of an engineered construct is not comparable to that of the surrounding native cartilage, gaps will develop along the interface between the two tissue compartments. These zones of discontinuity will interfere with fluid flow, the transport of nutrients, and intratissue signaling. Once formed, these zones are unlikely to be subsequently bridged.[4] The process of separation, albeit focal at first, will be expedited and exacerbated if the two tissue compartments are not well bonded at the time of implantation. Human partial-thickness defects, with a typical height of 2 to 4 mm, can easily be 1 to 2 cm or more in diameter. Hence, the interfacial area to be annealed is considerable.

Implanted matrices and engineered constructs adhere to the floor and walls of an articular cartilage defect, but adherence is poor because of the high aggrecan content of the native tissue. The adhesive properties of cartilage can be enhanced transiently, for several weeks, by enzymatically degrading superficially located proteoglycans with chondroitinase AC (Figure 3) or trypsin.[11] The use of these enzymes does not lead to cell loss or apoptosis nor to a reduction in the metabolic activity of chondrocytes bordering the defect[12] (Figure 3). Although this treatment may help to stabilize a construct during the initial phase of healing, it is unlikely to satisfy the more robust mechanical requirements of later stages.[13] Bonding could be strengthened by applying a biologic glue, such as tissue transglutaminase, which occurs naturally in cartilage tissue.[13] Fibrin glues could also be used, but because they need to be applied at much higher (nonphysiologic) concentrations than tissue transglutaminase to achieve comparable annealing

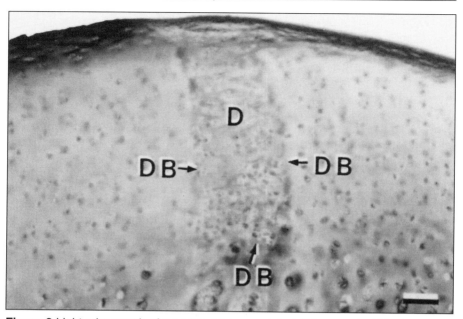

Figure 3 Light micrograph of a partial-thickness defect (D) created within the knee joint of articular cartilage in a mature Goettingen miniature pig. The defect was treated with chondroitinase AC and application of a fibrin matrix containing both free (4 ng/mL) and liposome-encapsulated (600 ng/mL) transforming growth factor-β_1. The defect is, 2 months after treatment, filled with repair tissue (D) that has a cartilage-like appearance. Cells have undergone transformation in the chondrocyte-like ones but are more isotropically distributed and present at a greater density than those within the surrounding native tissue. A gradient in cell size and shape is nonetheless apparent with chondrocytes in superficial regions being small and spindle-like, but those in deeper areas are larger with an oblate spheroid form. The chondrocyte-like cells have laid down a matrix that stains somewhat less homogeneously than that of native tissue, which circumstance indicates that it is not truly hyaline but still fibrous in nature. At the defect borders (DB), repair and native tissue are well integrated. Bar = 100 μm. (*Reproduced with permission from Hunziker EB: Growth-factor-induced healing of partial-thickness defects in adult articular cartilage.* Osteoarthritis Cartilage *2001;9:22-32.*)

results,[13] they are probably less suitable for use. The two tissue compartments could also be bonded chemically with cross-linking, but this reaction may generate artificial epitopes that could precipitate undesired immunoreactivity.

Biologic glues can also be used to affix tissue flaps in a lid-like fashion to the healthy articular cartilage surface surrounding defect voids. These materials are frequently used to prevent the loss of implanted material. Gluing is highly preferable to suturing in this situation, as will be discussed in the next section.

Chondrocyte Loss Associated With the Preparation of Defect Beds and With Suturing

The need for an intimate contact between native articular cartilage and the implanted construct is clearly established. A close association between these two tissue compartments will be facilitated during the press-fitting of a construct into a clinical lesion if the defect bed is first prepared by smoothing its walls and floor. This process involves the surgical excision of some healthy articular cartilage together with degenerated tissue. Both in vitro[14] and in vivo[12] studies have demonstrated that the surgical excision of healthy articular cartilage is associated with the loss of chondrocytes from tissue adjacent to the newly created wound edge. This loss of cells can be diminished if extremely sharp cutting instruments are used[15] and if tissue deformation is minimized by using a manual pressure that is as light as possible during the cutting action. Even routine suturing can be detrimental to the chondrocyte population. When inserted into articular cartilage, a suturing needle creates a canal, which, in effect, represents a small excavated lesion. Indeed, needle holes in cartilage tissue is associated with a loss of chondrocytes.[16]

In regions of articular cartilage tissue that have lost chrondrocytes, the surviving cells are not able to compensate for this loss by augmenting cell numbers or by altering their metabolic activity.[12] Because these cells must maintain and remodel a larger volume of matrix than before, the process of surgical "lesioning" could eventually result in the degeneration of healthy articular cartilage tissue.

Synovial Tissue Compartment

An engineered construct implanted within an articular cartilage defect is exposed directly to the synovial fluid and indirectly, via the joint cavity, to the synovial tissue compartment. The highly viscous synovial fluid is important not only for the lubrication and nourishment of the articular cartilage layer but also for its signaling activities and overall maintenance. Furthermore, it serves as a functional link between articular cartilage tissue and the synovial membrane, which provides a defense system against infection. Macrophages and mesenchymal cells originating from the synovial membrane are always present within the synovial fluid and may participate in either defense or repair activities.[4] Activation of the defense system, although important in counteracting infection, could jeopardize the engineered tissue and lead to its destruction. Thus there is a dilemma in trying to both sustain the beneficial activities of the synovium fluid and discourage adverse reactivity within the synovium. The physiologic rheology of the synovial compartment needs to be considered in any articular cartilage engineering approach. The degradation products of biologic or synthetic cell-carrier materials could, when released into the synovial fluid, stimulate an adverse reaction by the synovial membrane. Chondrogenic differentiation factors incorporated into the construct could, if released at too high a concentration into the synovial fluid, evoke inflammation of the synovium, joint effusion, pannus formation, destruction of articular cartilage tissue, and osteophyte formation within the subsynovial tissue space.[4,17,18]

Immunoreactivity

Both synthetic and biologic compounds can evoke an inflammatory response when implanted within an articular cartilage lesion. There is no physiologic situation in which an implanted material can be immediately accepted by the body and directly incorporated as a functioning unit. The implant must first be resorbed and then replaced. This problem needs to be confronted in both partial- and full-thickness articular cartilage defects. In partial-thickness defects, the absence of blood vessels and only indirect access to the immune system through the synovial fluid may lower the risk of immediate rejection of a construct. But all implanted materials will eventually be degraded and the products of degradation could elicit an adverse reaction at a later date. Clearly though, if the deposited material is exposed to the bone marrow and vascular spaces, as it will be in a full-thickness defect, the inflammatory response is likely to be extensive and chronic, with a high risk of rejection of the construct. The implanted material must be biocompatible and biocompatibility should be tissue-specific. Even autologous erythrocytes or autologous fibrin can elicit an inflammatory response and be rejected if they are deposited outside the blood compartment. For example, implantation of these materials within the connective tissue can cause rejection.[19]

A novel and elegant approach to the problem of immunoreactivity would be to enroll the activity of inflammatory cells in the neoformation of cartilage tissue. This concept has already been enacted in bone-formation processes. The set-up involved titanium-alloy implants bearing calcium-phosphate coatings into which an osteogenic signaling substance had been incorporated.[20]

Is the Aim to Recreate Ontogenetic Differentiation Processes During Articular Cartilage Repair the Right Approach to Tissue Engineering in Adult Human Patients?

The ideal course for successful articular cartilage repair is commonly held to be a recapitulation of embryologic, fetal, and early postnatal differentiation process. But I believe that this might not lead to success and that repair of articular cartilage in adult human patients requires a different approach. The reasons for this view are manifold. To begin with, the precursor cell populations functioning during ontogenesis are not present in the mature individual. Furthermore, precursor cells present in immature and adult organisms respond differently to the same growth factors. Moreover, the microenvironments within which these substances act is dramatically different in the embryologic state and in adults. The signaling agents operative during ontogenesis can undoubtedly be used to drive the differentiation of adult precursor cells in the desired direction. However, these substances will be acting within a different microenvironment and upon different populations of target cells whose responsiveness thereto will be unlike that in the embryologic state.

In addition, the structural and functional requirements of articular cartilage in the immature and mature organism differ. During embryologic, fetal, and early postnatal development, articular cartilage tissue has a high cellularity and an isotropic architecture. In contrast, adult articular cartilage has an extremely low cellularity[5] and an anisotropic structure with chondrocytes organized into morphologically distinct vertical columns and horizontal strata. The low cellularity and associated large matrix mass of adult articular cartilage tissue, together with the anisotropic ordering, are essential for its optimal biomechanical functioning. The high cellularity and isotropic organization of immature tissue cannot meet these biomechanical needs.[21] The change from an isotropic structure with high cellularity occurs during the early postnatal growth phase. It is based on a process of tissue resorption and substitution, and not on in situ remodelling (EB Hunziker, unpublished data). In the clinical situation, the maturation process cannot be allowed to proceed at the pace of postnatal development. Repair tissue that is both structurally competent and fully functional must be formed within a short period of time, usually within a few weeks. Although it is "unphysiologic" to form repair tissue this quickly, it is necessary to meet the patient's need for fully restored mobility as soon as possible after surgery.

It is thus neither realistic nor practical to approach the problem of articular cartilage repair in adult human patients by taking a pool of precursor cells and attempting to generate articular cartilage tissue by mimicking the ontogenetic differentiation process. The problem should be approached from an entirely new perspective that accounts for the peculiar set of conditions and needs of adult patients.

It is evident that an optimal solution to the engineering of articular cartilage tissue cannot be developed in the near future. Advances in this area will necessarily be laborious as success will be achieved by following a rational, systematic, step-by-step course. Only by proceeding in this manner can the numerous considerations of the microenvironment of tissue engineered products be addressed. Tissue engineering is a multidisciplinary art that involves a knowledge of chemical engineering, molecular biology, cell biology, physiology, pathology, and surgical technology. In addition, the concerted efforts of specialists in each of these fields are required for ultimate success in the generation of articular cartilage tissue.

Summary

There is a need for new methods to repair or replace cartilage in adult patients. The ultimate goal of tissue engineering is to produce articular cartilage with structural, compositional, mechanical, and endurance qualities similar to those of native tissue. All aspects of the vivo microenvironment need to be considered before tissue engineered constructs can be successfully used in clinical practice. For example, loss of chrondrocytes during preparation of the defect beds can have several negative effects. Implantation is also associated with a risk of immunologic responses that can adversely affect healing. The functioning of growth factors and osteoinductive factors in the microenvironment of the adult must also be considered. It must be recognized that precursor cell populations functioning during ontogenesis are not present in the mature individual. Therefore, attempting to mimic embryologic conditions may not be the ideal means of achieving success in adults. Only with a systematic and multidisciplinary approach can these issues be resolved.

References

1. Mankin HJ: The reaction of articular cartilage to injury and osteoarthritis (first of two parts). *N Engl J Med* 1974;291:1285-1292.

2. Furukawa T, Eyre DR, Koide S, Glimcher MJ: Biochemical studies on repair cartilage resurfacing experimental defects in the rabbit knee. *J Bone Joint Surg Am* 1980;62:79-89.

3. Hunziker EB: Biologic repair of articular cartilage: Defect models in experimental animals and matrix requirements. *Clin Orthop* 1999;367:S135-S146.

4. Hunziker EB, Rosenberg LC: Repair of partial-thickness articular cartilage defects: Cell recruitment from the synovium. *J Bone Joint Surg Am* 1996;78:721-733.

5. Hunziker EB, Quinn TM, Hauselmann HJ: Quantitative structural organization of normal adult human articular cartilage. *Osteoarthritis Cartilage* 2002;10:564-572.

6. Hunziker EB: From the preclinical model to the patient, in *Tissue Engineering of Cartilage and Bone*. John Wiley & Sons Ltd, in press.

7. Shapiro F, Koide S, Glimcher MJ: Cell origin and differentiation in the repair of full-thickness defects of articular cartilage. *J Bone Joint Surg Am* 1993;75:532-553.

8. Hunziker EB, Driesang IM, Saager C: Structural barrier principle for growth factor-based articular cartilage repair. *Clin Orthop* 2001;391:S182-S189.

9. Hunziker EB, Driesang IMK: Functional barrier principle for growth-factor-based articular cartilage repair. *Osteoarthritis Cartilage* 2003;11:320-327.

10. Aeschlimann D, Masterlark T, Hayashi K, Graf B, Vanderby R: Repair of cartilage defects with autogenous osteochondral transplants (mosaicplasty) in a sheep model. *Trans Orthop Res Soc*, 2000, 0183.

11. Hunziker EB, Kapfinger E: Removal of proteoglycans from the surface of defects in articular cartilage transiently enhances coverage by repair cells. *J Bone Joint Surg Br* 1998;80:144-150.

12. Hunziker EB, Quinn TM: Surgical removal of articular cartilage leads to loss of chondrocytes from cartilage bordering the wound edge. *J Bone Joint Surg Am* 2003;85:85-92.

13. Jurgensen K, Aeschlimann D, Cavin V, Genge M, Hunziker EB: A new biological glue for cartilage-cartilage interfaces: tissue transglutaminase. *J Bone Joint Surg Am* 1997;79:185-193.

14. Tew SR, Kwan APL, Hann A, Thomson BM, Archer CW: The reactions of articular cartilage to experimental wounding: Role of apoptosis. *Arthritis Rheum* 2000;43:215-225.

15. Redman S, Thompson B, Archer CW: The effects of blunt and sharp trauma on articular chondrocyte cell death and matrix metabolism. *Trans Orthop Res Soc*, 2002, 0419.

16. Hunziker EB, Driesang IMK: Surgical suturing of adult articular cartilage is associated with a loss of chondrocytes and an absence of wound healing. *Trans Orthop Res Soc*, 2003, 0192.

17. Hunziker EB: Growth-factor-induced healing of partial-thickness defects in adult articular cartilage. *Osteoarthritis Cartilage* 2001;9:22-32.

18. Elford PR, Graeber M, Ohtsu H, et al: Induction of swelling, synovial hyperplasia and cartilage proteoglycan loss upon intra-articular injection of transforming growth factor-beta-2 in the rabbit. *Cytokine* 1992;4:232-238.

19. Hunziker EB: Articular cartilage repair: Basic science and clinical progress. A review of the current status and prospects. *Osteoarthritis Cartilage* 2002;10:432-463.

20. Liu Y, Hunziker EB, de Groot K, Layrolle P: Introduction of ectopic bone formation by BMP-2 incorporated biomimetically into calcium phosphate coatings of titanium-alloy implants, in Ben-Nissan B, Sher D, Walsh W (eds): *Bioceramics-15*. Sydney, Australia, Trans Tech Publications, 2002, pp 667-670.

21. Wong M, Ponticiello M, Kovanen V, Jurvelin JS: Volumetric changes of articular cartilage during stress relaxation in unconfined compression. *J Biomech* 2000;33:1049-1054.

Chapter 19

Mesenchymal Stem Cells in Joint Therapy

Frank Barry, PhD

Abstract

Mesenchymal stem cells (MSCs) isolated from adult bone marrow have the capacity to differentiate into a variety of connective tissue cells including bone, cartilage, tendon, muscle, and adipose tissue. In the adult, stem cell populations may provide a reservoir of progenitor cells for the repair of damaged or diseased tissues. There is evidence to suggest that stem cell populations are depleted with age and that this decline contributes to a loss of bone mass in osteoporosis and the inability to regenerate new bone tissue following fracture. Stem cells may be functionally depleted in some degenerative diseases such as osteoarthritis (OA). In patients with advanced OA, MSCs in the bone marrow have a reduced proliferative capacity and a reduced ability to differentiate into chondrocytes and adipocytes. This is a systemic effect observed in cells from marrow from the site of surgery and from a remote site. The delivery of stem cells may have a therapeutic benefit by accelerating the repair of injured tissue or by slowing the degenerative process that occurs in OA. Stem cell therapy is an exciting and promising area of biomedical research. However, more work is needed to better understand the differentiation of stem cells and to devise methods for the delivery of stem cells to an injured joint. The selection of appropriate preclinical models to test the safety and efficacy of these new therapies is also critical.

Introduction

All bones of the axial skeleton are formed via a cartilage intermediate during endochondral ossification. During this process, mesenchymal cells condense and begin to differentiate along a pathway that ultimately leads to hypertrophic chondrocytes and mineralized tissue. Cartilage, ligaments, and tendons also originate from mesenchymal cells. In adult bone marrow, MSCs persist as a population of cells[1,2] with the capacity to differentiate into osteoblasts,[3] chondrocytes,[4,5] or adipocytes[2] both in vitro and when implanted in vivo.[6] These cells may be isolated from the marrow using standardized techniques and expanded in culture through many generations, while retaining their capacity to differentiate under appropriate culture conditions. This provides therapeutic potential for the treatment of lesions in mesenchymal tissues such as articular cartilage,[7] bone,[8] tendon,[9] meniscus,[10] and heart.[11] Much information about the physiologic role and potential therapeutic ben-

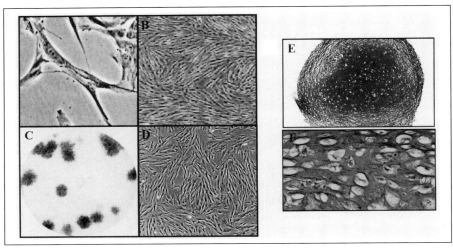

Figure 1 Isolation and culture of adult stem cells. An aspirate of marrow is harvested, usually from the superior iliac crest. Initial fractionation of the marrow is achieved by density gradient fractionation and cells are grown on culture dishes in the presence of fetal bovine serum. MSCs are cultured as the adherent cell population **(A)** but hematopoietic cells remain in suspension and are removed. The adherent cell population is expanded in culture **(B)** to form colonies **(C)**. Cells continue to proliferate through many passages **(D)** exhibiting a fibroblastic morphology. When cultured in the presence of a chondrogenic inducer such as TGF-β3 the cells differentiate into chondrocytes and express a cartilaginous matrix rich in glycosaminoglycan **(E)** and type II collagen **(F)**.

efit of stem cells can be derived from a study of the conditions that induce differentiation. This chapter discusses the control of stem cell differentiation and describes applications of stem cells in orthopaedic medicine.

Isolation and Differentiation of MSCs

In adults, MSCs are commonly isolated from an aspirate of bone marrow taken from the superior iliac crest of the pelvis. The cells are usually plated following initial fractionation on a density gradient. During the period of primary culture, the nonadherent hematopoietic cell fraction is removed. The adherent MSC population, which represents less than 0.001% of the nucleated cells in marrow, forms colonies that can be expanded through many generations. The specific signals required for induction of differentiation of bone marrow-derived cells have been well defined for the chondrogenic (Figure 1), osteogenic, and adipogenic pathways. In a high-density culture system and in the presence of specific factors such as transforming growth factor-β (TGF-β), MSCs undergo rapid differentiation to mature chondrocytes. These mature chondrocytes can form an abundant matrix rich in cartilage-specific proteins, such as type II collagen, aggrecan, and fibromodulin.[12] The synthesis of this matrix is maintained for a prolonged period in culture. When cultured in the presence of dexamethasone, ascorbic acid, and β-glycerol phosphate, MSCs can readily differentiate into osteoblasts. This is accompanied by the upregulation of alkaline phosphatase activity and the production of a mineralized extracellular matrix. There is evidence of the osteogenic nature of these cells in vivo and their probable role in normal bone growth and fracture repair. In the presence of dexamethasone and 3-isobutyl-1-methylxanthine, MSCs differentiate into adipocytes containing lipid-rich vacuoles. Adipogenesis can also be induced by long-chain fatty acids and inhibited by interleukin-1β (IL-1β), tumor necrosis factor-α (TNF-α), and TGF-β.[13] The decrease in bone density in osteopenia along with the increase in marrow adipose tissue may be a result of suppression of osteogenesis and activation of the adipogenic pathway.[13]

The differentiation potential of progenitor cells in the marrow may be broader than was originally anticipated and these cells may exhibit a high degree of plasticity.[14] A population of cells in rodent marrow possesses the ability to differentiate into cells with ectodermal, neuroectodermal, and endodermal characteristics. It appears that these marrow-derived cells contribute to most somatic tissues. These cells resemble embryonic stem cells in terms of the conditions required for their culture and they may represent a population of cells that persists after birth and into adult life.[14] These remarkable observations raise many intriguing questions about stem cell plasticity and the role of these cells in tissue growth and repair.

Stem Cells in Arthritic Diseases

Several principles of the physiology of stem cells in the adult guide our understanding of their therapeutic potential. The first of these suggests that resident populations of stem cells are the natural units of repair of tissues damaged by traumatic injury or disease. Several studies indicate that when stem cells are infused to a wounded host, they migrate to the site of injury. Furthermore, endogenous stem cell populations may be depleted or altered in certain degenerative diseases such as OA, and this may contribute to the lack of repair observed in these diseases. This hypothesis is supported by the observation that stem cells from the marrow of OA patients undergoing total joint arthroplasty surgery (either hip or knee) have reduced proliferative capacity and a reduced ability to differentiate into chondrocytes and adipocytes.[15] Alterations in stem cell activity may be a component of disease progression, or may simply reflect the disease environment, specifically the presence of inflammatory cytokines such as IL-1β and TNF-α. These areas need to be addressed more thoroughly. A direct corollary of these observations is that the delivery of stem cells will have a therapeutic benefit. Studies to define the efficacy of these approaches are ongoing.

Stem Cell Therapy in Joint Disease

The therapeutic potential of stem cells in stimulating the regeneration of meniscal tissue following injury was evaluated in a series of experiments in

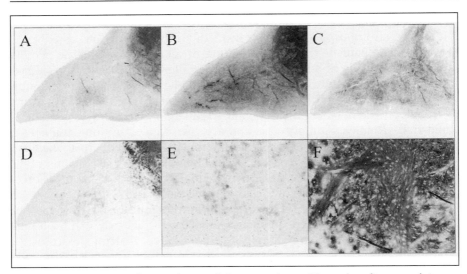

Figure 2 Appearance of posterior medial neomeniscus 26 weeks after complete medial meniscectomy and 20 weeks following injection of a preparation of MSCs delivered by intra-articular injection as a suspension in dilute HA. The figure shows sections of neomeniscal tissue from a treated joint stained with Safranin O **(A)**, Toluidine blue **(B)** and with an antibody specific for type I collagen **(C)** and type II collagen **(D through F)**. A through D: original magnification × 20; E and F: original magnification × 100.

goats.[16] Traumatic loss of meniscal tissue significantly increases the risk of developing OA, with degeneration of articular cartilage and changes to subchondral bone, ligaments, synovium, and the periarticular musculature. In this study, MSCs were retrovirally transduced to express enhanced green fluorescent protein and were injected into the joint as a suspension in sodium hyaluronan (HA). Following unilateral complete medial meniscectomy, the animals received an intra-articular injection of a suspension of 10^7 MSCs in 5 mL of HA 1 week (n=5) or 6 weeks (n=5) after injury. Animals used as controls received HA alone without cells. Regenerated neomeniscal tissue associated with the medial compartment was observed upon gross evaluation of joints (Figure 2). In goats injected with cells 1 week after surgery, the amount of regenerated neomeniscal tissue was significantly greater than those of goats treated with HA alone. Immunohistochemical and histologic staining of the regenerated neomeniscal tissue showed a dense, cellular, type I collagen–positive, fibrous network with small areas that were positive for proteoglycan. Damage to the cartilage in these areas was less in joints treated with MSCs compared with those of the control group. This finding suggests that cartilage in animals treated with MSCs may be protected from the abnormal mechanical forces associated with the loss of meniscus. These data demonstrate that the injection of allogeneic MSCs into destabilized joints results in regeneration of meniscal tissue. The time of injection of cells had a significant effect on the outcome as delivery of cells one week following injury was more beneficial than delivery at 6 weeks. Thus allogeneic MSCs may contribute to the natural repair process in joints after meniscectomy and impact the progression to OA.

Future Directions

It is generally accepted that three components must be supplied for tissue engineering to be successful: (1) cells, (2) extracellular matrix, and (3) signaling factors. Successful therapeutic applications of MSCs may require a specific combination of these three components. The ease with which stem cells can be transduced has been described[17] and the most effective treatments may involve the use of stem cells in combination with appropriate therapeutic genes or exogenous factors. Stem cell-based tissue engineering holds much promise for orthopaedic medicine. There are several challenging areas, however, including the need to optimize methods of cell delivery and fixation within the local environment. Although methods to treat focal defects in bone and cartilage continue to be successful in preclinical models, the greatest challenge will be in the application of stem cell therapies in conditions that are associated with a significant loss or destruction of tissue, such as OA, rheumatoid arthritis, and osteoporosis. The successful application of stem cell therapies to reduce joint degeneration would represent a significant advance in the treatment of OA and would have an enormous impact on this serious public health issue.

Summary

The ability of MSCs isolated from adult bone marrow to differentiate into a variety of connective tissue cells is the basis for the study of stem cell therapies to repair damaged or diseased bone tissue. Stem cell populations may be a natural mechanism to repair tissues in adults; however, these populations appear to decrease with aging and may become depleted in certain disease states. The delivery of stem cells has been demonstrated to facilitate the repair of damaged tissue in experimental models. Although initial results have been promising, additional research is needed to optimize the cells, matrices, and osteoinductive factors that would result in the best clinical outcomes.

References

1. Haynesworth SE, Baber MA, Caplan AI: Cell surface antigens on human marrow-derived mesenchymal cells are detected by monoclonal antibodies. *Bone* 1992;13:69-80.

2. Pittenger MF, Mackay AM, Beck SC, et al: Multilineage potential of adult human mesenchymal stem cells. *Science* 1999;284:143-147.

3. Jaiswal N, Haynesworth SE, Caplan AI, Bruder SP: Osteogenic differentiation of purified, culture-expanded human mesenchymal stem cells in vitro. *J Cell Biochem* 1997;64:295-312.

4. Johnstone B, Hering TM, Caplan AI, Goldberg VM, Yoo JU: In vitro chondrogenesis of bone marrow-derived mesenchymal progenitor cells. *Exp Cell Res* 1998;238:265-272.

5. Mackay AM, Beck SC, Murphy JM, Barry FP, Chichester CO, Pittenger MF: Chondrogenic differentiation of cultured human mesenchymal stem cells from marrow. *Tissue Eng* 1998;4:415-428.

6. Dennis JE, Haynesworth SE, Young RG, Caplan AI: Osteogenesis in marrow-derived mesenchymal cell porous ceramic composites transplanted subcutaneously: Effect of fibronectin and laminin on cell retention and rate of osteogenic expression. *Cell Transplant* 1992;1:23-32.

7. Wakitani S, Goto T, Pineda SJ, et al: Mesenchymal cell-based repair of large,full-thickness defects of articular cartilage. *J Bone Joint Surg Am* 1994;76:579-592.

8. Bruder SP, Kurth AA, Shea M, Hayes WC, Jaiswal N, Kadiyala S: Bone regeneration by implantation of purified, culture-expanded human mesenchymal stem cells. *J Orthop Res* 1998;16:155-162.

9. Young RG, Butler DL, Weber W, Caplan AI, Gordon SL, Fink DJ: Use of mesenchymal stem cells in a collagen matrix for Achilles tendon repair. *J Orthop Res* 1998;16:406-413.

10. Walsh CJ, Goodman D, Caplan AI, Goldberg VM: Meniscus regeneration in a rabbit partial meniscectomy model. *Tissue Eng* 1999;5:327-337.

11. Shake JG, Gruber PJ, Baumgartner WA, Senechal G: Mesenchymal stem cell implantation in a swine myocardial infarct model: Engraftment and functional effects. *Ann Thorac Surg* 2002;73:1919-1925.

12. Barry F, Boynton RE, Liu B, Murphy JM: Chondrogenic differentiation of mesenchymal stem cells from bone marrow: Differentiation-dependent gene expression of matrix components. *Exp Cell Res* 2001;268:189-200.

13. Nuttall ME, Patton AJ, Olivera DL, Nadeau DP, Gowen M: Human trabecular bone cells are able to express both osteoblastic and adipocytic phenotype: Implications for osteopenic disorders. *J Bone Miner Res* 1998;13:371-382.

14. Jiang Y, Jahagirdar BN, Reinhardt RL, et al: Pluripotency of mesendrymal stem cells derived from adult marrow. *Nature* 2002;418:41-49.

15. Murphy JM, Dixon K, Beck S, Fabian D, Feldman A, Barry F: Reduced chondrogenic and adipogenic activity of mesenchymal stem cells from patients with advanced osteoarthritis. *Arthritis Rheum* 2002;46:704-713.

16. Murphy JM, Kavalkovich KW, Young RG, Dodds RA, Barry FP: *Trans Orthop Res Soc* 2003;28:883.

17. Mosca JD, Hendricks JK, Buyaner D, et al: Mesenchymal stem cells as vehicles for gene delivery. *Clin Orthop* 2000;379:S71-S90.

Chapter 20

Autologous Chondrocyte Implantation of the Knee

Tom Minas, MD, MS

Abstract

In the past, patients with full-thickness chondral injuries had few treatment options. Autologous chondrocyte implantation (ACI) has been shown to provide encouraging and predictable short- and mid-term clinical results. The formation of a hyaline-like repair tissue is possible with ACI in young active patients who have large (> 2 cm²) lesions. Comorbid conditions such as axial alignment, ligamentous stability, and meniscal pathology need to be addressed to ensure optimal outcomes. Traditionally, indications for ACI have been restricted to isolated grade 3 to 4 lesions of the femoral condyle or trochlea of the knee. However, young active patients frequently have multiple lesions or early arthritic changes. Using a series of categories of clinical defects, namely, simple = 4 cm², complex = 7 cm², and salvage = 12 cm², an overall improvement of 87% was obtained after ACI. Prospective evaluation of the patients by a validated outcomes analysis suggests that the indications for ACI may be broadened to include patients with complex or salvage conditions. To date, more than 8,000 patients worldwide have been treated by this technique. In long-term histologic and mechanical evaluations, mechanically firm repair tissues have been correlated with a hyaline-like histologic appearance and a good clinical outcome. In contrast, repair tissues that are less firm mechanically are associated with a fibrocartilaginous appearance and a suboptimal clinical outcome. Preoperative counseling is imperative to ensure that patients understand the intensive rehabilitation process that is required and the potential complications. Future technological developments will render ACI a less invasive and a more universally applicable procedure.

Introduction

The excitement generated by ACI grafting since its introduction in 1994[1] has renewed interest in the field of cartilage repair. Prior to the development of ACI, there were no satisfactory or reproducible clinical methods to repair cartilage. This was the first clinical application of in vitro autologous cell cultivation in clinical orthopaedics. In essence, it was the introduction of tissue engineering to orthopaedic clinical practice. Instead of the classic use of an engineered scaffold, a suspension of autologous chondrocytes is implanted and held in place by a periosteal flap.

The incidence and natural history of articular cartilage lesions have not been well defined; however, a lesion in the weight-bearing surface of the knee may progress to osteoarthritis. This may be especially true for larger lesions that are unshouldered.[2] The fibrocartilage generated by a marrow-stimulating technique may be adequate to stabilize small lesions and to relieve symptoms of catching, but not to arrest disease progression.[3] When lesions covering an area of 4 cm² or more are so treated, the fibrocartilage is usually broken down within 24 to 36 months because of its poor mechanical properties and durability.[4]

The use of ACI has produced hyaline repair tissue in experimental models[5-9] and in biopsies from second-look arthroscopies.[1,10,11] This technique has also produced durable results at 2- to 9-year follow-up in patients.[12] In this trial, 96% of the autologous chondrocyte transplants that were doing well at 2 years remained durable at 10 years.[12] When measured with an arthroscopic mechanical indentation instrument, the mechanical viscoelastic properties of hyaline repair tissue, fibrocartilage, and fibrous tissue were well correlated with the histologic appearance of biopsies. Biopsies that demonstrated a hyaline-like repair tissue had mechanical properties similar to those of native articular cartilage. These results corresponded to good or excellent clinical outcome. Conversely, patients with repair tissue that was fibrous or fibrocartilaginous in appearance had a poorer clinical outcome and the mechanical properties of their repair tissue were less than half as good as those of native articular cartilage.

Encouraging results have also been obtained at short-term follow-up in a series of patients from the United States.[11] In this series, many of the patients who improved significantly after ACI had multiple or complex cartilage lesions. By resurfacing these areas of chondral injury with hyaline cartilage, this technique permitted a resolution of the symptoms and a return to an active lifestyle in young patients. Additional evaluations should reveal whether the progression of osteoarthritis was halted or delayed.

Many studies have explored the healing process of ACI. Studies with rabbits have demonstrated that the in vitro cultured chondrocytes are responsible for the majority of the repair tissue formed in vivo, and that this repair tissue surpasses periosteum alone in quality and quantity.[5,6] Canine studies[7-9] have revealed that the healing process has distinct phases. These phases have been categorized as a proliferative stage (0 to 6 weeks), a transition stage (7 to 12 weeks), and a remodeling and maturation stage that occurs over an extended period (13 weeks to 3 years). These stages of healing have been useful in guiding the clinical rehabilitation after ACI.

Preoperative Clinical Assessment

Patients with articular cartilage injuries may present with a variety of symptoms including pain, effusion, locking, and catching or giving way, with or without a prior traumatic injury. A thorough patient history and physical examination are essential to determine whether a patient is eligible for ACI.

Prior to or at the time of chondral repair, patients should be assessed for preexisting factors for chondral injuries, including axial malalignment, ligament stability, size of the chondral defect, status of the meniscus, and genetic predisposition to osteoarthritis.

Pain is a frequent symptom and its location can help localize the site of chondral injury. Locking or catching of the knee may represent the presence of a loose body or flap from an osteochondral injury. These symptoms also may reflect meniscal injury. The possibility of meniscal injury must be assessed before treatment of the chondral injury. If present, the meniscal injury should then be treated.

Axial malalignment may accompany full-thickness chondral weight-bearing injuries. Clinical evaluation may include the observance of genu varum or valgum and the presence of gait disturbance. However, radiographs of the long axial alignment should always be obtained to confirm or refute the clinical impression of alignment. As with the patellofemoral joint, any consideration of ACI must first include a correction of any axial malalignment to avoid exposing the immature chondrocytes to increased forces from a malaligned extremity.

Prior traumatic injuries are common, and in my experience, patients who have had anterior cruciate ligament (ACL) injuries are particularly predisposed to chondral injuries. Ligamentous stability may be just as important to the knee as correct axial alignment. Furthermore, full-thickness chondral injuries occur with an incidence of between 5% and 10% in patients who have experienced acute work-related or sporting injuries and present with an acute hemarthrosis.[13] Minas and Nehrer[2] have noted that full-thickness cartilage loss chronically develops in 20% of ACL-injured knees. Performing ACI without restoring joint stability may result in the graft site experiencing increased shear forces from recurrent episodes of instability and predispose the graft to failure.

Recurrent effusions are also associated with full-thickness chondral injuries. This is particularly true for trochlear lesions. Additional complaints that focus on the patellofemoral articulation are frequent. Patellofemoral pathology may be caused by prior dislocations or patellar maltracking, which result in full-thickness chondral injuries. Maltracking must be addressed if ACI is being considered for treatment of patients with this articulation. Early failures of ACI may have been related to the continual exposure of immature cells to the increased shear forces generated by maltracking.[10]

Radiographic Evaluation

Evaluation of radiographs is a useful screening tool to exclude patients who are not eligible for ACI. A standard protocol should include radiographs of weight-bearing anteroposterior and lateral views to assess joint space narrowing and osteophyte formation. Rosenberg's standing 45° angle posteroanterior views are also helpful in diagnosing tibiofemoral cartilage loss[14] or osteochondritis dissecans (OCD) lesions. Standing 54-in axial alignment views that include the center of the femoral head and talus are essential to assess the overall axial alignment of the extremity and to determine whether realignment osteotomy will be required. Merchant, or sunrise, views assess the patellofemoral articulation for evidence of joint space narrowing.

Patellofemoral maltracking is best assessed by clinical examination. CT scans of the knee with the leg in extension and with the quadriceps first contracted and then relaxed to better assess patellofemoral tracking may be required for some patients.

MRI allows visualization of the articular cartilage as well as the menisci, ligaments, and bone. Thus MRI is a valuable tool for evaluation of patients suspected of having articular cartilage injuries. However, technical issues, mainly related to spatial resolution and volume averaging, can make it difficult to determine the exact dimensions of cartilage defects and of repair tissues.[15] Newer imaging techniques and protocols have improved the accuracy of evaluating cartilage lesions and repair tissues. Winalski and Minas[15] noted that the techniques representing state-of-the-art clinical evaluation of articular cartilage include fat-suppressed three-dimensional T1-weighted gradient echo images, fast-spin-echo sequences, and magnetic resonance arthrography. These techniques have improved the sensitivity of detecting cartilage injury to more than 85% for moderate to severe cartilage lesions (Outerbridge grade: 2 to 4)[16] and maintain a specificity of more than 90%.[15] However, the availability of these techniques may be limited to specialized centers.

General Considerations

Patient education is critical for a successful outcome. Patients characterized as "salvage" may not have other viable treatment options and therefore, they are not excluded from consideration for ACI in my practice provided that they have a thorough understanding of the rehabilitation process and the potential complications. Extensive preoperative counseling is required with these patients to involve them in the decision-making process because they may have an increased risk for failure and complications such as graft hypertrophy and arthrofibrosis. If ACI does not appeal to a patient or if it fails, alternative approaches include unicondylar knee replacement, a custom patellofemoral arthroplasty, or salvage treatment. Total knee arthroplasty is reserved for tricompartmental disease states.

General medical comorbidities such as cigarette smoking and the use of medications that may impair cell proliferation, such as nonsteroidal or immunosuppressive drugs, should be investigated. The patient must be free from nicotine as this has been found to impair healing under certain conditions.[16,17] In addition, patients taking narcotic pain medications on presentation due to numerous prior surgeries must cease use before ACI to enhance postoperative pain management.

Arthroscopic Assessment

Arthroscopy has been the gold standard for evaluating a patient with a suspected articular cartilage injury, and it continues to play a critical role in patient evaluation and treatment selection. Arthroscopic evaluation and cartilage biopsy require a careful and systematic approach. The arthroscopic assessment should begin with an examination of the knee under anesthesia.

The range of motion and any ligamentous insufficiencies should be noted. An arthroscopic probe should be used to determine the extent of grade 3 and 4 chondromalacia. The opposing articular surface should also be examined for evidence of cartilage damage because ACI has traditionally been reserved for unipolar lesions with no more than grade 2 chondromalacia[18,19] on the opposing articular surface. The dimensions of the lesion(s) should be recorded for use in planning future treatments, especially to ensure that an appropriate volume of cultured chondrocytes is obtained. Biopsies may be derived from the superior medial edge of the trochlea, the superior transverse trochlear margin adjacent to the suprapatellar synovium, or the lateral intercondylar notch.[19] The biopsy should consist of 200 to 300 mg of articular cartilage (approximately 5 mm in width by 10 mm in length). This amount of biopsy tissue contains approximately 200,000 to 300,000 cells; this quantity is generally required for enzymatic digestion and cell culture.[19] The in vitro expansion of cells requires 3 to 5 weeks of cell culture. One vial of cultured autologous chondrocytes has a volume of 0.4 mL and contains approximately 12 million cells. This amount is sufficient to cover defects of approximately 4 to 6 cm^2.[19]

Indications

ACI has been used in Sweden since 1987.[20] Since the initial report on this technique in the *New England Journal of Medicine* in 1994,[20] ACI has become a recognized technique for the treatment of chondral injuries of the knee. Based on this initial study, the conventional indications for ACI include treatment of symptomatic patients between ages 15 and 55 years, with isolated Outerbridge grade 3 to 4 lesions of the femoral condyle or trochlea, and with no more than grade 1 to 2 chondromalacia on the opposing articular surface.[20] OCD of the medial or lateral femoral condyles has also been treated with ACI. However, ACI has not been commonly used for kissing lesions or for patients with radiographic evidence of joint space narrowing. Although ACI has not been generally accepted for transplants of the patella articulations, the results are good to excellent, approaching 80%, when maltracking is addressed. However, these isolated lesions are uncommon.[11,21] My experience has extended the indications to include treatment of patients with multiple lesions and those considered salvage who have early arthritic changes. The results have been encouraging at follow-up times of 2 to 7 years.[11]

Technique

An ACI is performed as a separate procedure after the removal of cartilage biopsies and cell expansion in vitro. Preoperative antibiotics are administered and a tourniquet is usually used for the stages of arthrotomy, the preparation of the defect site, and the harvesting of the periosteum. The tourniquet is then let down for periosteal microsuturing and the injection of the chondrocyte suspension.

The technical aspects of the procedure are documented elsewhere in more detail.[19,22] One technical aspect that is underemphasized and leads to early failure involves the radical débridement of the chondral lesion. Débridement is performed back to a stable edge of intact articular cartilage with vertical edges. In situations in which marrow-stimulating techniques have previously been performed, or when there is a stiff subchondral bone plate from early arthritic changes, intralesional osteophytes frequently exist and must be débrided. These débridements are best performed with the use of a high-speed burr. This can be accomplished without bleeding from the bone bed because the thickened bone is often avascular and does not allow good attachment of developing repair tissue.

After the lesion has been débrided, it is most easily templated with sterile paper. A marking pen is used to outline the size of the lesion on the paper and to mark the proper orientation of the lesion. A thrombin- and epinephrine-soaked sponge is then placed over the prepared defect to limit bleeding when the tourniquet is deflated.

The periosteum is harvested after a second incision over the medial tibia distal to the pes anserine insertion. The sharp incision in the periosteum should be 1 to 2 mm larger than the prepared template. With large lesions, multiple lesions, or revision surgeries, it may be difficult to obtain an adequate amount of periosteum from the medial aspect of the tibia. Additional periosteum can be harvested from the distal medial or lateral aspect of the femur; however, this may increase the risk of arthrofibrosis.

After harvesting the graft, the tourniquet is deflated and hemostasis achieved in the soft tissues and in the prepared defect. The graft is placed in its correct orientation with the cambium layer facing the defect floor. Suturing is then performed, at 3- to 5-mm intervals, using No. 6-0 vicryl thread that has been soaked in sterile mineral oil. An opening is left for the insertion of a catheter, through which saline is introduced to test graft integrity. Fibrin glue is placed over the edges of the graft to obtain a watertight seal. The saline is then aspirated out of the defect and the cell suspension injected into it through the catheter. Finally, the opening at the catheter insertion site is sealed by suturing and with fibrin glue.

Postoperative Protocol

A knee immobilizer is used to keep the leg in extension during the first postoperative day to permit the chondrocytes to adhere to the underlying subchondral bone and surrounding articular cartilage. Following this, postoperative rehabilitation is guided by the three healing phases of ACI, as identified in canine studies.[7-9] The principles of rehabilitation are to protect the graft, to retain knee motion, and to gradually increase activity and weightbearing mobility. The rehabilitation process is somewhat different for lesions of the patellofemoral joint and for those of the tibiofemoral joint.

The maturation phase of healing begins after 12 weeks and can extend up to 2 years postoperatively. During this time, assistive devices are discarded and activity levels are increased. Bicycling continues to be encouraged as well as the use of elliptical training machines and treadmill walking. Patients should be encouraged to take an active role in their own rehabilitation and can begin to undergo resistive training with the stationary bike and treadmill. Running is not permitted until graft hardness is similar to that of the surrounding cartilage, which may take 12 to 24 months.[19] Activity that fatigues the thigh musculature should be encouraged, but patients should be counseled to lessen their level of activity if they feel knee discomfort.

Complications

Periosteal graft hypertrophy, graft failure or delamination, and arthrofibrosis are the most common complications following ACI.[11] If a patient develops painful catching, new-onset pain, or recurrent effusions during the postoperative period, activity is decreased and MRI with gadolinium-enhancement is performed. Arthroscopy may then be performed in certain patients. In situations in which advanced MRI techniques are not available, diagnostic arthroscopy may be required on a more frequent basis.

In my experience, second-look arthroscopy for persistent symptoms is required in 25% of patients[7] with 20% being performed due to hypertrophy of the periosteum and 5% due to arthrofibrosis following the arthrotomy.

Graft failure occurs with an incidence of 7% to 13% in short- and midterm studies.[10,11] A MRI scan frequently reveals progressive or persistent subchondral edema under failing graft sites. Graft failure or delamination is heralded by the presence of a loose flap at the edge of the graft apparent with second-look arthroscopy. Patients are counseled about the possibility of performing marrow-stimulating techniques or an osteochondral graft-transfer procedure to address areas of possible failure when there is a small area of graft delamination (< 1.5 cm^2). If the area of graft failure is larger, consideration is given to revision ACI, fresh osteochondral allografting, or unicompartmental arthroplasty.

Results

The first clinical results of ACI came from Sweden and were published in 1994.[20] In this initial pilot study, 14 of the 16 patients treated by ACI for isolated lesions of the femoral condyle had good or excellent results. Individuals treated for lesions of the patella did not fare as well with only 2 of 7 patients achieving good or excellent results. These findings prompted further interest in ACI for full-thickness chondral injuries.

The durability of ACI has been confirmed in a 2- to 9-year follow-up study from Sweden that included the patients who were treated in the initial pilot study.[10] This mid-term retrospective assessment of 101 patients treated by ACI revealed good to excellent results in 92% of patients with isolated lesions involving the femoral condyle. The results were further divided according to diagnosis. Good to excellent results were obtained in 89% of patients treated for OCD, in 75% of those with concomitant ACL reconstruction, in 67% of individuals with multiple lesions, and in 65% of patients with patellar lesions. However, the fate of patellar lesions may be improved. Improvement, with good to excellent results in 11 of the 14 patients with patellar lesions, was noted when more attention was paid to adequate débridement and patellar realignment when needed.[10]

The results obtained for the first 169 patients treated by ACI at the Brigham and Women's Hospital in Boston, MA are also encouraging.[11] In this short-term follow-up study, 87% of the patients benefited, even though "simple" isolated lesions of the femoral condyle were unusual in this series. Patients were divided into simple, complex, and salvage groups on the basis of lesion size, and number and the presence of early arthritic changes in the salvage category. The treatment area per knee was large with treatment areas of 4 cm^2 in the simple category, 7 cm^2 in the complex category, and 12 cm^2 in the salvage category. Patients were on average age 35 years in the simple and complex categories and age 39 years in the salvage category.

Patients were prospectively evaluated and numerous outcome parameters were used to assess the results. Short Form-36 scores indicated clinical improvement in pain relief and were statistically significant in the complex and salvage groups compared with baseline values. Although statistical significance was not attained in the simple category, 10 of the 12 patients were free from pain and again participated in sporting activities. The Knee Society scores improved in all groups with statistical significance being achieved at 24 months and maintained at 4 years. Western Ontario McMaster University Osteoarthritis Index scores were statistically significant compared with baseline values at 24 months. Results in the salvage group were the most satisfactory, with 93% of patients (n = 15 patients at 24 months) reporting that they would choose the surgery again.

In general, there was a time-dependent improvement that was maximal after 24 months with femorotibial resurfacing. The majority of patients who underwent femorotibial resurfacing were able to resume sporting activities. Rehabilitation took longer (36 months) after patellofemoral resurfacing, and resumption of sporting activities was not always possible. Improvements were neither as significant nor as predictable when the patellofemoral joint, as opposed to the tibiofemoral joint, was involved.

Summary

Since the first report of its use in 1994, ACI has been shown to result in promising and predictable short- and mid-term clinical outcomes. In long-term evaluations, mechanically firm repair tissues have been correlated with a hyaline-like histologic appearance and a good clinical outcome. Conversely, repair tissues that are less firm mechanically are associated with a fibrocartilaginous appearance and a poorer clinical outcome. Preoperative patient evaluation and counseling are necessary to ensure the best possible clinical outcomes. Comorbid conditions such as axial alignment, ligamentous stability, and meniscal pathology need to be addressed before the ACI procedure. Patients must understand the intensive rehabilitation process that is required and the potential complications. The indications for ACI may be broadened to include patients with complex or salvage conditions. The use of a series categories of clinical defects, namely, simple = 4 cm^2, complex = 7 cm^2, and salvage = 12 cm^2, showed that an overall improvement of 87%

was obtained after ACI. With advances in technology, methods of ACI are expected to improve and its use in the clinical setting can be optimized.

References

1. Brittberg M, Lindahl A, Nilsson A, et al: Treatment of deep cartilage defects in the knee with autologous chondrocyte implantation. *N Engl J Med* 1994;331:889-895.

2. Minas T, Nehrer S: Current concepts in the treatment of articular cartilage defects. *Orthopaedics* 1997;20:525-538.

3. Messner K, Maletius W: The long-term prognosis for severe damage to weight-bearing cartilage in the knee: A 14-year clinical and radiographic follow-up in 28 young athletes. *Acta Orthop Scand* 1996;67(2):165-168.

4. Nehrer S, Spector M, Minas T: Histological analysis of failed cartilage repair procedures. *Clin Orthop* 1999;365:149-162.

5. Grande DA, Pitman MI, Peterson L: The repair of experimentally produced defects in rabbit cartilage by autologous chondrocyte transplantation. *J Orthop Res* 1989;7:208-218.

6. Brittberg M, Nilsson A, Lindahl A, et al: Rabbit articular cartilage defects treated with autologous cultured chondrocytes. *Clin Orthop* 1996;326:270-283.

7. Shortkroff S, Barone L, Hsu HP: Healing of chondral and osteochondral defects in a canine model: The role of cultured chondrocytes in regeneration of articular cartilage. *J Biomater Res* 1996;17:147-154.

8. Breinan H, Minas T, Hsu HP, et al: Effect of cultured articular chondrocytes on repair of chondral defects in a canine model. *J Bone Joint Surg Am* 1997;79:1439-1451.

9. Breinan HA, Minas T, Barone L, et al: Histological evaluation of the course of healing of canine articular cartilage defects treated with cultured autologous chondrocytes. *Tissue Eng* 1998;4:101-114.

10. Peterson L, Minas T, Brittberg M, Nilsson A, Sjogren-Jansson E, Lindahl A: Two to nine year outcome after autologous chondrocyte transplantation of the knee. *Clin Orthop* 2000;374:212-234.

11. Minas T: Autologous chondrocyte implantation for focal chondral defects of the Knee. *Clin Orthop* 2001;391S:S349-S361.

12. Peterson L, Brittberg M, Kiviranta I, et al: Autologous chondrocyte transplantation: Biomechanics and long-term durability. *Am J Sports Med* 2002;30:2-12.

13. Noyes FR, Bassett RW, Noyes FR, et al: Arthroscopy in acute traumatic hemarthrosis of the knee: Incidence of anterior cruciate tears and other injuries. *J Bone Joint Surg Am* 1980;62:687-695.

14. Rosenberg T, Paulos L, Parker R, et al: The forty-five degree posterior-anterior flexion weight bearing radiograph of the knee. *J Bone Joint Surg Am* 1988;70:1479-1483.

15. Winalski C, Minas T: Evaluation of chondral injuries by magnetic resonance imaging: Repair assessments. *Op Tech Sports Med* 2000;8(2):108-119.

16. Riebel GD, Boden SD, Whiteside TE, Hutton WC: The effect of nicotine on incorporation of cancellous bone graft in an animal model. *Spine* 1995;20(20):2198-2202.

17. Raikin SM, Landsman JC, Alexander VA, Froimson MJ, Plaxton NA: Effect of nicotine on the rate and strength of long bone fracture healing. *Clin Orthop* 1998;353:231-237.

18. Outerbridge RE: The etiology of chondromalacia patella. *J Bone Joint Surg Br* 1961;43:752-767.

19. Minas T, Peterson L: Autologous chondrocyte transplantation. *Op Tech Sports Med* 2000;8(2):144-157.

20. Peterson L: International experience with autologous chondrocyte transplantation, in Insall JN, Scott WN (eds): *Surgery of the Knee*, ed 3. Philadelphia, PA, Churchill Livingstone, 2001, pp 341-356.

21. Curl W, Krome J, Gordon S: Cartilage injuries: A review of 31,516 knee arthroscopies. *Arthroscopy Arthrosc Rel Surg* 1997;13:456-460.

22. Minas T, Peterson L: Advanced techniques in autologous chondrocyte transplantation. *Clin Sports Med* 1999;18:13-44.

Chapter 21

Manufacture of Cartilage Tissue In Vitro

Eugene J. Thonar, PhD
Robert L. Sah, MD, ScD
Koichi Masuda, MD

Abstract

Most attempts to engineer cartilage tissue have involved the seeding of cultured cells within a biologic or synthetic scaffold. We developed a novel two-step approach in which cartilaginous-like tissue is formed in vitro by adult bovine chondrocytes without the aid of a synthetic matrix. The first step consists of culturing chondrocytes in an alginate gel for 7 days using conditions that maintain the rounded shape and molecular phenotype of the cells. The second step consists of recovering the chondrocytes with their cell-associated matrix and culturing them for an additional 7 days on a tissue culture insert with a porous membrane. This approach results in the formation of a cartilaginous tissue with a relatively high proteoglycan-to-collagen ratio. This in vitro scaffold-free system appears ideal to study the development of transplantable cartilaginous tissue.

Introduction

Cell-laden cartilaginous tissues engineered in vitro may be useful as a graft material for cartilage repair. These types of tissues have been synthesized primarily using immature chondrocytes or chondroprogenitor cells in combination with various types of scaffolds. Some methods form implants composed only of cells and their products. Immature chondrocytes can form cartilaginous tissue when cultured as monolayers[1] or multilayers[2] in the absence of a scaffold at high density, a configuration that somewhat replicates the high density of cells in fetal cartilage. This approach has been less successful when using chondrocytes from adult articular cartilage as these cells form a much smaller amount of matrix under the same culture conditions.[1]

New methods to form cartilaginous tissue from cells derived from adults are needed for use in repair of tissue in adults.[3] Chondrocytes can be cultured in gels that promote the retention of their rounded shape, which appears to be critical for the maintenance of the phenotype that is capable of forming a cartilaginous matrix.[4] The chondrocytic phenotype of adult chondrocytes can be stabilized in an alginate culture for up to 8 months. Such cultures accumulate cartilage matrix components, including type II collagen and aggrecan. The matrix that rapidly forms around the cells in alginate has

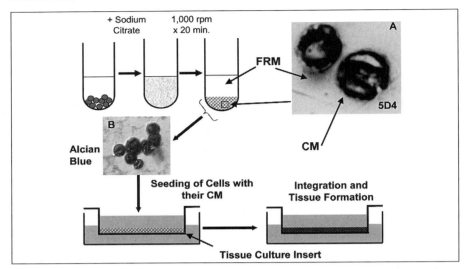

Figure 1 ARC method chondrocytes were cultured in alginate beads with complete medium (Dulbecco's Minimum Essential Media, Cellgro-Mediatech, Herndon, VA) containing gentamicin (50 μg/mL) and L-glutamine (360 μg/mL) and 10% or 20% FBS for 7 days. The beads were then dissolved to recover the chondrocytes with their CM by adding sodium citrate buffer. The CM was formed during culturing in alginate. A: Results of immunostaining for keratan sulfate (A: 5D4). After release from the beads, the CM structure was maintained without affecting cell viability. B: Results of staining with Alcian blue. The cells with their CM were seeded onto the tissue culture insert and cultured for an additional 7 days. Individual cells formed a cohesive matrix with time. *(Reproduced with permission from Masuda K, Sah RL, Hejna MJ, Thonar EJ: A novel two-step method for the formation of tissue engineered cartilage by mature bovine chondrocytes: The alginate-recovered-chondrocyte (ARC) method. J Orthop Res 2003;21:139-148.)*

been termed the cell-associated matrix (CM), and with time, a matrix known as the further removed matrix forms at some distance from the cells. The cells with their CM can be recovered from alginate beads cultured for 1 to 2 weeks by dissolving the alginate polymer with agents that chelate divalent cations.[5]

We report that bovine adult chondrocytes, grown in alginate and then recovered with their associated matrix, are subsequently capable of rapidly forming cartilaginous tissue with properties typical of articular cartilage.

The Alginate-Recovered-Chondrocyte Method of Forming Cartilaginous Tissue

We developed a two-step method of growing adult chondrocytes in culture (Figure 1). This method, known as the Alginate-Recovered-Chondrocyte

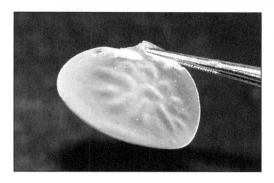

Figure 2 Gross appearance of the ARC tissue after 1 week of culture. Chondrocytes were cultured in alginate in the presence of 20% FBS for 7 days. The cells and their CM were then recovered, seeded onto a tissue culture insert, and grown in culture for an additional 7 days. The tissue formed using this method was white and opaque, and had a disk-like structure (diameter of tissue: 23 mm). *(Reproduced with permission from Masuda K, Sah RL, Hejna MJ, Thonar EJ: A novel two-step method for the formation of tissue engineered cartilage by mature bovine chondrocytes: The alginate-recovered-chondrocyte (ARC) method.* J Orthop Res *2003;21:139-148.)*

(ARC) Method, was developed[3] with the intent to produce tissue in vitro that could ultimately be used to form cartilaginous tissue in vivo.

STEP 1: After entrapping adult bovine articular chondrocytes (from animals aged 18 to 24 months) in alginate beads, the beads were incubated in 10-cm diameter Petri dishes without a tissue culture coating (Corning Costar, Cambridge, MA) in complete culture medium (300 beads in 12 mL) supplemented with 20% fetal bovine serum (FBS).

STEP 2: After 7 days of culture, the medium was collected, the beads (300 beads/Petri dish) were dissolved, and the cells with their CM were obtained by mild centrifugation.[5] The resulting pellet containing "alginate-recovered-cells" with their CM was then resuspended in complete medium (2.5 mL/300 beads) containing 20% FBS. In the meantime, the same complete medium containing FBS (3 mL) was added to the bottom portion of each well of a 6-well Falcon Cell Culture Insert Companion plate (Becton Dickinson, Franklin Lakes, NJ) and prewarmed to 37°C in an incubator in the presence of 5% CO_2 for 20 minutes. Next, a Falcon Cell Culture Insert bearing a transparent polyethylene terephthalate (PET) membrane, 23 mm in diameter, and with a pore size of 0.45 μm (Becton Dickinson), was placed within each well of the prewarmed companion plate. The suspension (2.5 mL) of alginate-recovered-chondrocytes and their associated matrix entrapped in 300 beads was plated onto each insert. After 7 additional days of culture, each insert was removed from the tissue culture plate and placed in a Petri dish. The PET membrane was cut along its circumference using a scalpel, the insert was removed, and the membrane was carefully peeled away, releasing the de novo ARC tissue shown in Figure 2.

Cartilaginous Tissue Formed From Adult Chrondrocytes

We observed that adult bovine chondrocytes can form a cartilaginous tissue de novo in as few as 14 days. These results extend those of previous studies on the fabrication of cartilage tissue. Although fetal articular chondrocytes maintained as a high-density monolayer appear to be capable of reforming a cartilage-like tissue within a few weeks, there is strong evidence that adult articular chondrocytes are much less effective in producing neocartilage.[1,2] Studies of cells isolated from steer articular cartilage cultured in alginate, a medium that promotes the retention of the chondrocytic phenotype, demonstrate that adult chondrocytes do not completely lose this capacity. Indeed, after 8 months of culture in alginate, these cells had reestablished around themselves a tissue that histologically and biochemically resembled the cartilage from which the cells were isolated.[6,7] The current findings demonstrate that adult chondrocytes cultured in alginate can be manipulated, even after release from the gel, to rapidly form a cohesive cartilaginous tissue laden with viable chondrocytes. This tissue is rich in keratan sulfate-bearing aggrecan and type II collagen, and contains only minimal amounts of type I collagen.

The cartilaginous tissue formed de novo is, like certain other cartilage constructs, devoid of a nonbiologic scaffold material that has to be degraded after implantation. Several strategies for forming such tissue have been used. Some involve the culture of immature chondrocytes on a surface that is subsequently mechanically separated from the cartilaginous tissue.[1,2] Others involve, in principle, an extended period of culture that results in the hydrolysis of resorbable materials and the diffusion of products out of the formed tissue.[8] The current approach actively removed the temporary scaffold. By means of cation chelation, the alginate gel is easily solubilized without affecting cell viability and fully washed out before the cells with their CM are used to form a cohesive cartilaginous tissue.

The first step of the ARC method allows the chondrocytes to surround themselves with a CM rich in proteoglycan aggregates that are turned over very rapidly. This constantly remodeled CM has been identified as the metabolically active compartment of the tissue.[5] Although it occupies less than 2% of the volume of adult human knee articular cartilage,[7] it plays a critical role in mediating the interactions of cells with the extracellular matrix as well as with neighboring cells. The mechanisms that allow the cells surrounded by a thin rim of CM to subsequently become rapidly integrated into a cohesive tissue are unclear. However, it appears that a specific amount or degree of maturity of the CM is essential, because prolonged culture in alginate does not lead to successful integration into a cartilaginous tissue. The cell-matrix organization of the tissue formed appears to mimic that of normal development and growth of articular cartilage. Both the in vitro tissue and fetal cartilage have a high density of cells and a low density of collagen relative to mature cartilage.[9]

214

Potential Future Applications

The ARC system to form cartilage-like tissue could be manipulated to address the mechanisms underlying the formation of the CM by the cells cultured in alginate gels or the integration of the cells with their CM to form a cohesive tissue. In addition, the process of subsequent growth and maturation of the newly formed tissue could be useful in several ways. Such a process could serve as a model system for analyzing not only cartilage growth, but for study of the relationships between tissue composition, structure, and biomechanical function. Preliminary results indicate that the tissue gains biomechanical competence with increasing time in culture.[10] The development of cartilaginous implants with biomechanical properties approaching those of native cartilage, at various stages of normal development and growth, may have practical utility. Such implants for cartilage defects may considerably reduce the time of rehabilitation compared with implants that consist of cells without a matrix. Additional work is needed to determine whether it is more important for the tissue to be immature and malleable, to ensure integration with the host cartilage tissue and remodeling,[11] or to be mature and stiff, to facilitate handling, rigid fixation, and the ability to withstand biomechanical forces after implantation.

Manipulation of the phenotype of the cells used to form cartilaginous constructs may be useful in several ways. Directing cells within tissue constructs to have a chondrocytic phenotype may be important in shortening the rehabilitation process if these constructs are used for cartilage repair. The ARC approach could also be applied using chondrocytes that have been amplified in number by culturing and passaging in monolayer cultures, as often done for autologous chondrocyte implantation procedures.[12] Indeed, such cells appear capable of regaining their chondrocytic phenotype when cultured in alginate gels.[13]

The application of the ARC cartilage tissue fabrication method to cells from other tissues or species, including humans, has not been fully evaluated. However, studies to determine whether human adult chondrocytes are capable of rapidly forming a cartilaginous tissue using the ARC methods are ongoing. Although the adult bovine cells used in this study produced an abundant matrix when stimulated by the growth factors present in FBS, additional growth factors might be used to induce human cells to successfully produce an equivalent amount of cartilaginous tissue. Whether growth-regulating factors, including defined growth factors or mechanical stimuli, can obviate the need for serum has yet to be determined.

Summary

We developed a novel two-step culture method, known as the ARC method, for the production of cartilaginous tissue by adult chondrocytes. The first step is to culture chondrocytes in alginate gel under conditions that maintain the normal phenotype of chondrocytes and modulate the formation of a CM rich in aggrecan molecules. After 1 week, the alginate gel is solubilized and the cells with their attached CM are recovered and cultured on a porous membrane. Within 1 week, the alginate-recovered chondrocytes become integrated into a cohesive cartilaginous tissue mass containing abundant amounts of aggrecan and type II collagen but minimal levels of type I collagen. Immunohistologic staining indicated that the glycosaminoglycan keratan-sulfate, present in high amounts in native bovine articular cartilage, was also detectable in the newly formed cartilaginous tissue. This scaffold-free system provides a novel means to study the development of transplantable cartilaginous tissue.

Acknowledgments

The authors wish to acknowledge that this work was supported, in part, by NASA grants NAG 8-1571 and NAG 9-1354 (RS), by NIH grants AG-04736 and 2 -P50-AR39239 (KM and ET), AG07996, AR44058, and AR46555 (RS), by NSF grant 9987353 (RS), and by a grant from the Rush Arthritis and Orthopaedics Institute.

References

1. Adkisson HD, Gillis MP, Davis EC, Maloney W, Hruska KA: In vitro generation of scaffold independent neocartilage. *Clin Orthop* 2001;391S:280-294.

2. Yu H, Grynpas M, Kandel RA: Composition of cartilaginous tissue with mineralized and non-mineralized zones formed in vitro. *Biomaterials* 1997;118:1425-1431.

3. Masuda K, Sah RL, Hejna MJ, Thonar EJ: A novel two-step method for the formation of tissue engineered cartilage by mature bovine chondrocytes: The alginate-recovered-chondrocyte (ARC) method. *J Orthop Res* 2003;21:139-148.

4. Benya PD, Shaffer JD: Dedifferentiated chondrocytes re-express the differentiated phenotype when cultured in agarose gels. *Cell* 1982;30:215-224.

5. Mok SS, Masuda K, Häuselmann HJ, Aydelotte MB, Thonar EJ: Aggrecan synthesized by mature bovine chondrocytes suspended in alginate: Identification of two distinct metabolic matrix pools. *J Biol Chem* 1994;269:33021-33027.

6. Häuselmann HJ, Fernandes RJ, Mok SS, et al: Phenotypic stability of bovine articular chondrocytes after long-term culture in alginate beads. *J Cell Sci* 1994;107:17-27.

7. Häuselmann HJ, Masuda K, Hunziker EB, et al: Adult human chondrocytes cultured in alginate form a matrix similar to native human articular cartilage. *Am J Physiol* 1996;40:C742-C752.

8. Vunjak-Novakovic G, Martin I, Obradovic B, et al: Bioreactor cultivation conditions modulate the composition and mechanical properties of tissue-engineered cartilage. *J Orthop Res* 1999;17:130-139.

9. Williamson AK, Chen AC, Sah RL: Compressive properties and structure-function relationships of developing bovine articular cartilage. *J Orthop Res* 2001;19:1113-1121.

10. Chen SS, Chen AC, Masuda K, et al: Tissue engineered cartilage from adult human chondrocytes: Biomechanical properties and function-composition relationships. *Trans Orthop Res Soc* 2003;28:945.

11. Schreiber RE, Ilten-Kirby BM, Dunkelman NS, et al: Repair of osteochondral defects with allogeneic tissue engineered cartilage implants. *Clin Orthop* 1999;367:382-395.

12. Brittberg M, Lindahl A, Nilsson A, Ohlsson C, Isaksson O, Peterson L: Treatment of deep cartilage defects in the knee with autologous chondrocyte transplantation. *N Engl J Med* 1994;331:889-895.

13. Flechtenmacher J, Mollenhauer J, Davies SD, et al: Human adult articular chondrocytes which no longer synthesize type II collagen regain the cartilage phenotype when cultured in alginate. *Orthop Trans* 1994;18:476-477.

Chapter 22

The Promise of Chondral Repair Using Neocartilage

Joseph Feder, PhD
H. Davis Adkisson, PhD
Neil Kizer, PhD
Keith A. Hruska, PhD
Regina Cheung, BS
Alan Grodzinsky, ScD
Yan Lu, MD
John Bogdanske, BS
Mark Markel, DVM, PhD

Abstract

A novel scaffold-free method for producing neocartilage in static culture has been developed. Using this model system, we demonstrate that juvenile chondrocytes are over 100-fold more active than adult chondrocytes in producing cartilage matrix. The characteristics of neocartilage formed from juvenile chondrocytes continue to be studied. For example, histologic methods have been used to confirm the hyaline nature of the tissue. Assays of the composition of neocartilage indicate that type II collagen is the predominant species of collagen synthesized. And, although the electrophoretic profile of the proteoglycans of neocartilage was similar to that of native tissue, there were some differences, including the extent of glycosylation of some proteoglycans. The biomechanical properties of the neocartilage have also been evaluated. The dynamic compressive stiffness of human neocartilage is 10% of that reported for adult human articular cartilage, and the tensile properties of neocartilage exceed dynamic compressive stiffness measurements by a factor of 10- to 50-fold. Most importantly, we have found that chondrocytes derived from neocartilage appear to be immune privileged. Dissociated chondrocytes cocultured with unrelated peripheral blood lymphocytes fail to induce proliferation of T lymphocytes. The immune privileged status of neocartilage is further supported by the fact that human neocartilage is not rejected following transplantation into experimental articular cartilage defects in animals. In summary, juvenile chondrocytes are naturally programmed to synthesize and organize cartilage matrix into neocartilage tissue at a significantly higher rate than adult chondrocytes. The enhanced biologic activity and the apparent lack of alloreactivity displayed by neocartilage allografts support the potential use of neocartilage for the repair and replacement of injured or diseased articular cartilage in humans.

Introduction

We developed a unique in vitro method for producing neocartilage, a viable biomechanically stable cartilage-like tissue, by culturing disaggregated chondrocytes derived from human, ovine, or rabbit articular cartilage. This neocartilage exhibits biochemical and morphologic properties similar to those of native tissue. In studies using a sheep model with surgically created lesions in cartilage, allografts of neocartilage integrated extremely well with host cartilage and retained their articular cartilage phenotype. These findings support the potential use of human neocartilage allografts for the repair and replacement of injured or diseased articular cartilage in humans. In this chapter, the characteristics of neocartilage are reviewed and its unique ability to repair surgically created chondral defects in the medial femoral condyle of sheep is discussed.

In Vitro Production of Neocartilage

Adkisson and associates[1] discovered that chondrocytes derived from juvenile articular cartilage could be grown under defined serum-free conditions to produce a cartilaginous tissue known as neocartilage. This neocartilage has many features that are similar to those of hyaline cartilage. The production of this neocartilage differs from other attempts to grow articular cartilage. The process exploits the unique potential of juvenile chondrocytes to form, in the absence of exogenous factors, a three-dimensional, structurally intact, hyaline-like cartilage tissue. The resulting neocartilage is similar to the tissue used as the source of the chondrocytes. Unlike some other tissue engineering techniques, the production of neocartilage does not require the use of polymeric scaffolds to form the three-dimensional tissue. The cells are grown in a defined serum-free medium that encourages the production of cartilage matrix molecules. The culture conditions used produce a biologic material that exhibits a tremendous potential for healing in the resurfacing of partial-thickness chondral defects in animal models.

Effect of Donor Age on Neocartilage Formation

To explore the effect of donor age on neocartilage formation, the ability of juvenile and adult human chondrocytes to produce cartilage matrix components, including sulfated glycosaminoglycans (S-GAGs) and hydroxyproline content of collagens, was assessed in vitro. Chondrocytes derived from the femoral and tibial condyles of individuals ranging in age from newborn to 72 years were grown under defined serum-free conditions as described previously.[1] Neocartilage harvested between days 44 and 58 of culturing was digested with papain and analyzed for its content of S-GAG, hydroxyproline, and DNA.[2-4] A cross-sectional analysis of more than 48 donors revealed chondrocytes derived from postpubertal individuals (> 20 years of age) to have lost their ability to produce amounts of S-GAG and hydroxyproline sufficient to support tissue formation (Figure 1). This loss of ability to produce neocartilage macromolecules manifested by chondrocytes derived from skeletally mature individuals follows an exponential decay curve. Juvenile

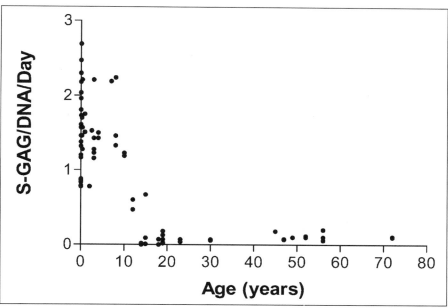

Figure 1 Effect of donor age on the synthesis of sulfated glycosaminoglycans (S-GAG) by neocartilage in vitro. Neocartilage was grown under defined serum-free conditions as previously described.[1] Matrix proteoglycan content was measured colorimetrically using dimethyl methylene blue between days 44 and 58 of culture after digestion with papain. Data are expressed as the average daily amount of proteoglycan deposited in the tissue matrix after normalization to DNA content. Juvenile chondrocytes produced nearly 100-fold greater levels of S-GAG than adult chondrocytes.

chondrocytes are over 100-fold more active than adult chondrocytes in producing cartilage matrix. These findings may have a significant impact on autologous approaches to cartilage repair because they suggest that in the absence of exogenous growth factors, adult chondrocytes cannot deposit matrix in sufficient quantities to repair large traumatic defects.

Characterization of Neocartilage

Identification of Neocartilage Collagens and Proteoglycans

When newly synthesized collagens were extracted from the matrix of neocartilage by limited pepsinization and subjected to sodium dodecylsulfate polyacrylamide gel electrophoresis, the pattern of distribution was virtually indistinguishable from that of native juvenile articular cartilage.[1] Type II collagen was the predominant type of collagen synthesized, but the minor collagens, type IX and XI, were also detected by staining with Coomassie Blue. Type I collagen was not detected by Western-blotting techniques. More recent studies assessed the long-term retention of the articular chondrocyte

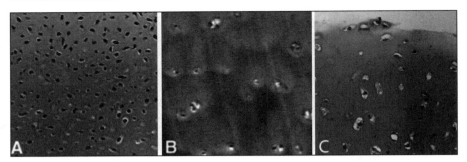

Figure 2 Microscopic appearance (original magnification × 400) of native juvenile (tibial) and adult (femoral condyle) articular cartilage and of engineered neocartilage (obtained after 90 days of culturing). As within juvenile and adult articular cartilage, chondrocytes within neocartilage have a rounded profile and are located within individual lacunoe. The cellular density of neocartilage is intermediate between that of juvenile tissue and that of adult articular cartilage.

phenotype in neocartilage with the use of enzyme linked immunosorbant assays to quantify type II and type I collagen epitopes. These studies showed that type II collagen accumulated in a linear fashion ($r^2 = 0.972$) during 1 year in culture. Nearly 1 mg of type II collagen was recovered from 6-mm diameter punches of intact neocartilage, but the level of type I collagen in the same samples was below the detection limits of the assay.

Newly synthesized proteoglycans were extracted from neocartilage. Using 1.2% agarose slab gels, the electrophoretic migration profile of proteoglycans from neocartilage was comparable to that of proteoglycans isolated from both juvenile and adult human articular cartilage.[1] Neocartilage contained an aggrecan that was more highly glycosylated than the adult species of this macromolecule, but which was characteristic of that in juvenile articular cartilage.[5] Surprisingly, the low-molecular-weight proteoglycan, decorin, was barely discernable in the neocartilage matrix after metabolic labeling with ^{35}S-sodium sulfate for 24 hours, and no biglycan was detected.

Morphologic and Ultrastructural Characterization of Neocartilage Matrix

Histologic staining of formalin-fixed and paraffin-embedded sections of neocartilage with Safranin O confirmed the hyaline nature of the tissue (Figure 2). Neocartilage chondrocytes have a rounded profile and are located within individual lacunae. The cellular density of neocartilage is intermediate between that of juvenile tissue and that of adult articular cartilage. Using transmission electron microscopy, the neocartilage matrix was observed to have an ultrastructural appearance that closely mimicked that of the transitional zone in adult articular cartilage, with a random distribution of 20-mm diameter type II collagen fibrils.[1]

Functional Biomechanical Properties of Neocartilage

The biomechanical properties of neocartilage have recently been evaluated in a pilot study designed to determine the relationship between dynamic compressive stiffness and biochemical composition. Human juvenile chondrocytes were seeded at various densities and grown for 45 to 150 days under static conditions in serum-free medium. The values for dynamic compressive stiffness for the neocartilage produced by these cells ranged from 0.080 MPa to 0.780 MPa and corresponded to S-GAG/DNA content ratios of 30 to 160. The human neocartilage used in our animal studies has a dynamic compressive stiffness of 0.30 MPa to 0.50 MPa. These values are approximately 10% of that reported for adult human cartilage derived from the distal femur (4.6MPa).[6] In contrast, the tensile properties of neocartilage are approximately 10- to 50-fold greater than its dynamic compressive stiffness. The ratio of tensile properties to dynamic compressive stiffness of neocartilage is characteristic of mature articular cartilage.[6] The dynamic compressive stiffness of human neocartilage compares favorably with that of native ovine cartilage derived from the distal femur (0.909 ± 0.540 MPa). This result suggests that neocartilage has sufficient strength to survive transplantation in our sheep studies. The dynamic compressive stiffness of ovine neocartilage ranges from 0.092 MPa to 0.40 MPa, depending on the duration of culture and the related S-GAG/DNA content.

Potential Alloreactivity of Neocartilage

The potential for graft rejection is a challenge to any engineered graft, particularly with allografts in which the cells used to produce the tissue are derived from an unrelated donor. To survive and integrate into the host, a neocartilage graft must evade the host's immune system. Once this biologic challenge is met, the graft can function as intended with potential for long-lasting repair.

An alloreactive response can be evolved when antigen-presenting cells that express human leukocyte antigen (HLA) class II (also known as major histocompatibility complex or MHC class II molecules) and at least one costimulatory signal (B7.1 or B7.2) interact with the appropriate receptors on naive T-cells.[7] The activated T-cells secrete interleukin-2 and undergo proliferation.

Jobanputra and associates[8] and other investigators[9,10] reported that chondrocytes have the ability to present class II HLA molecules, which are cell-surface bound glycoproteins. Although many cells have the potential to express class II HLA molecules, this does not necessarily render them alloreactive. Excellent examples of this premise are melanomas that constitutively express class II HLA but fail to stimulate the proliferation of primary T-cells.[11]

To explore the alloreactivity of neocartilage, we tested the ability of human chondrocytes derived from neocartilage to stimulate T cell proliferation in mixed lymphocyte cultures. Chondrocytes derived from neocartilage cultures that had been produced by human chondrocytes from eight different cadavers were used. In agreement with Jobanputra and associates,[8] cells from each of these sources failed to stimulate T cell proliferation when introduced into mixed lymphocyte cultures.[12] To further explore the absence of alloreactivity by these chondrocytes, we investigated their expression of co-stimulatory molecules. As previously reported,[12] polymerase chain reaction indicated that there was no expression of mRNA for either of the costimulatory molecules (B7.1 and B7.2) in any of the chondrocyte populations, irrespective of donor age.

Furthermore, when human neocartilage is implanted within ovine femoral condyle defects, it not only fails to be rejected, but there is no microscopic evidence of inflammatory cells 12 weeks after transplantation. These results strongly support the immune privilege of human neocartilage allografts. A low risk of alloreactive response with human neocartilage allografts would obviously be an important consideration for clinical use of these grafts.

Repair of Partial-Thickness Chondral Defects in the Ovine Knee

Several studies have examined the feasibility of allografted or xenografted neocartilage to induce cartilage repair in adult sheep. These studies evaluated initial graft fixation, determined the required period of immobilization and no weight bearing for graft integration, and assessed the effect of weight bearing on graft survival. In these studies, circular defects (5 mm in diameter and 300 to 500 µm in depth) were created in the weight-bearing region of the medial femoral condyle without violating the subchondral bone. Ovine or human neocartilage disks were cored to fit the defect, press-fitted in place, and sutured to the defect using resorbable thread. A grading system was established to evaluate the efficacy of the implant integration with native tissue and the repair of the surgically created lesion. The grading system included gross evaluation of graft survival, graft viability as measured by confocal microscopy, histologic evaluation for graft integration at the defect borders, and Safranin O staining to estimate the S-GAG content of the graft. In general, the survival and integration of neocartilage were excellent if the graft endured an 8-week period of immobilization. We also evaluated the efficacy of a tissue glue, tissue transglutaminase, to enhance graft fixation, integration, and survival.[13] One group of animals was studied after 8 weeks of no weight bearing and the second group after 1 to 3 weeks of weight bearing following the period of no weight bearing. The gross long-term survival of neocartilage allografts was excellent in all groups. The use of tissue transglutaminase did not significantly alter survival of the allografts. Integration determined by microscopic evaluation was also good, and histologic examination of surviving grafts revealed an excellent recovery of proteoglycan synthesis as assessed by staining with Safranin O in both the transplanted neocartilage and the surrounding host tissue (Figure 3). Chondrocyte viability, as determined by confocal microscopy using an ethidium homodimer and a calcein ester, was excellent in all of the surviving grafts.

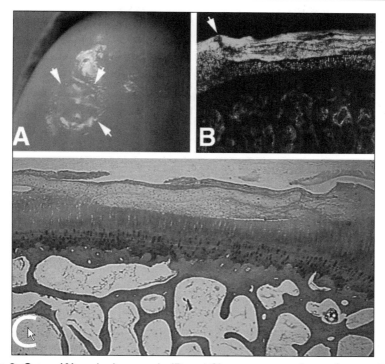

Figure 3 Gross **(A)** and microscopic **(B** and **C)** appearance of an ovine condyle that received a neocartilage allograft within a surgically created chondral lesion and was then subjected to 8 weeks of immobilization and 4 weeks of ambulation. **A,** Grossly, the graft (demarcated by arrows) has survived and appears to be well integrated with host articular cartilage. **B,** Low-power confocal photomicrograph of the same graft after staining for cell viability with a calcein ester (acetoxy-methylester) and an ethidium homodimer. Viable and nonviable chondrocytes stain green and red, respectively. One of the boundaries between the graft and native articular cartilage is indicated by an arrow (original magnification × 20). **C,** Safranin O staining of a formalin-fixed and paraffin-embedded section of the tissue represented in **(B).** The intensity of Safranin O staining within the neocartilage/repair tissue is similar to that in native articular cartilage. The original magnification was × 20.

Summary

We developed a novel in vitro method for producing neocartilage that involves culturing disaggregated juvenile chondrocytes derived from articular cartilage of various species. We show that juvenile chondrocytes are over 100-fold more active than adult chondrocytes in producing cartilage matrix. Proof-of-concept studies demonstrate that neocartilage holds great promise for the repair of articular cartilage defects arising from traumatic joint injury. The biochemical and morphologic features of neocartilage closely mimic those of juvenile articular cartilage. More importantly, neocartilage derived from juvenile chondrocytes appears to retain the autocrine growth factors that may be critical in facilitating integration of the transplanted graft with surrounding host articular cartilage.

Acknowledgments

The authors wish to acknowledge support from the National Institutes of Standards and Technology Advanced Technology Program Award (Grant #70NANB1H3027) and the National Institutes of Health Small Business Innovation Research Grant (Phase I: 1 R43 AR 46145-01; phase II: 2 R44 AR 46145-02).

References

1. Adkisson HD, Gillis MP, Davis EC, Maloney W, Hruska KA: In vitro generation of scaffold independent neocartilage. *Clin Orthop* 2001;391:S280-S294.

2. Farndale RW, Murray JC: Improved quantitation and discrimination of sulphated glycosaminoglycans by use of dimethylmethylene blue. *Biochim Biophys Acta* 1986;883:173-177.

3. Stegmann H, Stadler K: Determination of hydroxyproline. *Clin Chim Acta* 1967;18:267-273.

4. Labarca C, Paigen K: A simple, rapid, and sensitive DNA assay procedure. *Anal Biochem* 1980;102:344-352.

5. Cs-Szabo G, Roughley PJ, Plaas AH, Glant TT: Large and small proteoglycans of osteoarthritic and rheumatoid articular cartilage. *Arthritis Rheum* 1995;38:660-668.

6. Treppo S, Koepp H, Quan EC, Cole AA, Keuttner KE, Grodzinsky AJ: Comparison of biomechanical properties of cartilage from human knee and ankle pairs. *J Orthop Res* 2000;18:739-748.

7. Sharpe AH, Freeman GJ: The B7-CD28 superfamily. *Nat Rev Immunol* 2002;2:116-126.

8. Jobanputra P, Corrigall V, Kingsley G, Panayi G: Cellular responses to human chondrocytes: Absence of allogeneic responses in the presence of HLA-DR and ICAM-1. *Clin Exp Immunol* 1992;90:336-344.

9. Alsalameh S, Jahn B, Krause A, Kalden JR, Burmester GR: Antigenecity and accessory cell function of human articular chondrocytes. *J Rheumatol* 1991;18:414-421.

10. Jahn B, Burmester GR, Schmid H, Weseloh G, Rohwer P, Kalden JR: Changes in cell surface expression on human articular chondrocytes induced by gamma-interferon. Induction of Ia antigens. *Arthritis Rheum* 1987;30:64-74.

11. Brady MS, Lee F, Eckels DD, Ree SY, Latouche JB, Lee JS: Restoration of alloreactivity of melanoma by transduction with B71. *J Immunother* 2000;23:353-361.

12. Adkisson HD, Sung JH, Tippen MP, Maloney WJ, Hruska KA: Immune privilege permits neocartilage transplantation. *Trans Orthop Res Soc* 2000;25:612.

13. Lu Y, Bogdanske J, Kalscheur VL, et al: Effect of tissue transglutaminase on transplantation of ovine neocartilage allografts. *4th Symposium, International Cartilage Repair Society*, 2002, Toronto.

Chapter 23

Functional Assessment of Articular Cartilage

Jerry C. Hu, BS
Kyriacos A. Athanasiou, PhD, PE

Abstract

There is a great deal of interest in the application of tissue engineering techniques to the repair of damaged or diseased articular cartilage. Parallel with the advancement of tissue engineering techniques, there is a need for new methods to determine the functional capacity of tissue engineered cartilage. The most direct way to assess the function of cartilage is through the determination of its mechanical properties. Structural information also aids in the diagnosis of the pathophysiology of cartilage. Histologic methods, biochemical analyses, and noninvasive imaging techniques all provide valuable information about the structure and function of cartilage. However, these methods have not been optimized for use with tissue engineered constructs. In addition, the structural organization of tissues at the level of biochemical components cannot be examined in a noninvasive manner with currently available techniques. It is anticipated that new methods will continue to be developed to correlate the structure of cartilage grown in vitro to its function, and allow for more directed approaches to cartilage tissue engineering and repair.

Introduction

Tissue engineering to repair articular cartilage damaged by osteoarthrosis, injury, or other disease represents an area of intense interest and research. Degenerative joint diseases or injuries are associated with pain, swelling, joint stiffness, and reduction or loss of mobility. Cartilage often does not heal well given its avascular nature. In patients with osteochondral defects, mechanically inferior fibrocartilage is formed. Arthroplasty may be needed in patients with extensive degeneration of the articular surface.

Research of the use of tissue engineering techniques to repair or replace cartilage is ongoing. A particular challenge is the creation of repaired tissue with the appropriate biologic and mechanical properties. The function of cartilage is ultimately dictated by its mechanical properties, but structural information is also of key importance. This review will focus on methods to assess the functioning of cartilage. In particular, we focus on the unmet needs for methods that noninvasively determine the function and structure of cartilage.

Structure and Function of Articular Cartilage

Articular cartilage serves as a lubricated, wear-resistant, friction-reducing, and slightly compressible surface that evenly distributes forces onto bone. This aneural, avascular, alymphatic connective tissue that covers the articulating ends of bones may experience loads up to 18 megapascals (MPa) in the hip when rising from a chair.[1] The negative charges on the glycosaminoglycans (GAGs) in cartilage attract cations, thus creating a positive osmotic pressure, known as Donnan osmotic pressure that causes cartilage to swell.[2,3] When cartilage is compressed initially, the interstitial water becomes pressurized and supports a significant portion of the load.[4] The water is then forced out of the matrix, and friction between the water and the matrix dissipates the applied force.[5] The cartilage equilibrates as the load balances out the osmotic pressure. When the load is removed, fluid is imbibed back into the GAG network.

The need for functional assessment of cartilage arises from various forms of injury and pathology. The ability of cartilage to function is most directly assessed by its mechanical properties, although structural information also aids in the diagnosis of the pathophysiology of cartilage. Structural damage indicates a loss of function and is sometimes observed with arthroscopy or with the application of India ink. In the early stages, damage to cartilage is often imperceptible to the naked eye; however, cartilage in the later stages of osteoarthrosis or cartilage tears can be visualized with these methods. The gradual loss of proteoglycans compromises the mechanical properties of cartilage[6] and therefore its function, but this condition cannot be discerned arthroscopically. Although cartilage stiffness may be qualitatively assessed by probing the tissue, human touch is not sensitive enough to detect minute changes. Visualization of the structural damage of cartilage with histologic methods, biochemical analyses, and mechanical evaluations of cartilage not only extend the vision and touch of the clinician, but may also allow the damage to be quantified.

Cartilage Assessment Methods

Historically, the analysis of the structure of cartilage came before an evaluation of cartilage function because histologic and biochemical techniques were well developed. The basis for the current views of the structural and functional relationships of cartilage can be traced back to the early 1900s, when Benninghoff[7] postulated that the zonal variations in collagen orientation were a direct result of tissue function. It was not until the 1980s that a full model of the biomechanics of cartilage was developed by Mow and associates.[8] Functional assessment of cartilage should be directed toward its mechanical properties. However, structural changes also reflect changes in function and even basic techniques, such as histologic methods, and can provide important information with regard to the progression of functional loss or restoration of cartilage.

Structural Description of Articular Cartilage Function

Due to the structure-function relationship of cartilage, histologic methods can rapidly provide valuable information on the extracellular matrix (ECM). The presence or absence of a particular biochemical component, its relative amount, distribution, and, to some degree, organization within the tissue can be observed. Thus, histologic evaluation is often performed after gross inspection of the tissue when evaluating the function of normal, diseased, repaired, or tissue engineered cartilage.

Because GAGs and collagen are the two most abundant components of articular cartilage, they are often the first to be stained and observed with histologic methods. Stains for GAG target the anionic groups and include Safranin O/fast green, Alcian blue, and Toluidine blue O. Masson's and Gomori's trichrome, and Movat's pentachrome are often used to differentiate collagen from other components of the ECM. Immunohistochemistry can be used to detect the collagen type of interest. This is especially useful for evaluating repaired or tissue engineered cartilage. Unfortunately, type I, instead of type II, collagen is generally found in repaired fibrocartilage[9] and in engineered tissue containing cells that are no longer chondrocytic.

Staining is also useful in visualizing the organization of the tissue. Polarized microscopy can be used to observe collagen alignment.[10] Other techniques, such as scanning electron microscopy, offer direct visualization and quantification of collagen bundle organization and size, respectively.[11] Disorganized tissue is mechanically inferior and is expected to function poorly.

Histologic methods, though destructive, are indispensable for research. The field of tissue engineering has presented new problems in viewing tissues with histologic methods. Organic solvents, used to dry or clear slides, often dissolve the polymer scaffolds used in tissue engineering to house cells before tissue has fully developed. This obscures visualization of the polymer degradation process that plays an important role in tissue integration. Acids used to differentiate stains can also dissolve certain polymers. Uncharged polymers do not adhere to positively charged slides and thus sections can detach after the extended incubations often required for immunohistochemical analyses. Polymer networks, such as hydrogels, can retain stains. This increases the intensity of the background color and makes visualization of structures more difficult. Some of these problems can be addressed. Cryosectioning and aqueous mounting of polymer scaffolds can circumvent the use of organic solvents. Slides with adhesives that can be crosslinked can retain noncharged sections. The use of histologic methods in tissue engineering applications is still evolving as new polymeric scaffolds are developed. Histologic analyses may elucidate how tissue organization contributes to mechanical strength. Because tissue engineering of cartilage needs to overcome the lack of structural integrity in engineered constructs, a better understanding of structure-function relationships may lead to ways to direct cells to organize the matrix. Ultimately, structural analyses must be considered in relation to the function of the tissue.

Quantifiable Constituents of Articular Cartilage as Related to Function

Histology offers only a semiquantitative approach to the assessment of cartilage function because the relative intensity of staining depends on several variables in sample preparation and processing. Biochemical analyses are useful in situations when quantitative data are needed. For example, tissue function is indicated by the relative amounts of collagen, which imparts tensile strength to the tissue, and the fluid-filled GAG network that allows the compressive forces to dissipate.

In theory, assessment of the mechanical properties of cartilage does not necessarily have to involve physically touching the tissue. If data regarding the mechanical properties are unavailable, the performance of cartilage may be inferred through biochemical analyses as each component of the ECM serves a specific functional role. Certain techniques, such as MRI[12] and ultrasound,[13] represent noninvasive methods to assess tissue biochemistry. Data regarding biochemical properties of a tissue can then be used to predict mechanical properties with the use of models that correlate content with function. However, tissue function cannot be accurately predicted based on biochemical content alone. Understanding the structural organization of the tissue is also a key component in predicting tissue function. To date, it has been difficult to induce appropriate structural organization in tissue engineered constructs. Without the correct structural organization, the mechanical strength required for in vivo use cannot be achieved. Although MRI was used to determine GAG content as early as 1991,[14] at the time it was too strongly correlated to cartilage thickness to be a viable clinical tool. The recent development of contrast agents has renewed interest in the use of MRI[12] and results of modified MRI techniques have now been correlated to collagen orientation.[15] However, these imaging techniques are not yet fully developed. Furthermore, the structural organization of tissues at the level of biochemical components cannot be examined in a noninvasive manner. Thus mechanical function of cartilage cannot currently be determined with the use of biochemical assays or available imaging techniques alone.

Mechanical Evaluation of Articular Cartilage

The function of cartilage is ultimately dictated by its mechanical properties. From a mechanical point of view, cartilage can be modeled as a porous solid phase that contains and interacts with a fluid phase that fills the pores. This model, called the biphasic theory, models cartilage as consisting of an incompressible, porous-permeable solid and an incompressible viscous fluid.[8] The viscoelastic behavior of cartilage results from a viscoelastic dissipation within the collagen-proteoglycan matrix and a frictional resistance to flow of the interstitial fluid with respect to the permeable solid matrix.[16] Thus, when cartilage is loaded, the force imposed on the cartilage is counterbalanced by the fluid-flow drag of the interstitial fluid within the solid matrix.

Modeling with the biphasic theory yields three independent variables from an indentation test: the equilibrium compressive modulus, Poisson's ratio, and permeability. Articular cartilage has an aggregate modulus ranging from 0.47 MPa to 1.11 MPa. The Poisson's ratio of cartilage ranges from 0.0 to 0.4 and the permeability of cartilage ranges from 0.4×10^{-15} m⁴/Ns to 4.7×10^{-15} m⁴/Ns.[17-19] The size of the pores in the solid matrix has been estimated to range from 30 to 60 Å.[20]

Indentation testing is the technique most often used to assess the mechanical properties of cartilage.[17,18,21] Recently, a light scattering technique has also been used to measure the compressive modulus of cartilage by correlating the light scattering of collagen with properties of cartilage.[22] These devices work well in a research setting, but few in situ methods exist. To date, the biomechanical characteristics of cartilage cannot be determined in a noninvasive manner. The future of functional assessment is therefore to adapt current in vitro techniques to analysis in vivo, with minimal invasiveness. Once devices easily used in a clinical setting have been developed, cartilage evaluation can then be used in patients with diseases or conditions that are not usually or directly associated with diarthrodial joints. For example, it has been shown that patients with diabetes mellitus often have decreased cartilage stiffness,[23] but these patients may be more preoccupied with managing their insulin levels than with monitoring their cartilage health. Loss of bone density is a common problem for postmenopausal women and for women who have undergone total hysterectomy. Cartilage function is not routinely assessed in these women; however, the observation that cartilage stiffness is decreased in ovariectomized sheep suggests that evaluation of cartilage in women at risk for osteopenia and/or osteoporosis is indicated.[24] The development of hand-held devices, such as the Actaeon probe (OsteoBiologics Inc, San Antonio, TX)[25,26] and the Artscan (Artscan Oy, Helsinki, Finland),[27] may be useful to assess cartilage in these and other patient populations. For example, using the Actaeon, the stiffness of cartilage is measured arthroscopically in a fraction of a second and displayed as a numeric readout. Tests have shown that the readout does not depend on cartilage thickness, and studies on degenerated tissue have correlated probe readings to biochemical content.[25,26]

Summary

The ability to use tissue engineering techniques to repair or replace damaged cartilage would represent an important advance in medicine. As tissue engineering evolves, there is also a need for methods to accurately determine the functional characteristics of the repaired cartilage. Several histologic, biochemical, and imaging techniques are useful in the assessment of the properties of cartilage. However, to date, the biomechanical characteristics of cartilage cannot be assessed in a noninvasive manner. It is anticipated that currently available techniques used to assess cartilage function in vitro can be adapted for use in vivo. Methods to accurately and easily assess cartilage function in vivo with minimal invasiveness continue to be the evaluated.

References

1. Hodge WA, Carlson KL, Fijan RS, et al: Contact pressures from an instrumented hip endoprosthesis. *J Bone Joint Surg Am* 1989;71(9):1378-1386.

2. Maroudas A: Measurement of swelling pressure of cartilage, in Maroudas A, Kuettner K (eds): *Methods in Cartilage Research*. New York, NY, Academic Press, 1990.

3. Lai WM, Mow VC, Zhu W: Constitutive modeling of articular cartilage and biomacromolecular solutions. *J Biomech Eng* 1993;115:474-480.

4. Soltz MA, Ateshian GA: Experimental verification and theoretical prediction of cartilage interstitial fluid pressurization at an impermeable contact interface in confined compression. *J Biomech* 1998;31:927-934.

5. Ateshian GA, Warden WH, Kim JJ, Grelsamer RP, Mow VC: Finite deformation biphasic material properties of bovine articular cartilage from confined compression experiments. *J Biomech* 1997;30:1157-1164.

6. Kempson GE, Tuke MA, Dingle JT, Barrett AJ, Horsfield PH: The effects of proteolytic enzymes on the mechanical properties of adult human articular cartilage. *Biochim Biophys Acta* 1976;428:741-760.

7. Benninghoff A: Form und Bau der Gelenkknorpel in ihren beziechungen zur funktion: I. Die modellierenden und formerhalterden Faktoren des Knorpelreliefs. *Z ges Anat* 1925;76:43-63.

8. Mow VC, Kuei SC, Lai WM, Armstrong CG: Biphasic creep and stress relaxation of articular cartilage in compression? Theory and experiments. *J Biomech Eng* 1980;102:73-84.

9. Buckwalter JA: Articular cartilage: Injuries and potential for healing. *J Orthop Sports Phys Ther* 1998;28:192-202.

10. Arokoski JP, Hyttinen MM, Helminen HJ, Jurvelin JS: Biomechanical and structural characteristics of canine femoral and tibial cartilage. *J Biomed Mater Res* 1999;48:99-107.

11. Adams ME, Wallace CJ: Quantitative imaging of osteoarthritis. *Semin Arthritis Rheum* 1991;20:26-39.

12. Nieminen MT, Rieppo J, Silvennoinen J, et al: Spatial assessment of articular cartilage proteoglycans with Gd-DTPA-enhanced T1 imaging. *Magn Reson Med Oct* 2002;48:640-648.

13. Qin L, Zheng Y, Leung C, Mak A, Choy W, Chan K: Ultrasound detection of trypsin-treated articular cartilage: Its association with cartilaginous proteoglycans assessed by histological and biochemical methods. *J Bone Miner Metab* 2002;20:281-287.

14. O'Byrne EM, Paul PK, Blancuzzi V, et al: Magnetic resonance imaging of the rabbit knee: Detection of cartilage proteoglycan degradation. *Agents Actions* 1991;34:214-216.

15. Xia Y, Moody JB, Burton-Wurster N, Lust G: Quantitative in situ correlation between microscopic MRI and polarized light microscopy studies of articular cartilage. *Osteoarthritis Cartilage* 2001;9:393-406.

16. Mow VC, Ratcliffe A, Poole AR: Cartilage and diarthrodial joints as paradigms for hierarchical materials and structures. *Biomaterials* 1992;13:67-97.

17. Athanasiou KA, Rosenwasser MP, Buckwalter JA, Malinin TI, Mow VC: Interspecies comparisons of in situ intrinsic mechanical properties of distal femoral cartilage. *J Orthop Res* 1991;9:330-340.

18. Athanasiou KA, Agarwal A, Dzida FJ: Comparative study of the intrinsic mechanical properties of the human acetabular and femoral head cartilage. *J Orthop Res* 1994;12:340-349.

19. Athanasiou KA, Agarwal A, Muffoletto A, Dzida FJ, Constantinides G, Clem M: Biomechanical properties of hip cartilage in experimental animal models. *Clin Orthop* 1995;316:254-266.

20. Knudson CB: Hyaluronan receptor-directed assembly of chondrocyte pericellular matrix. *J Cell Biol* 1993;120:825-834.

21. Mow VC, Gibbs MC, Lai WM, Zhu WB, Athanasiou KA: Biphasic indentation of articular cartilage–II: A numerical algorithm and an experimental study. *J Biomech* 1989;22:853-861.

22. Kovach IS, Athanasiou KA: Small-angle HeNe laser light scatter and the compressive modulus of articular cartilage. *J Orthop Res* 1997;15:437-441.

23. Athanasiou KA, Fleischli JG, Bosma J, et al: Effects of diabetes mellitus on the biomechanical properties of human ankle cartilage. *Clin Orthop* 1999;368:182-189.

24. Turner AS, Athanasiou KA, Zhu CF, Alvis MR, Bryant HU: Biochemical effects of estrogen on articular cartilage in ovariectomized sheep. *Osteoarthritis Cartilage* 1997;5:63-69.

25. Niederauer GG, Schmidt DR, DeLee JC, et al: Applications and advantages of a hand-held indentation device. Paper presented at: 2002 International Cartilage Repair Society Symposium "Biophysical Diagnosis of Cartilage Degeneration and Repair", June 15-18, 2002, Toronto, Canada.

26. Niederauer GG, Niederauer GM, Cullen LC, Athanasiou KA, Thomas JB, Niederauer MQ: Sensitivity of a hand-held indentation probe for measuring the stiffness of articular cartilage. Paper presented at: The Second Joint Meeting of the IEEE Engineering in Medicine and Biology Society and the Biomedical Engineering Society, 2002, Houston, Texas.

27. Toyras J, Lyyra-Laitinen T, Niinimaki M, et al: Estimation of the Young's modulus of articular cartilage using an arthroscopic indentation instrument and ultrasonic measurement of tissue thickness. *J Biomech* 2001;34:251-256.

Future Directions

Tissue Engineering of Cartilage

The objective of the discussion, in which practicing orthopaedic surgeons, basic scientists, and experts in the fields of bioengineering and biomechanics participated, was to outline a realistic way forward in our endeavour to transpose articular cartilage engineering from the laboratory to clinical practice.

The tissue engineering approach to articular cartilage repair has been tackled with great enthusiasm in recent years. But the complexity of the task has been so grossly underestimated that we have been unable to realize our high expectations. In the future, it will be necessary to proceed in a more rational, stepwise fashion if we are to achieve ultimate success in the regeneration of articular cartilage tissue.

Choice and development of a clinically relevant experimental defect type

It is the general consensus that a partial-thickness defect type is of greater utility in a clinical context than a full-thickness one. In clinical practice, focal lesions are typically irregular in outline and are fibrillated, consequently, the ideal experimental model would thus reproduce this situation. However, since an experimental model should be reproducible, it must of necessity streamline the clinical situation. The most useful working defect models are deemed to be either a partial-thickness one, which does not penetrate the subchondral bone plate, or a very shallow full-thickness one, which just scratches the subchondral bone tissue so as to generate a few bleeding points. The deep, full-thickness defect type so frequently used in published studies is considered to be inappropriate and not representative of practical clinical needs.

Animal models

Many different experimental animals are used in articular cartilage repair studies, including rabbits, dogs, miniature pigs, sheep, goats, and monkeys. Furthermore, the genomes of several of these have been characterized, that should permit species-specific DNA sequence determination for the investigation of functional phenotypes. It was agreed that no single animal species ideally represents in all respects the human clinical situation, but that each may have its advantages or disadvantages according to the question to be tackled; the choice should reflect the specific requirements of the case in hand.

In an extremely large proportion of patients with articular cartilage defects, the lesions are caused by osteoarthritic degeneration. To adequately represent cases identified during the early stages of osteoarthritis, when lesions are still of a focal nature, a large animal defect model is required. However, a suitable model has not yet been established.

Cell sources for articular cartilage tissue engineering

A variety of autologous and allogeneic cell types have been used in articular cartilage tissue engineering studies, such as mature chondrocytes, bone-marrow-derived stromal cells, and chondroprogenitor cells derived from adipose, periosteal, or synovial tissue. It was agreed that there is a pressing need to specifically identify and characterize the chondroprogenitor cell pools obtained from adult or fetal sources and to determine the relative potential of each to produce hyaline-like articular cartilage tissue by elucidating the differentiation mechanisms and signaling pathways involved. Chondroprogenitor cells can produce several types of cartilage, including elastic, fibrous, and hyaline. But the factors governing a chondroprogenitor cell's commitment to a particular course of tissue differentiation are poorly understood. Only when we have a better comprehension of the differentiation process will we be in a position to rationally direct the engineered process as desired, namely, towards the production of a hyaline cartilage tissue.

Cell-carrier system

The choice of a suitable cell-carrier system for articular cartilage engineering is recognized to be an onorous task, in that an optimal matrix is not yet at our disposal. Indeed, such a matrix needs to fulfill so many requirements that it may be unrealistic to suppose that any one material will ever be satisfactory in all respects. Nevertheless, basic and applied research in this field is encouraged in an endeavour to combine within a single system as many as possible of the desired chemical, physical, and pharmacologic properties. Apart from such obvious features as biocompatibility and biodegradability (without yielding toxic by-products), the matrix should be preferably functionalized and easy to apply by arthroscopic intervention; it should have volume stability, appropriate biomechanical properties, and should adhere well to the walls and floor of the defect. This latter attribute, ie, the adequate fixation of a construct to the defect bed, is a prerequisite for the repair tissue's ultimate integration with native articular cartilage. As yet, there is no satisfactory solution to the problem. Neither surgical suturing nor the use of aggressive cross-linking agents is to be recommended, both being associated with adverse side effects. Physiologic bioadhesion appears to be the most promising approach to pursue in the future.

Signaling substances for articular cartilage repair

Many signaling substances with chondrogenic potential have now been identified, most of which are operative principally during embryologic development. It is a common supposition that these agents will trigger similar differentiation responses in an adult cartilage repair system, thereby effecting a recapitulation of the embryologic maturation process. However, the microenvironmental conditions pertaining, as well as the types of stem cells on which these substances will be acting, are quite different in the adult situation, and the therapeutic protocol elaborated cannot be based on an assumption of embryologic mimicry. We need to approach the problem of cartilage tissue engineering from a completely novel standpoint, drawing on more specific signaling agents and perhaps incorporating specific small peptide sequences into cell-carrier systems or transfecting cells with therapeutic genes. Needless to say, the concentration range and release kinetics of the signaling substances, as well as the target-cell effects, must be thoroughly characterized in order to yield optimal repair results. At this juncture, it is appropriate to mention the circumstance that most cell-based constructs are cultured in serum-containing media. With a view to generating a more reproducible system, we should now move away from this tradition. By culturing cell-based constructs in serum-free media, we will eliminate a variable and ill-defined source of signaling substances whose effects can neither be controlled nor monitored independently of those exerted by defined agents.

Structure, mechanical competence, and long-term survival of repair cartilage

The ultimate aim of articular cartilage engineering is to generate a type of tissue that is indistinguishable from the native hyaline one. This calls for a restoration of its anisotropic structure, the attainment of full mechanical competence, perfect integration with neighboring tissue, and the quality of endurance.

The mechanical properties of articular cartilage tissue are known to be closely coupled with its anisotropic structure. But we are still far from understanding the conditions that must be satisfied to achieve this organization, and this is an avenue that should be rigorously pursued in the future. There are indications in the literature that the physical properties of native articular cartilage need to be almost fully restored at a very early phase of healing if the repair tissue is to survive in the long run. Further research should now be conducted to precisely define the "minimal" mechanical needs during the initial stages of healing. A great many microenvironmental factors are known to influence the survival rate of cells, not only within an implanted construct but also within native tissue bordering the defect area, and this parameter naturally has a bearing on repair tissue integration and survival. A systematic investigation of these microenvironmental factors is of high priority in future cartilage tissue engineering endeavours.

Standardization of repair tissue quality assessments and of clinical outcome monitoring

The criteria chosen to gauge cartilage repair results in animal models and human patients vary considerably, which renders a comparison of different treatment strategies extremely difficult. A general consensus should be reached as to the critical defining parameters and attempts made to objectify, standardize, and preferably quantify evaluations in the future. In living patients, a similar approach should be adopted using noninvasive or minimally invasive technologies.

From the animal model to the patient

Cartilage tissue engineering with the human patient in mind calls for the consideration of a peculiar set of concerns that may be easily overlooked when working with animal models.

The components of an engineered construct, such as autologous or allogeneic stem cells and signaling substances, will all be subject to rigorous preclinical screening, and investigators should reflect on the likelihood of a proposed component being accepted for clinical use. A construct's potential for large-scale industrial production in bioreactors and its capacity for storage must also be considered.

The clinical practicability of a system must be constantly borne in mind. Simplicity of technology and ease of handling will weigh heavily in favor of a system being widely adopted in clinical practice. Developments in this particular direction may be costly and unprofitable from the point of view of investors and industry, but they are extremely important from the stance of public health services, and these conflicts of interest need to be resolved by representatives of the investment community, industry, and public health agencies.

Also of importance is the rapidity with which an implanted construct becomes established (if fully differentiated) or is transformed (if in a rudimentary form) into a well-integrated, mechanically competent hyaline-type of cartilage tissue. The postoperative rehabilitation phase must be kept to a minimum, not only to expedite a patient's social and professional reintegration but also to diminish the financial burden on health care systems. Although mechanical loading during this postoperative phase is known to have an impact on the outcome of the repair tissue, applied clinical studies are now required to precisely define this influence by carefully controlling the mobilization and physiotherapeutic protocols instigated.

Section Four

Tissue Engineering of Ligament, Tendon, Meniscus, Intervertebral Disk, and Muscle

Louis Almekinders, MD
Albert J. Banes, PhD
Don Bynum, MD
James Cook, DVM, PhD
Robert G. Dennis, PhD
David Dines, MD
C. Frank, MD
Jo Garvin, BS, MS
Daniel Grande, PhD
Mark Grynpas, PhD
D. Hart, PhD
Mark Hurtig, DVM
Brian Johnstone, PhD
Rita Kandel, MD

Melissa Maloney, BE, MS
James Mason, PhD
Robert Pilliar, BASc, PhD
Jie Qi, PhD
Pasquale Razzano, MS
Cheryle Seguin, MSc
David Tuckman, MD
Stephen Waldman, BASc, MSc, PhD
Neal Watnik, MD
Jung U. Yoo, MD

Chapter 24

Clinical Application of Tissue Engineered Tendon and Ligaments

C. Frank, MD
D. Hart, PhD

Abstract

There are a variety of clinical circumstances that require surgical repair of either tendons or ligaments. The natural healing ability of tendons and ligaments is variable and not all structures are able to truly heal. Suture apposition of torn ends of tendons and ligaments in combination with appropriate rehabilitation is a common method of improving their repair and can result in reasonable restoration of function in many situations. However, some ligament injuries have continued to have relatively poor functional clinical outcomes with available therapies. Several areas are being explored to enhance repair of tendons and ligaments. These include ways to exploit the plasticity of scar tissue to improve healing, possibly in conjunction with tissue engineering techniques. The use of intact tissues as grafts is one method being evaluated. A second major approach to tendon and ligament bioengineering has been the ex vivo creation of bioengineered struts that can be surgically implanted into tendon or ligament gaps or to replace the entire structure right into bone. For example, ligament cells can be isolated and implanted into lattices, gels, or tendon grafts. Understanding the inflammatory mechanisms responsible for changes to grafts and engineered tissues in the postimplantation local environment may be critical if these techniques are to be successful in the clinical setting. Basic studies to advance our understanding of the maturation processes essential to achieve fully functional tendons and ligaments are also needed.

Introduction

It is important to begin the discussion of replacing damaged tendons and ligaments by clearly defining the normal structures and functions of each of these unique dense fibrous connective tissues. As with many musculoskeletal tissues, tendons and ligaments look fairly similar, simple, homogeneous, and biologically inert. However, these tissues are surprisingly different, complex, heterogeneous, and dynamic.[1-6] To successfully replace these tissues, their complex functions must be duplicated (or at least substituted). Replacement of function is likely to require that their anatomic structures be reproduced. It is expected that at least some of their composition and microscopic substructures will need to be recreated. Therefore, the subtle aspects of these structures and their composition need to be appreciated.

Normal Tendons

Tendons are generally characterized as dense collagenous tissues with parallel fibers. Tendons contain a relatively small number of elongated tenocytes arranged in chains along the length of the fibers that run from muscular junction to bony attachment. The complex nature of these attachment sites at the muscular and bony ends has been studied in detail.[4,6] The complexity of the attachment sites clearly represents one of the challenges of tendon replacement. There are transitional cells and materials at the interfaces that bond tendon to muscle at the muscular end of the tendon and presumably help to distribute stresses more gradually at the tendon-bone interface at the opposite end. The body of the tendon may have different forms but generally appears as a cord-like structure that is either round or somewhat flattened as it runs from one end to the other. These cords may or may not be covered by synovial tissue or be contained within sheaths. In the hand, these cords run through fibrous pulleys, but at the ankle they can wrap around bony grooves. In the shoulder, rotator cuff tendons merge into a sheet that surrounds part of the humeral head under the acromion and represents a common area of injury, tearing, and/or impingement of the tendinous cuff. The morphology, physiology, and cell biology, including their healing and regenerative capacities, of different tendons have shown to vary slightly with different anatomic locations.[1,6]

Normal Ligaments

Ligaments are also dense, collagenous tissues that contain slightly plump but still elongated fibroblasts or fibrocytes between their parallel fibers. These cells generally run along the length of the fibers within the structure and between its bony insertions at both ends. There are some important differences in cell shapes and cellular behaviors between different ligaments. Midsubstance cells from normal anterior cruciate ligaments (ACLs) grow and migrate more slowly in vitro than cells derived from medial collateral ligaments (MCLs). Cells from ACLs and MCLs produce slightly different matrices. Similar to tendons at bone insertions, ligaments are morphologically complex. The zones of ligaments transition from soft midsubstance, through zones of stress-distributing fibrocartilage and calcified fibrocartilage, before merging into bone. The gross morphology of ligaments is more variable than that of tendons. Ligaments have structures ranging from discrete cord-like bands that are covered with synovium (eg, collateral and cruciate ligaments in the knee) to ill-defined capsular ligaments that encapsulate virtually all diarthrodial joints.[1,6]

Mechanical Properties of Tendons and Ligaments

Both tendons and ligaments have some key mechanical characteristics that must be duplicated if replacement is to be successful. Structural properties or load-deformation behaviors are structure specific. These are typically nonlinear, with both ligaments and tendons being relatively stiff and strong compared with other soft connective tissues such as skin. Different tendons and ligaments range in strength from hundreds to thousands of Newtons and structural strength is typically a function of their size. These tissues tend to be very stiff structurally with increasing stiffness as a function of size. When normalized to cross-sectional area and to changes in elongation (material properties or stress-strain behaviors), tendons and ligaments generally have tensile strengths of 60 to 80 Mpa and failure strains of less than 10% of their lengths. In addition, tendons and ligaments exhibit nonlinear viscoelastic behaviors that result in small amounts of stress-relaxation and creep under fixed deformations or stresses, respectively. These characteristics likely contribute to resistance to fatigue damage.[7] Both tendons and ligaments must resist tensile damage over a lifetime and are likely cyclically loaded in various degrees of tension more than other physiologic structures. Each type of tendon or ligament must have quite specific lengths to allow specific but minimal tensile displacement before the fibers in that structure carry and then share loads.[7,8] This load sharing among increasingly recruited collagen fibers is typical of tensile loading of both tendons and ligaments and contributes to their nonlinearity and possibly to their fatigue resistance. In addition to potential proprioceptive roles of embedded nerve endings,[9] these mechanical characteristics are likely critical for successful long-term replacement of either type of structure.

The Clinical Problem

A variety of clinical circumstances require some type of surgical repair of either tendons or ligaments. These may include pain, instability, or loss of normal full function. Tendons in the extremities are frequently lacerated or completely severed by penetrating sharp injuries. Segments of tendon can also be crushed or lost completely in degloving injuries, after traumatic losses of limb segments, or due to infections. Degenerative diseases including osteoarthritis and bursitis, and inflammatory disorders such as rheumatoid arthritis can damage adjacent tendons. In the worst cases of rheumatoid arthritis, tendons can rupture. Tendons might also become impinged, as with the rotator cuff tendons in the shoulder, causing them to fail from a combination of ischemic and physical damage. Repair of rotator cuff tendon injuries and/or defects is an important indication for bioengineered tendon tissue.

Ligaments are most often torn in traumatic joint injuries. They can be sprained, partially disrupted, or completely disrupted. The instability or laxity of the damaged joint tends to increase with increasing severity of the injury. Severe injuries are the most likely to become symptomatic. Combined injuries that affect multiple ligaments around a joint can make the joint unstable and less likely to heal spontaneously compared with isolated injuries. The functional outcomes of combined injuries tend to be worse than those of isolated ligament injuries and are more likely to require surgical repair. Replacement of injured cruciate ligaments in the knee, especially the ACL, are frequently required to address functional instabilities, particularly when cruciate injuries are combined with collateral ligament injuries.[10] Thus, improvements in the healing ability of both the cruciate and collateral ligaments of the knee are important. Inadequate healing or subsequent additional injury and failure of damaged ligaments are common in and around other joints, such as some finger and intercarpal ligaments, wrist ligaments, and ankle ligaments.

Suture apposition of torn ends of tendons and ligaments is a common method of improving their repair.[11] When used in combination with appropriate rehabilitation, this technique results in reasonable functional success in many situations, including repair of ruptured biceps tendons in the arm, quadriceps tendon ruptures in the leg, and repair of flexor or extensor tendon lacerations in the hand. Although healing is relatively slow with a prolonged rehabilitation in these situations, the ultimate clinical outcomes of such repairs are fairly good and rates of reinjury are quite low.

Based on their presumed similarity to tendons and success of acute surgical repairs of tendons, suture apposition of torn ends of ligaments was used in the mid-twentieth century to repair most acute ligament ruptures. The basis for use of this technique was that suture apposition of torn ligament ends should lead to better quality repair than nonsurgical approaches. However, it has become obvious that the natural history of healing of some tendon and ligament injuries, when combined with appropriate physiotherapy, is not detectably different from that of suture repairs.[12] Thus the need for suture repair of several tendon and ligament injuries has more recently been abandoned in favor of aggressive nonsurgical rehabilitation. It is not certain whether nonsurgical approaches work by supporting adequate healing of the damaged structures or through compensatory changes in complementary elements around each joint. Nonetheless, many nonsurgical approaches achieve adequate healing in a relatively safe and less invasive manner than with some surgical approaches.

Some ligament injuries, however, have continued to have relatively poor functional clinical outcomes with both natural healing plus aggressive and appropriate therapy, as well as with attempted suture repairs plus therapy. Suboptimal clinical outcomes are a particular concern in certain categories of patients such as young aggressive athletic males with ACL injuries. This observation suggests that other surgical or biologic approaches are needed to successfully address some injuries in certain individuals.[13] The ACL is probably the best example of this failure of natural and/or suture-only repairs in high-risk individuals.[4,14-16] Failure of rotator cuff injuries and their attempted suture repair in the shoulders of high risk middle-aged to older individuals are another example of suboptimal clinical outcomes. Various types of reconstructive surgery have evolved to treat these tendon and ligament lesions and other chronic tendon or ligament deficiencies of other joints.

Healing Processes, Healing Deficiencies, and Demands on Replacements

Accumulating evidence indicates that the natural healing ability of tendons and ligaments is variable.[17,18] Some structures have adequate compensatory mechanisms in response to injury and some can truly heal adequately to restore function. Other tissues lack these mechanisms and adequate, natural healing cannot be achieved. Physical, anatomic, and biologic reasons might explain these differences. In general, retraction of damaged ends leaves a gap that must be bridged in the most severe complete injuries of either tendons or ligaments. Gap size is an important factor in the ability to be bridged by formation of scar tissue. Gaps that are minimal have the potential to be successful bridged by local repair responses. Bigger gaps are structurally weaker and appear to increase the likelihood that a scar to bridge the gap will be flawed.[3] Some gaps are insurmountable, with inches between retracted muscle and tendon insertion sites onto bone. An example of such an injury is rupture of the biceps tendon in the upper arm.

Physical and anatomic barriers to the healing of tendon and ligament injuries are compounded by local biologic deficiencies that remain ill-defined, but include inadequate cellular division within hypocellular tissues, lack of local blood supply, local nerve supply, and the presence of inhibitors that interfere with local healing. For these reasons, certain structures have poor healing potential and require attempts at repair or restoration. Some of these problems, such as lack of local blood supply, are universal challenges for the use of bioengineered tissue replacements. However, the specific problems for a given tissue replacement will be somewhat unique and one approach may not work in all circumstances.

There are significant challenges in healing both tendons and ligaments. There is considerable evidence from animal models suggesting that both tendons and ligaments heal naturally by scar formation.[19] Even the human ACL expresses some scar-like substances.[20] Scar formation to repair tissues may start effectively but may be subsequently inhibited in the intra-articular environment of the ACL. This concept is supported by studies in sheep that compared healing in the ACL and MCL at the molecular level. In either situation, scar is a unique connective tissue that forms in the adult, apparently as a consequence of bleeding and inflammation surrounding the damaged connective tissue in the area of injury. Similar to scars of the skin, tendon and ligament scars are collagenous matrices built on fibrin clots with early cellular (fibroblastic) division and progressive deposition of an increasingly dense fibrous matrix. This matrix then remodels slowly to align along the main tensile axis of the structure. There are several morphological, ultrastructural, biochemical, and biomechanical changes that occur within the scar as it forms and remodels.[18] Some of these changes likely occur partly as a function of myofibroblast-mediated matrix contraction[21] with slowly progressive, but ultimately limited improvement toward normal tissue qualities. However, scar turnover can continue for months to years, creating a period of time to potentially intervene.

Some structures form better scars than others. Collateral ligaments, for example, form better scars than those of ACLs. Reasons for observed differences may include differences in intrinsic cell division and migration, the presence or absence of synovial cells around the structure to form a barrier to cellular and vascular ingrowth, local chemical or mechanical conditions, or other factors.[3,19] Reasons for ligament scar weakness in the skeletally mature host include flaws within the matrix,[22] abnormal collagen types, inadequate collagen cross-linking,[23] and abnormally small collagen fibril sizes[24] perhaps related to abnormal production of proteoglycans.

Whenever a scar forms, it has generally been shown to be weaker than normal tendons and ligaments. Scars of tendons and ligaments apparently stop improving after reaching roughly 30% to 50% of normal tissue quality, or roughly 25 to 40 Mpa of ultimate tensile strength.[2,25] This means that the only way such scars can replace normal ligament structural properties is to have cross-sectional areas that are two- to threefold larger than that of the normal structure. Fortunately, this appears to be possible for many tendons and ligaments. Viscoelastic behaviors of ligament scars (creep and load-relaxation behaviors) are also abnormal, but recovery of viscoelastic behaviors is apparently better than that of high load properties, at least in extra-articular healing environments. This has been the subject of relatively few investigations and the fatigue behaviors of such scars are not yet known.

Scar as a Pluripotential Substrate

There is considerable evidence to suggest that scars have the potential to be improved. Scars are hypermetabolic, hypercellular, and are responsive to local conditions in some important ways.[26,27] Cells in scars are connected abnormally to each other[28] and this may alter what they sense as the scar is loaded.[29] Research in animal models suggests that some joint motion, if appropriately controlled to stay below loads that cause failure or damage, improves the quality and quantity of scar formation of healing tendons and ligaments.[3] The optimal loads have not been defined, but it appears that some amount of tensile loading appears to improve scar quality and quantity. However, the appropriate loads are likely to be variable depending on the tissue and injury. A very unstable joint, such as with a combined ACL and MCL injury, may have too much motion and this may lead to an altered healing process with a prolonged inflammatory phase.

As with skin wound healing, biochemical methods of improving tendon and ligament scar quality have been examined quite extensively with evidence that scar can be modified by several extrinsic factors, including anti-inflammatory compounds,[30] growth factors,[31,32] and cytokines. Platelet derived growth factor appears to be an early stimulator of ligament scar formation.[33] Tissue growth factor-β1 and insulin-like growth factor-1 have been shown to have slightly later stimulatory effects on collagen synthesis and remodeling.[25,34] Hyaluronic acid may stimulate scar formation and type III collagen synthesis in ACLs in vivo.[35]

Adult scar tissue has some pluripotent capabilities. Application of compressive forces to the scar are apparently able to upregulate some cartilage-like molecules in vitro.[27] This has important implications for scar remodeling in vivo. This observation also suggests that scar tissue may be harvested, expanded, and modified ex vivo for subsequent reimplantation.[36-38]

Gene therapy for modification of ligament scar has been developed[39] and tested in animal models with some modest early success.[40-42] Decorin antisense therapy delivered by one injection of fusigenic Sendai virus (HVJ)-liposomes[42] was associated with some significant increases in early rabbit MCL scar strength and appeared to increase the size of some of the scar collagen fibrils.[41] However, the interpretation of these results was confounded by the alteration of many genes and variables within and between animals. Thus the mechanism by which changes occurred in this model are not fully understood.[43] In a similar trial of gene therapy for ligament healing, adenoviral-transfected myoblasts[40] and fibroblasts were injected into joints and shown to be capable of infiltrating both synovium and ACLs in the joint. Both of these animal studies suggest that gene-therapy mediated protein alterations are possible in normal ligament tissues, synovium, and scar tissue. These are important concepts for any injury because scarring is apparently a ubiquitous component of virtually any biologic tendon or ligament replacement.[18] The ability to modify scar tissue in combination with other healing tissues or surrounding uninjured structures could have important functional implications.

Tendon Grafts and Other Bioengineered Replacements

Intact Tissues as Grafts

Based on clinical and basic investigations that date back more than 100 years,[4,13,44-52] grafting of dense connective tissues from other sources, either within the same individual (autografting) or harvested from another individual of the same species (allografting), have become the gold standards for tendon and ligament replacement.[53-56] Grafting of animal tissues (xenografting) has received less attention[57] mainly because these grafts are often rejected by the recipient. Tendons and fascia are two major sources of harvestable tissues because these structures are potentially expendable, reasonable in terms of donor morbidity, and do appear to heal into bone tunnels over time.[58] Unlike tendon and fascial tissues, there are no nonessential ligaments that can be transplanted to another site without significant potential loss of function at the donor site.

Based on clinical studies during the past 50 years,[4] hamstring tendon grafts and partial patellar tendon grafts have become the two gold standard tendons for autograft in the lower extremity. Palmaris longus tendons, if present, are a low functional-risk graft source from the arm. Fascial strips from the lower extremity (the fascia lata) have also been harvested as tendon and ligament graft materials. Given the lack of concern for donor site morbidity in allografting, there are additional tendon and even potential ligament graft options such as the Achilles, tibialis anterior, ACL, and others. There have been considerable studies on these grafts during the past 25 years to determine their efficacy and potential shortcomings in a variety of situations.[48]

Key information regarding autografts and allografts and their healing processes is briefly summarized in the following sections.

Both patellar tendon grafts and multiply looped hamstring grafts have become nearly equal in terms of current use based on similar successes and relatively modest morbidities. However, there is some morbidity and potential risks with the harvesting of either type of tendon autograft, including pain, swelling, numbness caused by local nerve damage, scarring, stiffness, potential weakness, atrophy, tendinitis, and infection. These grafts involve significant surgical procedures and the tissues recover slowly over many months; however, their ultimate clinical outcomes are quite good.[52,54] The success rate of grafts in restoring ACL stability is estimated to be more than 80%, at least short-term. Although only a small proportion of these chronic reconstructions create normal knees from a symptomatic point of view, the majority restore functional stability for a significant period of time. Long-term clinical success rates likely decrease slightly, but these data are not yet firmly established.

Although both autograft and allograft tendons appear to have reasonable clinical success in terms of functional stability, there is considerable evidence that their structures and functions are altered over time. The cellularity and collagen matrix architecture of tendon grafts is likely altered shortly after transplantation.[49,52,59] Cells and vessels grow onto and into grafts, penetrating their dense matrix and contributing to local and generalized remodeling. Normally large collagen microfibrils are either replaced by new small scar-like fibrils,[3,19] or the normal fibrils become eroded to the smaller size.[60] Either way, the normal tendon tissue apparently becomes less stiff, weaker, and more prone to damage than at the time of transplantation. Grafts may also be softened by increased water and proteoglycan contents, potentially contributing to altered viscoelastic properties. Although not proven, there is some evidence from an animal model of collateral ligament grafting that even a supportive healing environment in the extra-articular space can cause fresh ligament grafts to change quite significantly.[26,61,62] Both autograft and allograft ligament grafts[50,63] have increased potential to stress-relax under fixed deformations and to creep under fixed loading conditions.[64] These changes could contribute to both altered load carrying ability of grafts at low functional loads and increases in relative laxity in vivo. Thus, these changes may not cause symptoms from instability but may contribute to altered joint mechanics, remodeling, and degradation of cartilage. This theory is currently being investigated in animal models.

There may be some common final pathways by which the original autografts or allografts undergo tissue remodeling and at least some partial replacement. Evidence from studies of biologic and biomechanic properties suggests that the changes in grafts are similar to those observed in natural tendon and ligament healing with some graft infiltration and partial replacement by scar. This finding implies that normal tendon and ligament grafts become somewhat scar-like as a natural consequence of their surgical placement when transplanted into a healing environment. Thus virtually any bioengineered tissue replacement surgically transplanted into the damaged area by methods that invoke bleeding and inflammation will almost certainly induce some scar formation in and around the replacement graft. In addition, there will almost certainly be some tissue infiltration[65] if not actual replacement of the bioengineered graft.[2] Therefore, strategies for successful bioengineered replacements of tendons and ligaments should consider ways to optimize the bioengineered graft-scar composite in situ during and after conditions in which scar tissue is formed. Principles that improve scar quality and quantity will likely be required to achieve successful bioengineered tendons and ligaments.

Lattices, Cells, and Composites

Another approach to the bioengineering of tendon and ligament replacements has been the ex vivo creation of bioengineered struts that can be surgically implanted into gaps in the tendon or ligament or be used to replace the entire structure. The struts that have been tested include many nonabsorbable substances such as carbon fibers[65] and Gortex,[4,66] ultra-high molecular weight polyethylene, silk,[17] and biodegradable polymers.[67-69] Substrates such as polylactide coglycolide, calcium pyrophosphate fibers,[70] collagen-chitosan composites,[71] natural or cross-linked collagen lattices without cells,[72-78] and fibroblast- or mesenchymal stem cell (MSC)-seeded collagen lattices[79,80] have also been tried. Ligament cells can be isolated[37,81] and implanted into lattices, gels, or tendon grafts; however, DesRosiers and associates[82] have shown that isolated ligament cells alter their molecular expression to become slightly more scar-like. Recreation of normal ligament cell behavior may require first recreating a normal ligament matrix (such as an autograft, allograft, or xenograft matrix) on which to grow these cells. Even if a repopulated matrix is generated, it will be difficult to recreate the communication potential of the cell networks that are present in normal ligament tissue but absent from scar tissue.[21]

Some materials created ex vivo have shown promising results in replacing tendons or ligaments.[43,79,83,84] Germain and associates[67] showed that a fibroblast-seeded, biomechanically-stimulated construct grown in vitro before implantation into ACL defects in dogs creates relatively good early restoration of ACL stiffness and strength. In a similar manner, Butler[85,86] and Awad and associates[87,88] used MSC-seeded gels to produce 18% to 33% stimulation of some healing properties of scars within rabbit patellar tendons 4 weeks after injury. In these studies, additional increases in the properties of healing rabbit Achilles tendons were observed at 12 weeks of healing. Although totally normal tissue properties have not been restored, these approaches seem very promising. More work is required to determine the long-term natural history of different constructs as compared with natural healing tissues and with normal tissues.

Potential Improvements in Bioengineered Tissues

Intact grafted tendon tissues deteriorate to some extent in vivo and both scar tissues and composites that have been implanted appear to have limited potential for total restoration of full tendon or ligament function. Some of the problems with restoration of function with grafts or implants of various materials may be associated with common pathways of degeneration and/or other processes that hinder repair. Two areas of investigation may lead to improvements in the efficacy of tendon and ligament replacements. The first is to develop methods to maintain graft quality after transplantation. Causes of the success or failure of repair of tendons and ligaments that are specific for structures or locations need to be elucidated. Studies of these factors may suggest solutions using bioengineering techniques. Infiltrating grafts with resistant cells is one such option. The second area of investigation is to develop ways to obtain or maintain grafts by building them from lattices with cells that can be manipulated. Alternatively, there may be methods to maintain the best possible autografts, allografts, or xenografts. The biochemical and biomechanical regulators of tissue synthesis, degradation, and remodeling need to be further elucidated and perhaps exploited to develop better methods of replacing tendons and ligaments.

Regulation of Inflammatory Processes: A Key Factor in Graft and Bioengineered Tissue Survival and Function

Other than some aspects of fetal wound healing,[12,36] injuries to tissues such as ligaments evoke an inflammatory response that is critical to repair processes. Surgical procedures used to place autografts, allografts, or bioengineered tissues also induce an inflammatory response. These inflammatory processes may be needed for healing. However, they may also play a role in the subsequent infiltration of a graft or engineered tissue by host cells. Some of these inflammatory processes can disrupt the integrity of the implanted material and result in formation of a scar-like material. Such changes are likely associated with diminished mechanical and biologic properties. Inhibition of inflammatory processes is one potential solution but may impair the healing of a graft or engineered tissue if the inhibition is not selective. Studies of the inflammatory mechanisms responsible for changes to implanted grafts and engineered tissues may be critical to develop specific interventions to maintain survival and maintenance of function of the implanted tissues. Perhaps bioengineered tissues could be generated with intrinsic anti-inflammatory or anti-angiogenic functions. Another solution would be to use methods to minimize specific inflammatory process extrinsically.

Recapitulation of Normal Maturation of Ligaments: Critical Factors in Improving Engineered Tissues

Ligaments undergo a temporally linked series of steps leading from a very cellular tissue at birth to a hypocellular tissue with a highly organized matrix at the time of skeletal maturation. Ligaments and tendons mature from a tissue with a unimodal distribution of small collagen fibrils to a larger tissue with a bimodal system of large and small diameter collagen fibrils. This maturation process has been described, but the molecular mechanisms responsible have yet to be fully elucidated. These steps have been the subject of intense investigations, and it will be important to consider this information in the design of engineered tissues. The maintenance of a bimodal distribution of collagen fibrils that are appropriately cross-linked and interact with a suitable set of proteoglycans will be essential for conveying the mechanical properties required for proper function of the tissue. This information can only come from basic investigations but is needed to optimize the engineering of functional replacement tissues for use in the clinical setting.

Summary

The natural healing ability of tendons and ligaments is variable and not all structures are able to truly heal. Suture apposition of torn ends of tendons and ligaments in combination with appropriate rehabilitation is a common method of improving their repair but has limitations. Although autografts and allografts may be used to replace tendons and ligaments, they are not always successful. In addition, there is evidence that their structures and functions are altered over time. Thus, successful engineering of tissues to replace tendons and ligaments in the clinical setting would represent an important advance in medicine. Additional data regarding healing of these tissues and the physiologic responses to grafts or implants are needed. It is especially important to understand factors affecting the maintenance of mechanical properties required for proper function of the tissue. It is anticipated that understanding factors critical to healing and maintaining tendons and ligaments will facilitate the engineering of replacement tissues for use in the clinical setting.

References

1. Amiel D, Billings E Jr, Akeson WH: Ligament structure, chemistry and physiology, in Daniel DM, Akeson WH, O'Connor JJ (eds): *Knee Ligaments: Structure, Function, Injury and Repair*. New York, NY, Raven Press, 1990, pp 77-91.

2. Frank C, McDonald D, Shrive N: Collagen fibril diameters in the rabbit medial collateral ligament scar: A longer term assessment. *Connect Tissue Res* 1997;36(3):261-269.

3. Frank C, Shrive N, Hiraoka N, Nakamura N, Kaneda Y, Hart D: Optimisation of the biology of soft tissue repair. *J Sci Med Sport* 1999;2(3):246-266.

4. Jackson DW, Simon TM, Kurzweil PR, Rosen MA: Survival of cells after intra-articular transplantation of fresh allografts of the patellar and anterior cruciate ligaments: DNA-probe analysis in a goat model. *J Bone Joint Surg Am* 1992;74:112-118.

5. Thornton GM, Boorman RS, Shrive NG, Frank CB: Medial collateral ligament autografts have increased creep response for at least two years and early immobilization makes this worse. *J Orthop Res* 2002;20:346-352.

6. Tomasek JJ, Gabbiani G, Hinz B, Chaponnier C, Brown RA: Myofibroblasts and mechano-regulation of connective tissue remodeling. *Nat Rev Mol Cell Biol* 2002;3(5):349-363.

7. Shrive NG, Chimich D, Marchuk L, Wilson J, Brant R, Frank C: Soft tissue "flaws" are associated with the material properties of the healing rabbit medial collateral ligament. *J Orthop Res* 1995;13:923-929.

8. Shino K, Inoue M, Nakamura H, Hamada M, Ono K: Arthroscopic follow-up of anterior cruciate ligament reconstruction using allogeneic tendon. *Arthroscopy* 1989;5:165-171.

9. Koski JA, Ibarra C, Rodeo SA: Tissue-engineered ligament: Cells, matrix and growth factors. *Orthop Clin North Am* 2000;31:437-452.

10. Cartmell JS, Dunn MG: Effect of chemical treatments on tendon cellularity and mechanical properties. *J Biomed Mater Res* 2000;49:134-140.

11. Thornton GM, Shrive NG, Frank CB: Altering ligament water content affects ligament pre-stress and creep behaviour. *J Orthop Res* 2001;19(5):845-851.

12. Peled ZM, Phelps ED, Updike DL, et al: Matrix metalloproteinases and the ontogeny of scarless repair: The other side of the wound healing balance. *Plast Reconstr Surg* 2002;110:801-811.

13. Dahners LE, Gilbert JA, Lester GE, Taft TN, Payne LZ: The effect of nonsteroidal anti-inflammatory drug on the healing of ligaments. *Am J Sports Med* 1988;16:641-646.

14. Arnold JA, Coker TP, Heaton LM, Park JP, Harris WD: Natural history of anterior cruciate tears. *Am J Sports Med* 1979;7:305-313.

15. Bosch U, Krettek C: Tissue engineering of tendons and ligaments: A new challenge. *Unfallchirurg* 2002;105:88-94.

16. Jackson DW, Simon TM: Tissue engineering principles in orthopaedic surgery. *Clin Orthop* 1999;367:31-45.

17. Altman GH, Horan RL, Lu HH, et al: Silk matrix for tissue engineered anterior cruciate ligaments. *Biomaterials* 2002;23(20):4131-4141.

18. Hildebrand KA, Woo SL-Y, Smith DW, et al: The effects of platelet-derived growth factor-BB on healing of the rabbit medial collateral ligament: An in vivo study. *Am J Sports Med* 1998;26:549-554.

19. Frank CB: The pathophysiology of ligaments, in Arendt EA (ed): *Orthopaedic Knowledge Update: Sports Medicine 2*. Rosemont, IL, American Academy of Orthopaedic Surgeons, 1999.

20. Lo IKY, Chi S, Ivie T, Frank CB, Rattner JB: The cellular matrix: A feature of tensile bearing dense soft connective tissues. *Histol Histolopathol* 2002;17:523-537.

21. Sun ZY, Zhao L: Feasibility of calcium pyrophosphate fiber as scaffold material for tendon tissue engineering in vitro. *Zhongguo Xiu Fu Chong Jian Wai Ke Za Zhi* 2002;16(6):426-428.

22. Sabiston P, Frank C, Lam T, Shrive N: Transplantation of the rabbit medial collateral ligament: I. Biomechanical evaluation of fresh autografts. *J Orthop Res* 1990;8:35-45.

23. Dunn MG, Liesch JB, Tiku ML, Zawadsky JP: Development of fibroblast-seeded ligament analogs for ACL reconstruction. *J Biomed Mater Res* 1995;29(11):1363-1371.

24. Frank CB, McDonald DB, Wilson JE, Eyre DR, Shrive NG: Rabbit medial collateral ligament scar weakness is associated with decreased collagen pyridinoline crosslink density. *J Orthop Res* 1995;13(2):157-165.

25. Thornton GM, Frank CB, Shrive NG: Ligament creep recruits fibres at low stresses and can lead to modulus-reducing fibre damage at higher creep stresses: A study in a rabbit medial collateral ligament model. *J Orthop Res* 2002;20(5):967-974.

26. Lo IKY, Ou Y, Rattner JP, Hart DA, Frank CB, Rattner JB: The cellular networks of normal ovine medial collateral and anterior cruciate ligaments are not accurately recapitulated in scar. *J Anat* 2002;200:283-296.

27. Lo IKY, Marchuk LL, Leatherbarrow KE, et al: The healing response of ACL and MCL injuries is initially similar but diverges with time. 4th World Congress for Biomechanics, Calgary, Alberta, August 4-9, 2002.

28. Lo IKY, Randle JA, Majima T, et al: New directions in understanding and optimizing ligament and tendon healing. *Curr Opin Orthop* 2000;11:421-428.

29. Lin VS, Lee MC, O'Neal S, McKean J, Sung KL: Ligament tissue engineering using biodegradable fiber scaffolds. *Tissue Eng* 1999;5(5):443-452.

30. Cunningham KD, Musani F, Hart DA, Shrive NG, Frank CB: Collagenase degradation decreases collagen fibril diameters: An in vitro study of the rabbit medial collateral ligament. *Connect Tissue Res* 1999;40(1):67-74.

31. Daniel DM: Selecting patients for ACL surgery, in Jackson DW, Arnoczky SP, Woo SLY, Frank CB, Simon TM (eds): *The Anterior Cruciate Ligament: Current and Future Concepts*. New York, NY, Raven Press, 1993, pp 251-258.

32. Laurencin CT, Ambrosio AM, Borden MD, Cooper JA Jr: Tissue engineering: Orthopaedic applications. *Annu Rev Biomed Eng* 1999;1:19-46.

33. Batten ML, Hansen JC, Dahners LE: Influence of dosage and timing of application of platelet-derived-growth factor on the early healing of the rat medial collateral ligament. *J Orthop Res* 1996;14:736-741.

34. Wiig ME, Amiel D, VandeBerg J, Kitbayashi L, Harwood FL, Arfors KE: The early effect of high molecular weight hyaluronan (hyaluronic acid) on anterior cruciate ligament healing: An experimental study in rabbits. *J Orthop Res* 1990;8:425-434.

35. Tan W, Krishnaraj R, Desai TA: Evaluation of nanostructured composite collagen-chitosan matrices for tissue engineering. *Tissue Eng* 2001;7(2):203-210.

36. Bleacher JC, Adolph VR, Dillon PW, Krummel TM: Fetal tissue repair and wound healing. *Dermatol Clin* 1993;11:677-683.

37. Hildebrand KA, Frank CB: Scar formation and ligament healing. *Can J Surg* 1998;41(6):425-429.

38. Jackson DW, Arnoczky SP, Woo SL-Y, Frank CB, Simon TM: (eds): *The Anterior Cruciate Ligament: Current and Future Concepts*. New York, NY, Raven Press, 1993.

39. Winters SC, Seiler JG, Woo SL-Y, Gelberman RH: Suture methods for flexor tendon repair: A biomechanical analysis during the first six weeks following repair. *Ann Chir Main Memb Super* 1997;16:229-234.

40. Lo IK, Marchuk L, Hart DA, Frank CB: Messenger ribonucleic acid levels in disrupted human anterior cruciate ligaments. *Clin Orthop* 2003;407:249-258.

41. Majima T, Lo IKY, Randle JA, et al: ACL transection influences mRNA levels for collagen type I and TNF-alpha in MCL scar. *J Orthop Res* 2002;20:520-525.

42. Menetrey J, Kasemkijwattana C, Day CS, et al: Direct-, fibroblast- and myoblast-mediated gene transfer to the anterior cruciate ligament. *Tissue Eng* 1999;5:435-442.

43. Goulet F: Tissue-engineered ligament, in Lanza R, Langer R, Chick WL (eds): *Principles of Tissue Engineering*. San Diego, CA, Academic Press Ltd, 1997, pp 633-644.

44. Aglietti P, Buzzi R, Zaccherotti G, DeBiase P: Patellar tendon versus doubled semi-tendinosus and gracilis tendons for anterior cruciate ligament reconstruction. *Am J Sports Med* 1994;22:211-217.

45. Arnoczky SP, Matyas JR, Buckwalter JA, Amiel D: Anatomy of the anterior cruciate ligament, in Jackson DW, Arnoczky SP, Woo SLY, Frank CB, Simon TM (eds): *The Anterior Cruciate Ligament: Current and Future Concepts.* New York, NY, Raven Press, 1993, pp 5-22.

46. Beynnon BD, Johnson RJ, Fleming BC: The mechanics of anterior cruciate ligament reconstruction, in Jackson DW, Arnoczky SP, Woo SLY, Frank CB, Simon TM (eds): *The Anterior Cruciate Ligament: Current and Future Concepts.* New York, NY, Raven Press, 1993, pp 259-272.

47. Beynnon BD, Fleming BC, Johnson RJ, Nichols CE, Renstrom PA, Pope MH: Anterior cruciate ligament strain behaviour during rehabilitation exercises in vivo. *Am J Sports Med* 1995;23:24-34.

48. Butler DL, Goldstein SA, Guilak F: Functional tissue engineering: The role of biomechanics. *J Biomech Eng* 2000;122:570-575.

49. King GJ, Edwards P, Brant RF, Shrive NG, Frank C: Intraoperative graft tensioning alters viscoelastic but no failure behaviours of rabbit medial collateral ligament autografts. *J Orthop Res* 1995;13:915-922.

50. Milthorpe BK: Xenografts for tendon and ligament repair. *Biomaterials* 1994;10:745-752.

51. Murray MM, Spector M: The migration of cells from the ruptured anterior cruciate ligament into collagen-glycosaminoglycan regeneration templates in vitro. *Biomaterials* 2001;17:2393-2402.

52. Nakamura N, Timmermann SA, Hart DA, et al: A comparison of in vivo gene delivery methods for antisense therapy in ligament healing. *Gene Ther* 1998;5:1455-1461.

53. Hodde J: Naturally occurring scaffolds for soft tissue repair and regeneration. *Tissue Eng* 2002;8(2):295-308.

54. Nakamura N, Hart DA, Boorman RS, et al: Decorin antisense gene therapy improves functional healing of early rabbit ligament scar with enhanced collagen fibrillogenesis in vivo. *J Orthop Res* 2000;18(4):517-523.

55. Rodeo SA, Arnoczky SP, Torzilli PA, Hidaka C, Warren RF: Tendon-healing in a bone tunnel: A biomechanical and histological study in the dog. *J Bone Joint Surg Am* 1993;75:1795-1803.

56. Ross SM, Joshi R, Frank CB: Establishment and comparison of fibroblast cell lines from the medial collateral and anterior cruciate ligaments of the rabbit. *In Vitro Cell Dev Biol* 1990;26(6):579-584.

57. Majima T, Marchuk LL, Shrive NG, Frank CB, Hart DA: In vitro cyclic tensile loading of an immobilized and mobilized ligament autograft selectively inhibits mRNA levels for collagenase MMP-1. *J Orthop Sci* 2000;5:503-510.

58. Ng GY, Oakes BW, Deacon OW, McLean ID, Lampard D: Biomechanics of patellar tendon autograft for reconstruction of the anterior cruciate ligament in the goat: Three-year study. *J Orthop Res* 1995;13:602-608.

59. Amiel D, Kleiner JB, Akeson WH: The natural history of the anterior cruciate ligament autograft of patellar tendon origin. *Am J Sports Med* 1986;14:449-462.

60. Chen EH, Black J: Materials design analysis of the prosthetic anterior cruciate ligament. *J Biomed Mater Res* 1980;14(5):567-586.

61. Kato YP, Dunn MG, Zawadsky JP, Tria AJ, Silver FH: Regeneration of Achilles tendon with a collagen tendon prosthesis: Results of a one-year implantation study. *J Bone Joint Surg Am* 1991;73(4):561-574.

62. Noyes FR, Barber-Westin SD: Reconstruction of the anterior cruciate ligament with human allograft: Comparison of early and later results. *J Bone Joint Surg Am* 1996;78:524-537.

63. Noyes FR, Butler DL, Paulos LE, Grood ES: Intra-articular cruciate reconstruction: I. Perspectives on graft strength, vascularization and immediate motion after replacement. *Clin Orthop* 1983;172:71-77.

64. Shino K, Oakes BW, Horibe S, Nakata K, Nakamura N: Collagen fibril populations in human anterior cruciate ligament allografts: Electron microscopic analysis. *Am J Sports Med* 1995;23:203-208.

65. Gisselfalt K, Edberg B, Flodin P: Synthesis and properties of degradable poly (urethane ureas) to be used for ligament reconstructions. *Biomacromolecules* 2002;3(5):951-958.

66. Casteleyn PP, Handelberg F: Non-operative management of anterior cruciate ligament injuries in the general population. *J Bone Joint Surg Br* 1996;78:446-451.

67. Germain L, Goulet F, Muolin V, Berthod F, Auger FA: Engineering human tissues for in vivo applications. *Ann N Y Acad Sci* 2002;961:268-270.

68. Li WJ, Laurencin CT, Caterson EJ, Tuan RS, Ko FK: Electrospun nanofibrous structure: A novel scaffold for tissue engineering. *J Biomed Mater Res* 2002;60(4):613-621.

69. Majima T, Marchuk LL, Sciore P, Shrive NG, Frank CB, Hart DA: Compressive compared with tensile loading of medial collateral ligament scar in vitro uniquely influences mRNA levels for aggrecan, collagen type II and collagenase. *J Orthop Res* 2000;18(4):524-531.

70. Sabiston P, Frank C, Lam T, Shrive N: Allograft ligament transplantation: A morphological and biochemical evaluation of a medial collateral ligament complex in a rabbit model. *Am J Sports Med* 1990;18:160-168.

71. Shelbourne KD, Nitz P: Accelerated rehabilitation after anterior cruciate ligament reconstruction. *Am J Sports Med* 1990;18:292-299.

72. Donnelly NI, Hart DA, Frank CB: Matrix mRNA levels in ligament tissue versus cells. *In Vitro Cell Dev Biol Anim* 1998;34:617-618.

73. Dunn MG, Tria AJ, Kato YP, et al: Anterior cruciate ligament reconstruction using a composite collagenous prosthesis: A biomechanical and histologic study in rabbits. *Am J Sports Med* 1992;20(5):507-515.

74. Dunn MG, Avasarala PN, Zawadsky JP: Optimisation of extruded collagen fibers for ACL reconstruction. *J Biomed Mater Res* 1993;27(12):1545-1552.

75. Hildebrand KA, Jia F, Woo SL-Y: Response of donor and recipient cells after transplantation of cells to the ligament and tendon. *Microsc Res Tech* 2002;58:34-38.

76. Kannus P, Järvinen M: Conservatively treated tears of the anterior cruciate ligament: Long-term results. *J Bone Joint Surg Am* 1987;69:1007-1012.

77. Kleiner JB, Amiel D, Roux RD, Akeson WH: Origin of replacement cells for the anterior cruciate ligament autograft. *J Orthop Res* 1986;4:466-474.

78. Krauspe R, Schmidt M, Schaible HG: Sensory innervation of the anterior cruciate ligament: An electrophysiological study of the response properties of single identified mechanoreceptors in the cat. *J Bone Joint Surg Am* 1992;74:390-397.

79. Bellincampi LD, Closkey RF, Prasad R, Zawadsky JP, Dunn MG: Viability of fibroblast-seeded ligament analogs after autogenous implantation. *J Orthop Res* 1998;16(4):414-420.

80. Letson AK, Dahners LE: The effect of combinations of growth factors on ligament healing. *Clin Orthop* 1994;308:207-212.

81. Ng GY, Oakes BW, Deacon OW, McLean ID, Lampard D: The long-term biomechanical and viscoelastic performance of repairing anterior cruciate ligament after hemitransection injury in a goat model. *Am J Sports Med* 1996;24:109-117.

82. DesRosiers EA, Yahia L, Rivard CH: Proliferative and matrix synthesis response of canine cruciate ligament fibroblasts submitted to combined growth factors. *J Orthop Res* 1996;14:200-208.

83. Frank CB, Hart DA, Shrive NG: Molecular biology and biomechanics of normal and healing ligaments: A review. *Osteoarthritis Cartilage* 1999;7(1):130-140.

84. Goodship AE, Wilcock SA, Shah JS: The development of tissue around prosthetic implants used as replacements for ligaments and tendons. *Clin Orthop* 1985;196:61-68.

85. Buckley SL, Barrack RL, Alexander AH: The natural history of conservatively treated partial anterior cruciate ligament tears. *Am J Sports Med* 1989;17:221-225.

86. Butler DL, Awad HA: Perspectives on cell and collagen composites for tendon repair. *Clin Orthop* 1999;367:S324-S332.

87. Awad HA, Butler DL, Boivin GP, et al: Autologous mesenchymal stem cell-mediated repair of tendon. *Tissue Eng* 1999;5(3):267-277.

88. Awad HA, Butler DL, Harris MT, et al: In vitro characteristics of mesenchymal stem cell-seeded collagen scaffolds for tendon repair: Effects of initial seeding density on contraction kinetics. *J Biomed Mater Res* 2000;51(2):233-240.

Chapter 25

Bioartificial Tendons: An Ex Vivo Three-Dimensional Model to Test Tenocyte Responses to Drugs, Cytokines, and Mechanical Load

Albert J. Banes, PhD
Jie Qi, PhD
Melissa Maloney, BE, MS
Jo Garvin, BS, MS
Louis Almekinders, MD
Don Bynum, MD

Abstract

The development and use of suitable constructs for tendon replacement or augmentation have been limited by issues with autogenous or allogeneic graft materials. Carbon fiber and polytetrafluorethylene scaffolds have been used without success. Most often, a tendon repair is performed by apposition and suturing of the tendon ends without the need for additional lengthening material. However, allografts sequentially weaken with time and even primary repair sites can fail. Various cell-populated matrix constructs, consisting of endogenous cells or stromal cells in a cell-assembled or collagen matrix, have been developed. We describe a model system for the investigation of tendon cells in a linear gel matrix. This system can be uniformly and reproducibly fabricated as an artificial tendon with custom compositions of matrix and cells, populated with animal or human cells, and mechanically loaded with an automated regimen. The ends may have a specialized composition for bone and muscle attachment. This system can easily be seeded with genetically engineered cells, cytokines, growth factors, or other ligands. These bioartificial tendons (BATs) have linearly arranged cells that have a tendon-like gene expression profile, histologic appearance of human flexor tendon, and remodel their matrix. These cells respond to ascorbate, adenosine triphosphate, norepinephrine, and interleukin 1-β as do whole tendons or isolated cells. The currently available BATs are relatively weak, but their strength can be increased by treatment with ascorbate. Fabricating BATs by populating matrix gels with autogenous tenocytes grown in mechanically loaded cultures may be a way to augment tendon healing and to grow a structurally competent bioartificial tendon for replacement of damaged tissue with integration into the bony insertion and origin in mus-

cle. Five areas of additional research include: (1) building a tendon segment to unite torn or ruptured tendons, (2) building an entire digit replete with bone and muscle attachments, (3) developing a construct to withstand the forces applied to small and large tendons, maintain correct architecture, cell types, blood supply, innervation, and attachment points to bone and muscle, (4) evaluating suturability, scalability, maintenance of gliding characteristics, strength, potential for adhesions, and response to cytokines and growth factors, and (5) integration of an artificial tendon into the native tendon matrix. The integration of the artificial tendon represents a challenge, whether the construct is designed as a void filler, to deliver cells, or as a tension-bearing construct that can be sutured or otherwise fixed between two points such as apposing ruptured tendon ends. It is hoped that these issues can be overcome to develop tendon replacements for clinical use.

Introduction

In the United States, more than 100,000 patients per year undergo surgery to repair tendon or ligament injuries.[1] The current gold standard for surgical repair of these injuries is the use of autogenous tendon. However, during repair, the mechanical strength and structural characteristics of the host tissue are permanently altered. During anterior cruciate ligament reconstruction, often with the use of patellar tendon, an initial loss of strength occurs with a gradual increase that never reaches the original level of strength.[2,3] Therefore, new treatments to provide long-term tendon or ligament repair are needed. Natural scaffold materials into which marrow stem cells are seeded have been tested in a rabbit patellar tendon implant model.[4-7] Acellular synthetic materials such as Dacron,[8-9] polytetrafluoroethylene,[10] polypropylene,[11] and carbon fibers[12] have been tried and have failed. Most of the synthetic materials do not approximate the properties of tendon or ligament and thus result in stress shielding in the natural tissue and production of wear debris or toxic degradation products. Fibrillar collagen scaffolds allow cells to integrate themselves and the matrix into surrounding tissue. This material can be formulated to approximate the collagen types of the host tissue that consists of 92% type I collagen in the tensile load bearing compartment with lesser amounts of type III, type XII, and other collagens.[13] Material properties of fibrillar collagen scaffolds have not matched those of tendon or ligament because of the low strength of collagen gel materials.[4,5,14,15] Collagen can be antigenic but is often minimally immunogenic even if derived from xenogeneic sources. The use of bovine tendon instead of bovine skin as a source of collagen reduces some of the collagen type mixing, potential for contamination with hair and other organelles, and bioburden. In addition, it would be advantageous to use a material seeded with native tendon cells because these cells are responsible for normal tissue maintenance, remodeling, and metabolism. Thus the use of native tendon cells adds components to the culture medium that stimulate collagen synthesis, collagen hydroxylation, and crosslinking. Together, these ideas form the basis for the hypothesis that mechanically conditioned tendon internal fibroblasts grown in a tethered, three-dimensional collagenous matrix can mimic native tendon in appearance, genetic expression, and, eventually, biomechanical strength.

Technology and Methods

Cell Culture

Avian tendon internal fibroblasts (ATIFs) were isolated from the flexor digitorum profundus tendons of 52-day-old White Leghorn chickens. Chicken feet were obtained from a Perdue processing plant (Robbins, NC). Legs were washed with soap and cold water before tendon isolation. The flexor digitorum profundus tendons were removed from the middle toes after transection at the proximal portion of the metatarsal and distal portion of the tibiotarsus. Using a sterile technique, tendons were dissected from their sheath and placed in a sterile dish of phosphate buffered saline (PBS) with 20 mM N-2-hydroxyethylpiperazine-N'-2-ethanesulfonic acid (HEPES) (pH 7.2) with 1X penicillin/streptomycin (100 units penicillin/100 mg streptomycin per mL) (1x PS). Cells were subsequently isolated by sequential enzymatic digestion and mechanical disruption.[16,17] Cells were cultured until confluent in Dulbecco's Minimum Essential Media-high glucose with 10% fetal calf serum (FCS), 20 mM HEPES, pH 7.2, 100 µM ascorbate-2-phosphate, and 1x PS.

Fabrication of a Three-Dimensional Bioartificial Tendon

The ATIFs were enzymatically disaggregated, washed in PBS, then plated in a type I collagen gel (Vitrogen, Cohesion Technologies, Palo Alto, CA) mixed with growth media and FCS, and neutralized to pH 7.0 with 1M sodium hydroxide. A total of 200 k cells/170 µL of the collagen mixture were dispensed to each well of TissueTrain (Flexcell International Corporation, McKeesport, PA) culture plate[15,18-21] (Figure 1). Linear, tethered, bioartificial tendons as three-dimensional cell populated matrices were formed by placing the TissueTrain culture plate on top of a 4-place gasketed baseplate with planar-faced cylindrical posts with centrally located, rectangular cut-outs (6-place BioFlex Loading Station with TroughLoaders, Flexcell International Corporation, McKeesport, PA) beneath each flexible well base. The TroughLoaders have vertical holes in the floor of the rectangle through which a vacuum can be applied to deform the flexible membrane into the trough. The trough provides a space for delivery of cells and matrix. The baseplate was transferred into a humidified incubator with 5% CO_2 at 37°C. The construct was held in position in the incubator under vacuum for 1.5 hours until the cells and matrix formed a gelatinous material connected to the anchor stems.[18,19] The BATs were then covered with 3 mL per well growth medium. The cultures were then digitally scanned and returned to the incubator.

Mechanical Loading

The BATs were uniaxially loaded by placing Arctangular (Flexcell International Corporation) loading posts (rectangle with curved short ends) beneath each well of the TissueTrain plates in a gasketed baseplate and

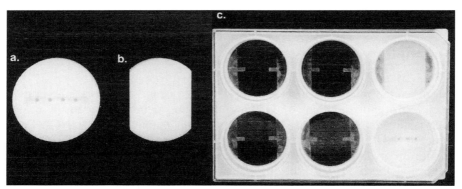

Figure 1 Panel a shows a 35-mm diameter Delrin Troughloader (Flexcell International Corporation) insert that is filling the space beneath a flexible well substrate of a TissueTrain (Flexcell International Corporation) culture plate. The trough is 25 mm × 3 mm × 3 mm. The four holes are 1 mm in diameter and communicate with the reservoir beneath the culture plate so that vacuum can draw the overlying rubber membrane into the trough creating a space into which cells and gel can be cast. Once the gel is cast, the Troughloader is removed. To mechanically load the bioartificial tissue (BAT), an Arctangle (Flexcell International Corporation) loading post (panel b) is placed beneath the TissueTrain well so that the linear sides correspond to the east and west poles of the anchors to which the linear gel is attached. Vacuum draws the flexible but inelastic anchors downward resulting in uniaxial strain on the BAT. Panel c shows a TissueTrain culture plate with linear anchors in each well and two wells with a TroughLoader and Arctangle loading post. *(Courtesy of Flexcell International Corporation, McKeesport, PA.)*

applying vacuum to deform the flexible membranes downward at east and west poles (Figures 2 and 3). The flexible, but inelastic, anchors deformed downward along the long sides of the Arctangular loading posts thus applying uniaxial strain along the long axis to each of the BATs. The loading regimen was 30 minutes per day at 1% elongation and 1 Hz using a Flexercell (Flexcell International Corporation) Strain Unit to control the regimen.

The effects of applying cyclic mechanical strain on biomechanical strength and moduli of the BATs were evaluated. We also tested the effects of an anabolic steroid, 100 nM nandrolone, in conjunction with cyclic load.[15]

Outcome Measures: Contraction Index, Histology, Gene Expression Profile, Material Properties of BAT Constructs

Each plate of BATs was imaged daily and the area was quantified with the use of IMAQ VISION software by National Instruments (Austin, TX). The BATs were fixed in situ and stained with hematoxylin and eosin. Sections were imaged at 10× and 40× magnification using an Olympus BH61 light microscope (Melville, NY). Other BATs were stained with rhodamine phalloidin for actin and 4´,6´-diamidino-2-phenylindole hydrochloride (DAPI)

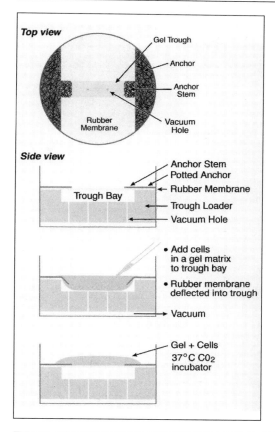

Figure 2 Diagram of one well of a TissueTrain (Flexcell International Corporation) 6-well culture plate (top view) shown from above, the gel trough into which the rubber membrane is drawn by vacuum, the nonwoven nylon mesh anchor bonded to the rubber in the sector portion, and the anchor stem with collagen bonded to it. On the side view, the anchor stem is shown free of the rubber bottom connected to the potted nylon anchor. Vacuum drawn through the trough loader holes pulls the rubber membrane downward to closely conform to the trough bay dimensions. Cells in a collagen gel are then added to the trough bay and the constructs are gelled at 37°C in a CO_2 incubator. After gelation, the vacuum is released and the cultures receive culture medium. *(Courtesy of Flexcell International Corporation, McKeesport, PA)*

for nuclear staining. Comparative gene expression profiles for cells grown in two-dimensional monolayer cultures, three-dimensional BATs, and native whole tendon were created using a quantitative reverse transcriptase polymerase chain reaction (n=3/group). Expression levels for collagens I, III, XII, decorin, tenascin, fibronectin, prolyl and lysyl hydroxylase, lysyl oxidase, connective tissue growth factor, β-actin, and 18 s ribosomal RNA were quantified. Tensile strength tests were performed at 7 and 14 days using an ElectroForce 3200 mechanical tester by EnduraTEC Systems Corporation (Minnetonka, Minnesota).

Results and Conclusions

Contraction Index

The ATIFs in a linear collagen gel attached to matrix-bonded anchor ends to form a three-dimensional "tendinous" construct (n = 6/group). The BATs were cultured for up to 30 days and initially assumed a rectangular to cylindrical shape. As the cells reorganized the collagen matrix, macroscopic radial contraction of the construct was evident. Over an 8-day period, ATIFs con-

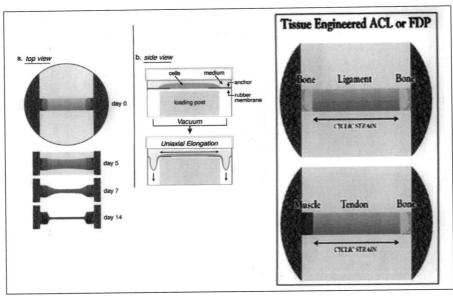

Figure 3 Panel A (top view) shows the dimensions of a typical bioartificial tendon (BAT) from the initial molding on day 0 through contraction phases on days 5, 7, and 14. The BAT transforms from a rectangular to an hourglass shape (days 5 and 7) and finally a cylindrical shape (day 14). Panel B (side view) shows one well of a TissueTrain (Flexcell International Corporation) culture plate with a molded linear BAT immersed in culture medium. The rubber membrane faces an apposing lubricated Arctangular (Flexcell International Corporation) shaped loading post (rectangle with curved short ends). When vacuum is applied to the well bottom, the rubber membrane deforms downward at east and west poles resulting in uniaxial elongation of the BAT. Panel C shows how a BAT may include muscle and bone at opposing ends to simulate a tendon or have bone at either end to simulate a ligament. *(Courtesy of Flexcell International Corporation, McKeesport, PA)*

tracted the overall area of the construct by 82% (mean ± SD; $P < 0.001$), with a reduction in midsection width of 89% ($P < 0.001$).

Histologic Evaluations

The BATs stained with hematoxylin and eosin appeared "tendon-like" with a multicellular top layer resembling an epitenon and deeper cells aligned in the direction of the long axis of the BAT. Staining with rhodamine phalloidin (for filimentous actin) and DAPI (for nuclei) showed elongated cells and nuclei stacked throughout the matrix. Numerous cell-to-cell contacts were observed with staining.

Gene Expression Profile

The ATIFs cultured as BATs retained their phenotypic expression profiles for the predominant collagens found in whole tendon as the expression of fibronectin, tenascin, decorin, lysyl oxidase, and lysyl hydroxylase remained unchanged. Cells grown in two-dimensional monolayers with a collagenous substrate also retained the genetic expression of the predominant collagens found in tendon cells and did not vary from the expression levels observed in BATs. In these cells, the expression of collagen XII was increased 60% and expression of tenascin was 10% less compared with levels of expression in whole tendon ($P < 0.001$). Mechanical loading increased the levels of mRNA for collagen XII at day 3 by 33% ($P < 0.05$). The level of mRNA for prolyhydroxylase was increased by 61% at day 3 and by 33% on day 5 ($P < 0.05$). Treatment of BATs with ascorbate increased the expression of CTGF dramatically and increased strength sixfold in the first week.[14,20]

Mechanical Properties

The average modulus for control BATs was 0.49 MPa on day 7 and 0.96 MPa on day 14. The average modulus for mechanically conditioned BATs was 1.8 MPa on day 7 and 4.3 MPa on day 14. BATs subjected to cyclic mechanical load of 1% elongation at 1 Hz for 1 h per day for 7 days had a 2.9-fold greater ultimate tensile strength compared with nonloaded controls ($P < 0.22$). At 2 weeks, the ultimate tensile strength of nonloaded BATs strength increased 6.9-fold compared with values at 1 week and that of loaded BATs increased twofold ($P < 0.36$) compared with values at 1 week. There was no significant difference in ultimate tensile strength between load and no load groups at week 2. Ascorbate increased strength sixfold in the first week of culture. Nandrolone and load increased strength by 60% in the first week.

Table 1 provides a comparison of results for tendon cells grown in two-dimensional culture or three-dimensional culture as bioartificial tendons to those of a gold standard, whole, native tendon.

Discussion

A three-dimensional tenocyte-populated linear bioartificial tendon was created using a novel molding process.[15,18,19,21] The goal was to use a three-dimensional cell culture to create a tissue replacement that mimicked the biologic behavior and physical properties of native tendon. The tenocytes had mitotic ability, remodeled their surrounding matrix, and retained their intrinsic phenotypic mRNA expression patterns and appearance. However, loading increased the expression of type XII collagen and prolylhydroxylase. Increased hydroxylase activity could be responsible for the greater stability in the collagen fibrils that resulted in greater ultimate tensile strength. These findings were based on BATs that were maintained in culture for 7 days. Lysyl oxidase expression did not change, suggesting that aldehyde creation from epsilon amino groups of lysine or hydroxylysine and subsequent formation of Schiff base crosslinks was not likely to be the cause of increased matrix strength. Collagen XII associates with fibrillar collagens to enhance the binding of cells, proteoglycans, or other extracellular matrix proteins to the fibrillar collagen network.[22] The biomechanical strength and moduli of the BATs was increased by applying cyclic mechanical strain in

Table 1 Outcome Measures Comparing Bioartificial Tendons (BATs) to Control Tissues

Strength	2D Culture	BATs		Whole Tendon
		- load	+ load	
Ultimate Tensile Strength	ND	+	++	++++
Strain Energy	ND	+	++	++++
Fatigue Resistance	ND	+	++	++++
Remodeling				
Matrix Contraction	ND	++	++++	ND
Cell Alignment	ND	+	+++	+
Actin Polymerization	ND	+	+++	+
Metalloproteinase ELISA				
MMP 1	+	++	+	+
MMP 2	+	++	+	+
MMP 3	+	++	+	+
Gene Expression Profile				
Collagen I	+	+	+	+
Collagen III	+	+	+	+
Collagen XII	+	+	++	++++
Collagen XIV	+	+	+	+
Fibronectin	+	+	+	+
Tenascin	+	+	++	+
Prolyl Hydroxylase	+	+	++	+
Lysyl Hydroxylase	+	+	+	+
Lysyl Oxidase	+	+	+	+
CTGF	+	+	+++	+

BATs are bioartificial tendons. Remodeling includes indices of how well cells contract the matrix, polymerize actin, align cells with the long axis of the tendon, and remodel the matrix via matrix metalloproteinases. The gene expression profile includes RNA expression of the listed genes by a quantitative PCR technique.
ND not done
+ indicates the parameter is detected and more + indicates a more robust effect.
MMP = matrix metalloproteases

vitro. In addition, it appears that use of an anabolic steroid, nandrolone, in conjunction with cyclic load can increase the strength of BATs populated with human supraspinatus tenocytes.[15] Application of daily, cyclic mechanical strain can enhance the biomechanical properties of bioartificial tendons.

Summary

There are several drawbacks to currently available methods to repair tendon or ligament injuries. Thus, there is a need for new treatments to provide a long-term solution for tendon or ligament repair. Our results indicate that a method to develop a three-dimensional tenocyte-populated linear bioartificial tendon may be useful in tissue engineering applications. Constructs may be used as tendon replacements or as cell or gene delivery systems. It is anticipated that this technology can be developed for future clinical applications.

Acknowledgments

We would like to thank Flexcell International Corporation for the use of the TissueTrain (Flexcell International Corporation, McKeesport, PA) 3D culture system, their research laboratory facilities, equipment, and supplies.

References

1. Goulet F, Germain L, Rancourt D, Caron C, Normand A, Auger FA: Tendons and ligaments, in Lanza R, Langer R, Chick W (eds): *Principles of Tissue Engineering*. San Diego, CA, Academic Press Inc, 1997, pp 631-644.

2. Jackson DW, Grood ES, Arnoczky SP, Butler DL, Simon TM: Cruciate reconstruction using freeze dried anterior cruciate ligament allograft and a ligament augmentation device (LAD). *Am J Sports Med* 1987;15:528-538.

3. Woo SL, Ritter MA, Ameil D, et al: The biomechanical and biochemical properties of swine tendons-long term effects of exercise on the digital extensors. *Connect Tissue Res* 1980;7:177-183.

4. Awad HA, Butler DL, Boivin GP, et al: Autologous mesenchymal stem cell-mediated repair of tendon. *Tissue Eng* 1999;5:267-277.

5. Awad HA, Butler DL, Harris M, et al: In vitro characterization of mesenchymal stem cell-seeded collagen scaffolds for tendon repair: Effects of initial seeding density on contraction kinetics. *J Biomed Mater Res* 2000;51:233-240.

6. Huang D, Chang TR, Agarwal A, Lee RC, Erlich PH: Mechanisms and dynamics of mechanical strengthening in ligament-equivalent fibroblast-populated collagen matrices. *Ann Biomed Eng* 1993;21:289-305.

7. Kleiner JB, Amiel D, Roux RD, Akeson WH: Origin of replacement cells for the anterior cruciate ligament autograft. *J Orthop Res* 1986;4:466-474.

8. Andrish JT, Woods LD: Dacron augmentation in anterior cruciate ligament reconstruction in dogs. *Clin Orthop* 1984;183:298-302.

9. Park JP, Grana WA, Chitwood JS: A high strength dacron augmentation for cruciate ligament reconstruction. *Clin Orthop* 1985;196:175-185.

10. Bolton CW, Bruchman WC: The GORE-TEX expanded polytetrafluoroethylene prosthetic ligament: An in vitro and in vivo evaluation. *Clin Orthop* 1985;196:202-213.

11. Kennedy JC, Roth JH, Mendenhall HV, Sanford JB: Presidential address: Intra-articular replacement in the anterior cruciate ligament-deficient knee. *Am J Sports Med* 1980;8:1-8.

12. Jenkins DH, Forster IW, McKibbin B, Ralis ZA: Induction of tendon and ligament formation by carbon implants. *J Bone Joint Surg Br* 1977;59:53-57.

13. Tsuzaki M, Yamauchi M, Banes AJ: Tendon collagens: Extracellular matrix composition in shear stress and tensile components of flexor tendons. *Connect Tissue Res* 1993;29:141-152.

14. Evans G, Garvin J, Qi J, Maloney M, Banes A: The Effects of Uniaxial Cyclic Loading on the Biomechanical Properties of Bioartificial Tendon Stimulates (BATs). Geargia Insititute of Technology, Tissue Engineering meeting, Nov, 2002. Atlanta, GA.

15. Triantafillopoulis IK, Banes AJ, Bowman KF, Garrett WE, Karas SG: Nandrolone decanoate and load increase remodeling and strength in human supraspinatus bioartificial tendons. *Am J Sports Med*, in press.

16. Banes AJ, Tsuzaki M, Hu P, et al: Cyclic mechanical load and growth factors stimulate DNA synthesis in avian tendon cells: Special issue on cytomechanics. *J Biomech* 995;28:1505-1513.

17. Banes AJ, Donlon K, Link GW, et al: Cell population of tendon: A simplified method for isolation of synovial cells and internal fibroblasts. Confirmation of origin and biological properties. *J Orthop Res* 1988;6:83-94.

18. Banes AJ: Loading station assembly and method for tissue engineering. US Patent # 6,472,202, October 29, 2002.

19. Banes AJ: Method and apparatus to grow and mechanically condition cell cultures. Patent Pending. US Provisional Patent # 60/254,144, October 28, 2002. International Patent Application # PCT/US01/47745.

20. Banes AJ, Maloney M, Evans G, Qi J: Ascorbate stimulates CTGF expression and increases strength in avian tendon cell-populated bioartificial tendons (BATs). 1st Cold Spring Harbor Symposium on Tissue Engineering, Nov, 2002.

21. Garvin J, Qi J, Maloney M, Banes A: Novel system for engineering bioartificial tendons and application of mechanical load. *J Tissue Eng* 2003;9:967-979.

22. Sugrue SP, Gordon MK, Seyer J, Dublet B, Van der Rest M, Olsen BR: Immunoidentification of type XII collagen in embryonic tissues. *J Cell Biol* 1989;109:939-945.

Chapter 26

Meniscus Repair Through Tissue Engineering

Brian Johnstone, PhD
Jung U. Yoo, MD

Abstract

There are several challenges in the repair and healing of the injured meniscus and currently available therapies have several limitations. Given the need for new strategies to repair or regenerate meniscus tissue, tissue engineering is being evaluated for its potential in healing the meniscus. Although animal models are useful in developing these methods, differences in biomechanical and biochemical characteristics of the meniscus tissue of various species must be considered. In particular, differences in the types of collagen must be addressed and there are difficulties in prompting cells to produce highly collagenous extracellular matrices in vitro. Another challenge is the selection of the most appropriate cell types to be used in tissue engineered constructs. Meniscal fibrochondrocytes and other types of cells are currently being evaluated. At this time, tissue engineering for meniscus repair or replacement is still in its infancy. However, it is hoped that additional studies will advance our understanding of the physiology of the meniscus and facilitate the development of new interventions to repair or replace injured meniscus tissue.

Introduction

The meniscus is a semilunar fibrocartilaginous structure that is located in the knee joint between the tibial and femoral articular cartilage surfaces. Of all the structures within the knee joint, the meniscus is the most frequently injured.[1] Thus, surgical procedures related to problems with the meniscus of the knee joint are among the most common in orthopaedic practice. In addition, the meniscus is susceptible to degenerative joint disease in older persons. The repair and healing of the injured meniscus presents a unique challenge for surgeons. Although very effective in the outer region of the meniscus, repair is not easily accomplished in the inner regions. Many techniques have been tried, but there are no obviously superior solutions. Thus, there is a need for new strategies to repair or regenerate meniscus tissue. Tissue engineering is a strategy that is currently being studied in a small but growing number of laboratories worldwide. As with other musculoskeletal tissues, the first attempts to engineer meniscus tissues have been relatively crude. However, as the field of tissue engineering expands, experimental procedures have become more sophisticated. This review discusses the challenges of developing tissue engineering techniques for meniscus repair with use in the clinical setting as the ultimate goal.

Meniscus Healing

The outer third of the meniscus has the potential to heal because of the perimeniscal capillary plexus that allows a normal wound healing reaction, formation of fibrovascular scar, and remodeling of the tissue after injury.[2,3] Tears and lesions in this region are referred to as red-red tears because there is blood supply on both sides. The capillaries do not extend into the meniscus much beyond the outer third. Tears in the middle third of the meniscus are referred to as red-white tears as there can be blood vessels on the capsular side. Tears in the inner third of the meniscus (white-white tears) will not heal because there is no blood supply.[4] The contrast in healing capability between the vascular and avascular regions has led to the development of techniques to bring the blood supply to wounds in the avascular region. Channels have been created from the wound site to the periphery of the meniscus to allow vascular ingrowth;[5,6] however, these vascular access channels were large and disrupted the normal peripheral architecture of the meniscus. Alternative strategies include stimulation of the synovium by abrasion to produce a vascular pannus and bringing a synovial flap to the injury site.[6-8] These can be performed in conjunction with addition of a fibrin clot to the wound site as first developed by Arnoczky and associates[5] and subsequently used in clinical studies. In a review of current clinical methods, Koski and associates[9] noted that preliminary evidence suggests that successful repair of simple lesions in the avascular region, either directly or with fibrin clot, appears possible. However, other methods are still needed for complex tears and more substantial loss of meniscus tissue.

Challenges for Tissue Engineering Meniscus Tissue

The morphology, biochemistry, and biomechanical properties of the meniscus differ between species (Figure 1). As with other musculoskeletal tissues, it is not obvious which animal model is best suited for use in studies of engineering of meniscus tissue. In an interspecies comparison, the compressive biomechanical properties of the meniscus of humans most closely resembled those of sheep and pigs and differed greatly from those of the monkey. Properties of canine and bovine meniscus tissues were intermediate between those of the other studied species.[10] The size of the animal can limit its use as a model for humans. Rabbits, for example, are an appropriate model for studies of block tissue resection. However, it is very difficult to mimic tears of the type occurring in humans and to determine biomechanical characteristics of repaired tissue in rabbits and smaller animals.

The composition of the meniscus also varies in different species. Both type I and type II collagen are present in the meniscus so it is categorized as a fibrocartilage. The extent of type II collagen expression differs between species and between regions within the meniscus. Type II collagen is mini-

Figure 1 Comparison of the morphology of the inner avascular region of the center of (A) rabbit, (B) dog, and (C) human medial meniscus (toluidine blue stain, 64 × magnification). Note the greater cellularity and more rounded phenotype of the rabbit cells. There are differences in metachromatic staining, which is greatest in the rabbit and least in the human meniscus in this region. These differences in staining reflect the different biochemical composition of the meniscus in each species.

mal in the normal human meniscus but is more extensively expressed in smaller animal species, especially in the middle region of the meniscus. These differences in biomechanical and biochemical parameters must be taken into account for designing tissue engineered constructs and when analyzing their success for repair in animal models.

Tissue engineering involves the use of cells, usually in combination with biodegradable scaffolds, to create the tissue of choice. The length of time in which cells are combined with the scaffold before implantation in the injury site ranges from minutes in the operating room to days or weeks in a cell culture environment. Tissue engineering does not produce the exact tissue it is meant to replace, but instead produces some kind of neotissue that ideally has some of the properties of the native tissue that requires repair. A particular problem in engineering the meniscus is the difficulty in prompting cells to produce highly collagenous extracellular matrices in vitro.

Early studies indicate that meniscal implants would possibly need a level of structural organization and biomechanical properties close to that of the native tissue. However, this may lead to a lower degree of host-implant integration, as demonstrated by in vitro studies with articular cartilage.[11] The challenge is to produce tissue with mechanical strength that allows both fixation and survival and that provides protection of the other tissues of the joint. As tissue remodeling to form a true meniscus structure is desired, the initial implant would need to facilitate this too. Biodegradable scaffolds would work best in this situation, but they require optimization to produce the appropriate timing of degradation that is coincident with the production of the meniscus tissue regenerate.

Another challenge for meniscal tissue engineering is selection of the most appropriate cell type or types to be used. Ibarra and associates[12] have experimented with meniscal fibrochondrocytes for tissue engineering, implanting autologous cell-seeded scaffolds subcutaneously in sheep for 4 weeks and then transferring them to defects created in the sheep knee meniscus. There was some production of meniscus-like tissue, but tissue shrinkage was evident. Although meniscal cells can be isolated and grown with relatively simple technique, there are few other studies of their use in meniscus repair. Another challenge is the relative lack of studies of meniscal cells as there is little knowledge of the stimuli that are required to promote and maintain meniscus tissue production in vivo. Despite extensive investigation into joint formation and much new information on the signaling factors that control it, the meniscus is rarely discussed in that literature.

As an alternative to meniscal fibrochondrocytes, cells with differentiation potential may be used for seeding in tissue engineering constructs. Autologous bone marrow-derived cells in an autologous fibrin clot were used in a goat model of a full-thickness lesion in the avascular zone.[13] The cell-loaded implant fared no better than the repair observed when a fibrin clot was used alone. However, the content of marrow mesenchymal progenitor cells in the cell population used was not determined. Walsh and associates[14] used more characterized autologous, bone marrow-derived mesenchymal stem cells, seeded in a type I collagen sponge, to regenerate tissue in a large meniscal defect in rabbits. In comparison to empty sponges, the cell-seeded implants had a greater percentage of fibrocartilage in some specimens.

To produce a cell-based fibrocartilage implant, Angele and associates[15,16] seeded autologous bone marrow-derived mesenchymal progenitor cells into a biodegradable composite matrix of hyaluronan and gelatin and cultured it in a chondrogenic medium before implantation. The cell-loaded, precultured implants integrated well with the surrounding host tissue, and meniscus-like fibrocartilage was noted in the majority of implants after 12 weeks. However, the protection of the adjacent articular cartilage was minimal and degeneration occurred. The study demonstrated the feasibility of creating a large tissue engineered implant and its use in the repair of a significant meniscal defect. However, it highlighted the need for meniscal implants to be of reasonable strength from the outset to provide protection for the surrounding articular cartilage during the period when the implant is remodeling into meniscal tissue.

The replacement of the injured or degenerated meniscus with allograft meniscus has been attempted. In a review of meniscus transplantation, Rodeo[17] described initial efforts as potentially useful in a subset of patients. He noted that there are indications that transplantation should be performed early, before significant joint degeneration has occurred. However, it is difficult to justify such a procedure in patients without symptoms given that there are no long-term studies demonstrating efficacy. Long-term efficacy is a particularly important consideration for tissue engineered meniscus replacements. Animal studies indicating long-term success will be needed before such replacements could be justified for use in patients with meniscectomies who have otherwise normal joints. One implant that has been developed in animal models and studied in clinical trials is a collagen-based scaffold for use as a meniscus replacement.[18-21] However, although described as a meniscal replacement, some native meniscus tissue is retained for attachment purposes. Suitable attachment regions to merge with or replace the ligamentous or fibrous attachments of the horns of the meniscus have not been developed.

Summary

The use of a tissue engineering approach for meniscus repair or replacement is still in its infancy, and there are deficits in our knowledge of normal meniscus biology, biochemistry, and biomechanics. Understanding the physiology of the meniscus could provide clues to optimize tissue engineering strategies. Furthermore, we do not have well-characterized animal models or analysis schemes. Although the limited studies completed thus far have been promising, it is clear that additional research is needed.

References

1. Renstrom P, Johnson RJ: Anatomy and biomechanics of the menisci. *Clin Sports Med* 1990;9:523-538.

2. Cabaud HE, Rodkey WG, Fitzwater JE: Medical meniscus repairs: An experimental and morphologic study. *Am J Sports Med* 1981;9:129-134.

3. Arnoczky SP, Warren RF: The microvasculature of the meniscus and its response to injury: An experimental study in the dog. *Am J Sports Med* 1983;11:131-141.

4. King D: The healing of semilunar cartilages. *J Bone Joint Surg Am* 1936;18:333-342.

5. Arnoczky SP, Warren RF, Spivak JM: Meniscal repair using an exogenous fibrin clot: An experimental study in dogs. *J Bone Joint Surg Am* 1988;70:1209-1217.

6. Gershuni DH, Skyhar MJ, Danzig LA, Camp J, Hargens AR, Akeson WH: Experimental models to promote healing of tears in the avascular segment of canine knee menisci. *J Bone Joint Surg Am* 1989;71:1363-1370.

7. Henning CE, Lynch MA, Clark JR: Vascularity for healing of meniscus repairs. *Arthroscopy* 1987;3:13-18.

8. Ghadially FN, Wedge JH, Lalonde JM: Experimental methods of repairing injured menisci. *J Bone Joint Surg Br* 1986;68:106-110.

9. Koski JA, Ibarra C, Rodeo SA, Warren RF: Meniscal injury and repair: Clinical status. *Orthop Clin North Am* 2000;31:419-436.

10. Joshi MD, Suh JK, Marui T, Woo SL: Interspecies variation of compressive biomechanical properties of the meniscus. *J Biomed Mater Res* 1995;29:823-828.

11. Obradovic B, Martin I, Padera RF, Treppo S, Freed LE, Vunjak-Novakovic G: Integration of engineered cartilage. *J Orthop Res* 2001;19:1089-1097.

12. Ibarra C, Koski JA, Warren RF: Tissue engineering meniscus: Cells and matrix. *Orthop Clin North Am* 2000;31:411-418.

13. Port J, Jackson DW, Lee TQ, Simon TM: Meniscal repair supplemented with exogenous fibrin clot and autogenous cultured marrow cells in the goat model. *Am J Sports Med* 1996;24:547-555.

14. Walsh CJ, Goodman D, Caplan AI, Goldberg VM: Meniscus regeneration in a rabbit partial meniscectomy model. *Tissue Eng* 1999;5:327-337.

15. Angele P, Kujat R, Nerlich M, Yoo J, Goldberg V, Johnstone B: Engineering of osteochondral tissue with bone marrow mesenchymal progenitor cells in a derivatized hyaluronan-gelatin composite sponge. *Tissue Eng* 1999;5:545-554.

16. Angele P, Johnstone B, Kujat R, Nerlich M, Goldberg V, Yoo J: Meniscus repair with mesenchymal progenitor cells in a biodegradable composite matrix. *Trans Orthop Res Soc* 2000;25:605.

17. Rodeo SA: Meniscal allografts: Where do we stand? *Am J Sports Med* 2001;29:246-261.

18. Stone KR, Rodkey WG, Webber RJ, McKinney L, Steadman JR: Future directions: Collagen-based prostheses for meniscal regeneration. *Clin Orthop* 1990;252:129-135.

19. Stone KR, Rodkey WG, Webber R, McKinney L, Steadman JR: Meniscal regeneration with copolymeric collagen scaffolds: In vitro and in vivo studies evaluated clinically, histologically, and biochemically. *Am J Sports Med* 1992;20:104-111.

20. Stone KR, Steadman JR, Rodkey WG, Li ST: Regeneration of meniscal cartilage with use of a collagen scaffold: Analysis of preliminary data. *J Bone Joint Surg Am* 1997;79:1770-1777.

21. Rodkey WG, Steadman JR, Li ST: A clinical study of collagen meniscus implants to restore the injured meniscus. *Clin Orthop* 1999;367:S281-S292.

Chapter 27

Gene Enhanced-Tissue Engineered Repair of the Meniscus

Daniel Grande, PhD
David Tuckman, MD
James Cook, DVM, PhD
Neal Watnik, MD
James Mason, PhD
Pasquale Razzano, MS
David Dines, MD

Abstract

The repair of meniscal injuries is a challenge for orthopaedic surgeons. Tears within the avascular zone of the meniscus often do not heal well. The local delivery of peptides using genetically engineered cells is a strategy that may be suitable for meniscus repair. We assessed whether integrative repair of the meniscus can be stimulated by the introduction of a cell construct engineered to synthesize insulin-like growth factor-1 (IGF-1). Meniscal fibrochondrocyte cell lines were established from menisci of adult bovine knee joints. Oligonucleotide primers were manufactured and used to generate cDNA fragments of the human IGF-1 genes by reverse transcriptase polymerase chain reaction (RT-PCR). Additional subcloning generated the retroviral vector plasmid LNCX-IGF-1. When cells reached 25% to 50% confluence, transductions with supernatants were performed. A selection process was used to expand the transduced cell line. Additional menisci were harvested from bovine knee joints and cut into segments 1.5 cm wide at the outer periphery. A bucket-handle tear was made in the inner one third (white zone). The tear was either left empty (as a control), filled with a cell construct, or filled with a cell construct transduced with the gene encoding for IGF-1. Menisci were then xenografted subcutaneously onto the dorsum of athymic (Nu/nu) rats and allowed to incubate in vivo for either 2 or 4 weeks at which time they were fixed in formalin and processed for histologic evaluation. Meniscal cells were easily cultured and transduced with the LNCX-IGF-1 gene construct. Northern blot analyses confirmed that the IGF-1 message was being upregulated. The active IGF-1 peptide was synthesized into the medium as determined by enzyme-linked immunosorbant assay (ELISA) ($6.25pg/1 \times 10^6/24$ h). Nontransduced meniscal cells had no detectable IGF-1 in the media supernatant. Histologic evaluations showed that menis-

ci constructs left empty (as a control) were not repaired. Implantation of cells without the IGF-1 gene typically filled the gap between the tear but did not demonstrate cellular bonding. The graft of the cells tranduced with the IGF-1 gene construct completely healed the tear and demonstrated integrative repair with significant collagen deposition and remodeling of the repair site. There was also an increase in cellular infiltration from the host meniscus to the repair site. In summary, biologic or integrative repair occurred in meniscal defects receiving a meniscal cell-based construct with the IGF-1 gene. Additional development of these types of constructs may lead to the ability to restore normal biomechanics and prevent osteoarthritis in patients who have sustained meniscal tears.

Introduction

Injuries to the knee joint are the most common injury in athletes. All tissues within the joint are at risk of being traumatized and the meniscus is the most commonly injured intra-articular structure.[1] Basic and clinical research has shown that avascular tears of the meniscus do not heal.[2,3] Meniscal fibrochondrocytes lack an adequate scaffolding in which to mount an effective wound repair.[4] Inadequate healing of the meniscus is also related to vascular access.[4-8] In the past, complete excision of the injured meniscus was used.[3] At present, the most commonly used surgical approach is arthroscopic examination and partial menisectomy of the injured portion of the meniscus.[9-13] However, the patient is subjected to a chronic course of degenerative changes secondary to altered biomechanics. Encouraged by the success with articular cartilage resurfacing in preclinical studies,[14] similar tissue engineering and gene therapy approaches are now being applied to the meniscus.

Recent evidence has shown that IGF-1 is a potent anabolic mediator and is mitogenic for cartilage.[15,16] We have demonstrated the successful use of transfected cell lines to deliver growth factor genes to damaged articular cartilage.[17] This technology has the potential to promote local, sustained delivery of a growth factor to enhance regional regeneration. In this study, we evaluated whether the use of transduced meniscal cells to introduce IGF-1 at the site of an avascular tear mediates functional and/or integrative repair.

Methods

Tissue Culture

Meniscal fibrochondrocyte cell lines were established from menisci of adult bovine knee joints. Tissue explants were rinsed in Ringers buffer and cut into 1-mm pieces before placement in six-well dishes containing D10 media (high glucose Dulbeccos modified Eagle medium; Gibco, Grand Island, NY) with 10% heat inactivated fetal bovine serum (Gibco) and supplemented with 1X concentration of antibiotic and antimycotic (Gibco). Cells were cultured at 37°C in 5% CO_2 and serial passaged no more than two times before transduction with retroviral vectors.

Human IGF-1 cDNA Cloning and Expression

Total RNA was isolated from human embryonic lung cells (Hel 299) using a commercially available RNeasy Kit (Qiagen, Germantown, MD). First strand synthesis was performed using the Reverse Transcription System (Promega, Madison, WI). The RT-PCR was performed using a Gene AMP PCR kit (Perkin-Elmer, Wellesley, MA) with oligonucleotide primers NS172 and NS173 to generate a 492 base pair (bp) PCR product of the human IGF-1 cDNA (NS172 = 5′ aaaaagcttgccgccgcaatgggaaaaatcagcag 3′; NS173= 5′ tttatcgatttactacatcctgtagttcttgtttcc 3′). The hIGF-1 PCR product was cloned into plasmid pT7Blue-3 (Novagen, Madison, WI) and sequenced to confirm that the hIGF-1 cDNA sequence was complete and correct. Plasmid pT7Blue3-hIGF-1 was double-digested with HindIII/ClaI and the hIGF-1 fragment isolated and cloned into HindIII/ClaI double-digested plasmid pLNB-BMP7 with the hIGF-1 gene replacing the gene for bone morphogenetic protein-7. The resulting retroviral vector plasmid pLNB-hIGF-1 contains the hIGF-1 gene driven off of the rat β-actin promoter and the neomycin resistance gene driven off of the retroviral long terminal repeat.

Transduction and Selection of Bovine Meniscal Chondrocytes

Primary bovine meniscal chondrocytes were cultured in S-DMEM (Dulbeccos modified Eagle medium with serum) in T-75 primaria flasks (BD Falcon, Franklin Lakes, NJ). Cells were grown to 50% confluence and transduced overnight at low passage number (P2 or P3) using 2 mL of amphotropically packaged retroviral vector LNB-hIGF-1 clone No. 7 (approximately 4 E 5 neo colony forming units/mL) in the presence of 8 µg/mL polybrene. Selection of transduced cells in the presence of 400 µg/mL active G418 (GibcoBRL, Invitrogen, Carlsbad, CA) was complete in approximately 10 days. Selected gene-enhanced cells were then used for subsequent experiments.

RT-PCR Analysis of hIGF-1 Expression in Primary Bovine Meniscal Chondrocytes

Total RNA (2 µg) was isolated from primary bovine meniscal chondrocytes, LNCX control transduced chondrocytes, and LNB-hIGF-1 transduced chondrocytes and used as template in RT-PCR using oligonucleotide primers NS172 and NS173. Approximately 5% to 10% of the reaction was electrophoresed on 1.5% agarose gels and stained with ethidium bromide for visualization of the ~492 bp PCR fragment that is diagnostic for hIGF-1 mRNA. A gel lane with water only and one with 1mg of plasmid pLNB-hIGF-1 template served as additional controls.

Brd-U Labeling

Cells were pulsed during log phase of in vitro growth with 10 µM bromo-deoxyuridine (Brd-U) for subsequent localization studies. After fixation of tissue samples and processing for histologic evaluations, sections were stained with fluorescein conjugated F(ab')$_2$ fragments of a monoclonal anti-Brd-U antibody (In situ Cell Proliferation Kit: FLUOS; Roche Molecular Biochemicals, Indianapolis, IN).

Surgical Procedure

Menisci were harvested from bovine knee joints and segments approximately 1.5 cm wide were cut in a radial fashion. A bucket-handle tear 1.0 cm long was made in the inner one third (white zone). The tear was either left empty (control), filled with a cell construct, transduced with the empty cassette (LNCX), or filled with a cell construct containing IGF-1 transduced cells. All meniscal tears were repaired with a meniscal arrow (Bionx Implants, Bluebell, PA) and then placed in nylon mesh pouches (0.45-µm pore size) to exclude host cells. Menisci were then xenografted subcutaneously onto the dorsum of athymic rats and allowed to incubate in vivo for either 2 or 4 weeks at which time they were fixed in formalin and processed for histologic evaluations.

Histological Preparation

Menisci were examined macroscopically for evidence of healing, photographed, and fixed in neutral buffered formalin. Samples were then dehydrated and embedded in paraffin. Sections 5-µm thick were either stained with hematoxylin and eosin or left unstained for immunoassay for Brd-U.

Results

Culture and Transduction

Bovine meniscal cells were easily subcultured from explants and expanded. Transduction of the cells with IGF-1 gene did not result in a decrease in doubling time or result in apoptosis. Northern blot analysis confirmed IGF-1 message being upregulated. Results from ELISAs demonstrated that there was synthesis of active IGF-1 peptide (6.25pg/1E6 cells/24 h) into the medium. Nontransduced meniscal cells had no detectable IGF-1 in the media supernatant.

Macroscopic Evaluation

At the time of harvest, in control defects, conspicuous lines at the site of the original bucket-handle tear could still be appreciated. The cells alone or cells plus IGF-1 gene groups showed an incremental improvement in apparent repair with less of the original line visible where the tear was reduced by placement of the arrow. Although there was no discernable difference between these two groups, they both demonstrated significantly improved annealing of the original defect than empty controls. These results were consistent for both the 2- and 4-week time points with no improvement or change by the later 4-week time point.

Microscopic Evaluations

Control menisci repaired with a meniscal arrow only demonstrated no repair at a microscopic level. There was no evidence of cellular migration to the defect site and a well-delineated gap was present (Figure 1). Implantation of transduced cells minus IGF-1 gene (LNCX) filled the gap between the tear bridging the opposing sides. By 4 weeks, the cells in this group began organizing into longitudinal rows parallel to the plane of the tear (Figure 2) with

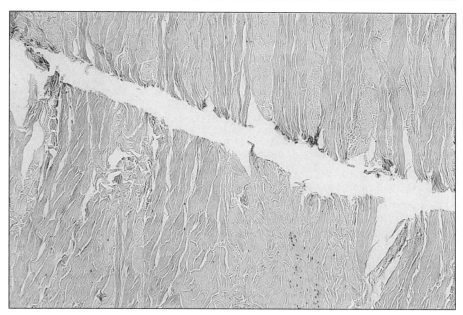

Figure 1 Light micrograph of an empty control group (meniscus repair arrow only) at 4 weeks following xenotransplantation. Note prominent defect gap with lack of repair. ×100 original magnification. H+E stain.

an apparent loss in stainable nuclei. The IGF-1 transduced cell group did not display this characteristic. The experimental IGF-1 grafts completely bridged the tear and demonstrated integrative repair with significant collagen deposition and extensive reorganization of the repair site (Figure 3). Additionally, there was extensive cellular migration within the host tissue to the site of the defect. The contribution of host cells could be readily appreciated when sections were stained for Brd-U. Cells that were implanted, although abundant within the space of the defect, did not migrate extensively into the host meniscus tissue from the tear gap site. Within the xenografted tissue repair there was extensive cellular infiltration to the site of the original bucket-handle tear. This cellular migration or proliferation within the implanted meniscus tissue to the repair site was not seen in any of the control groups.

Discussion

King[3] provided evidence that menisci will have a varied response to injury. From this early knowledge, intense debate and research have advanced efforts to repair and regenerate the meniscus. Although initially viewed as a useless remnant of tissue within the synovial cavity, the meniscus is now considered a vital tissue that imparts load-bearing properties, supplies nutri-

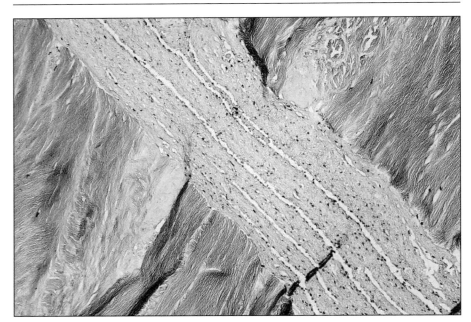

Figure 2 Light micrograph of cells alone (IGF-1 gene) group 4 weeks after xeno-transplantation. The defect gap is now bridged by grafting of cells. Note the organization of the cells into laminar sheets. ×100 original magnification. H+E stain.

ents, and provides crucial joint stability to preserve joint biomechanics.[1-3] Any loss or disruption of the meniscus leads to irreversible degeneration of the articular cartilage. Injuries to the meniscus often do not heal well and currently available treatments are less than ideal. Thus there is a need for novel methods to repair the meniscus and prevent the onset of degenerative changes.

To our knowledge, this study is the first to evaluate an intervention that combines tissue engineering methods and gene therapy to repair the meniscus. The presence of IGF-1 within the defect site, even at picomolar levels, significantly enhanced cellular activity in both the graft and the surrounding meniscal tissue. The increased cellular infiltration into the site from the meniscal implants could be inferred to be a result of chemotactic properties of IGF-1. The type of cellular infiltration observed in the group that received IGF-1 was not observed in the two groups that served as controls.

Meniscal fibrochondrocytes proved to be an excellent cell type for this application based on their ease of culture and high proliferative index.[18-22] These characteristics make meniscal fibrochondrocytes an excellent candidate for gene therapy that uses retroviral vectors. Although we selected the gene for IGF-1, several genes from a panel related to fibrin clot may also prove effective.[23-25] These genes include those for transforming growth factor β and platelet derived growth factor.

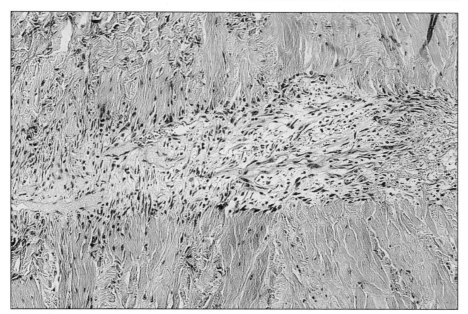

Figure 3 Light micrograph of experimental IGF-1 transduced cell graft. Note complete bridging of defect gap and extensive remodeling. Compare the extent of cellular infiltration from native meniscal tissue between Figures 1 and 2. ×100 original magnification. H+E stain.

The work reported here is similar to the approach taken by Perretti and associates,[26,27] which uses a cell-based approach for articular cartilage repair. A major difference between our study and other work[26,27] is that we enhanced repair by introducing the gene for IGF-1. In addition, we used mensical fibrochondrocytes instead of articular chondrocytes. Meniscal fibrochondrocytes may be a better match for the type of matrix synthesized.

The nude rat offers the potential to use xenografts to study integrative repair. This model also allows a large meniscus to be used, but is more cost effective than studies with large animal models. However, the lack of mechanical perturbation normally associated with weight-bearing knee motion is a limitation of the nude rat model.[28] Thus the athymic rat model is useful as an initial screening system, but studies of new interventions eventually need to be conducted in large animal models so that effects of weight bearing can be fully evaluated.

In this initial study, organization of the integrative repair tissue into discrete lacunae was not observed. This may be attributed to the timing of the evaluations and our study may not have been of adequate duration to observe such changes in the repaired tissue. Long-term studies are needed to more fully characterize tissue repair in this model.

Summary

Injuries to the knee joint are common among athletes and the meniscus is the most frequently injured intra-articular structure. Avascular tears of the meniscus do not heal well and current interventions are suboptimal. Thus, tissue engineering approaches are being explored as potential therapy for repair of the meniscus. The local delivery of IGF-1 using genetically engineered meniscal fibrochondrocytes is a strategy that may be suitable for meniscus repair. Our initial studies demonstrated biologic or integrative repair in mensical defects receiving a gene enhanced cell-based construct. For these experiments, we selected meniscal fibrochondrocytes based on characteristics such as their ease of culture and high proliferative index. We used IGF-1 based on its properties as an anabolic mediator and its mitogenic ability in cartilage tissue. The experimental IGF-1 grafts completely healed the tear and demonstrated integrative repair with significant collagen deposition and remodeling of the repair site. The clinical implication of these results would be the restoration of normal biomechanics and a prevention of osteoarthritis in patients having meniscal tears.[28]

Acknowledgments

The authors wish to thank Ana Duka for her help in conducting the histologic evaluations for this study.

References

1. Cooper DE, Arnoczky SP, Warren RF: Meniscal repair. *Clin Sports Med* 1991;10(3):529-548.

2. Ghadially FN, Wedge JH, Laland JMA: Experimental methods of repairing injured menisci. *J Bone Joint Surg Br* 1986;68:106-110.

3. King D: Healing of semilunar cartilages. *J Bone Joint Surg* 1936;18:333-342.

4. Arnoczky SP, Warren RF: Microvasculature of the human meniscus. *Am J Sports Med* 1982;10(2):90-95.

5. Danzig L, Goncalves MR, Resnick D, et al: Human meniscal blood supply. *Trans Orthop Res Soc* 1981;6:338.

6. Davies DV, Edwards DAW: The blood supply of the synovial membrane and intra-articular structures. *Ann R Coll Surg Eng* 1948;2:142-156.

7. McDevitt CA, Webber RJ: The ultrastructure and biochemistry of meniscal cartilage. *Clin Orthop* 1990;252:8-18.

8. Scapinelli R: Studies on the vasculature of the human knee joint. *Acta Anat (Basel)* 1968;70:305-331.

9. Cox JS, Cordell LD: The degenerative effects of the medial meniscus tears in dogs' knees. *Clin Orthop* 1977;125:236-242.

10. Fairbank TJ: Knee joint changes after menisectomy. *J Bone Joint Surg Br* 1948;30:664-670.

11. Kurosawa H, Fukubayashi T, Nakajima H: Load-bearing mode of the knee joint: Physical behavior of the knee joint with or without menisci. *Clin Orthop* 1980;149:283-290.

12. Lynch MA, Henning CE, Glick KR: Knee joint surface changes: Long-term follow-up meniscus tear treatment in stable anterior cruciate ligament reconstructions. *Clin Orthop* 1983;172:148-153.

13. Sonne-Holm S, Fledelius I, Ahn N: Results after menisectomy in 147 athletes. *Acta Orthop Scand* 1980;51:303-309.

14. Freed LE, Grande D, Emmanual J, et al: Joint resurfacing using allograft chondrocytes and synthetic biodegradable polymer scaffolds. *J Biomed Mat Res* 1994;28:891-900.

15. Nixon AJ, Fortier LA, Williams J, Mohammed H: Enhanced repair of extensive articular defects by insulin-like growth factor-I-laden fibrin composites. *J Orthop Res* 1999;17(4):475-487.

16. Nixon AJ, Lillich JT, Burton-Wurster N, Lust G, Mohammed HO: Differentiated cellular function in fetal chondrocytes cultured with insulin-like growth factor-I and transforming growth factor-beta. *J Orthop Res* 1998;16(5):531-541.

17. Mason J, Breitbart AS, Barcia M, Porti D, Grande D: Cartilage and bone regeneration using gene enhanced tissue engineering. *Clin Orthop* 2000;379S:171-178.

18. Webber RJ: In vitro culture of meniscal tissue. *Clin Orthop* 1990;252:114-120.

19. Webber RJ, Aubrey AJ Jr: Culture of rabbit meniscal fibrochondrocytes II: Sulfated proteoglycan synthesis. *Biochimie* 1988;70:193-204.

20. Webber RJ, Harris MG, Hough AJ Jr: Cell culture of rabbit meniscal fibrochondrocytes: Proliferative and synthetic response to growth factors and ascorbate. *J Orthop Res* 1985;3(1):36-42.

21. Webber RJ, York JL, Vandershilder JL, Haugh AJ: An organ culture model for assaying wound repair of the fibrocartilaginous knee joint meniscus. *Am J Sports Med* 1989;17(3):393-400.

22. Webber RJ, Zitaglio T, Hough AJ Jr: In vitro proliferation and proteoglycan synthesis of rabbit meniscal fibrochondrocytes as a function of age and sex. *Arthrit Rheum* 1986;29(8):1010-1016.

23. Arnoczky SP, Warren RF, Spivak JM: Meniscal repair using an exogenous fibrin clot. *J Bone Joint Surg Am* 1988;70:1209-1217.

24. Hashimoto T, Kurosaka M, Yoshiya S, Hirohata K: Meniscal repair using fibrin sealant and endothelial cell growth factor: An experimental study in dogs. *Am J Sports Med* 1992;20(5):537-541.

25. Henning CE, Yearsant KM, Vequist SW, Stallbaumer RJ, Decker KA: Use of the fascia sheath coverage and exogenous fibrin clot in the treatment of complex meniscal tears. *Am J Sports Med* 1991;19(6):626-631.

26. Peretti GM, Bonassar LJ, Caruso EM, Randolph MA, Trahan CA, Zaleske DJ: Biomechanical analysis of a chondrocyte-based repair model of articular cartilage. *Tissue Eng* 1995;4:317-326.

27. Peretti GM, Randolph MA, Villa MT, Buragas MS, Yaremchuk MJ: Cell-based tissue-engineered allogeneic implant for cartilage repair. *Tissue Eng* 2000;6(5):567-576.

28. Renstrom P, Johnson RJ: Anatomy and biomechanics of the menisci. *Clin Sports Med* 1990;9(3):523-538.

Chapter 28

Nucleus Pulposus Tissue Can Be Formed In Vitro

Rita Kandel, MD
Cheryle Seguin, MSc
Mark Grynpas, PhD
Stephen Waldman, BASc, MSc, PhD
Mark Hurtig, DVM
Robert Pilliar, BASc, PhD

Abstract

The intervertebral disk (IVD) is a specialized structure consisting of the anulus fibrosus (AF) and the nucleus pulposus (NP) that merge with the cartilage end plate. Studies of the regulation of NP regeneration have been limited in part because of the absence of an appropriate in vitro model. In addition, the development of a biological approach to treat disk degeneration has been hindered by an inability to develop engineered tissue with appropriate biomechanical properties to withstand the forces placed on it immediately after implantation. Our results show that it is possible to form NP tissue in the absence of a scaffold when the cells are placed on the top of an appropriate porous substrate. Histologic evaluation confirmed that tissue formed in vitro has some features similar to those of native NP. When bovine caudal NP cells are used to form the tissue, it has compressive properties similar to those of the native NP. This was unexpected as the NP tissue formed in vitro has only a quarter of the collagen content of the native NP. It is not clear why the relative paucity of collagen did not affect the mechanical properties of the tissue. This study describes our investigations that will help develop new methods to form NP tissue in vitro. Biochemical evaluation of NP from different sources showed that NP from bovine and sheep lumbar spines had similar cellularity and collagen content. The age of the animal seemed to affect cellularity as the fetal bovine tissue was more cellular than the young adult NP. Proteoglycan content appeared to be more variable and the proteoglycan to collagen ratio varied from 0.4 to 1.4 in the studied tissues. This study demonstrates the differences in NP tissue from bovine and sheep samples, and these differences should be considered when conducting additional studies of NP regeneration.

Introduction

Intervertebral disks anchor adjacent vertebral bodies and by doing so allow for spinal stabilization, load bearing, and movement. The IVD is a specialized structure consisting of two interdependent tissues, the AF and the NP, that merge with the cartilage end plate.[1,2] The AF is responsible for withstanding circumferential stresses resulting from radial and torsional forces acting primarily on the IVD. The NP resists compressive stresses during normal activity.[1-4] The compositions of AF and NP vary with the anatomic site in the tissue and the individual's age.[1-5]

The normal function of the disk is dependent on maintenance of the composition, organization, and integrity of its different components.[4,5] In an autopsy study, 97% of individuals 50 years or older showed disk degeneration.[6] It is not known why this condition is so common but may be due in part to the relative avascularity of the tissue,[7] mechanical factors,[8] the absence of notochordal cells,[9] and/or genetic factors.[10] As the disk is relatively avascular (only the outer portion of the anulus contains blood vessels), it has limited capacity for repair.[4] Although back pain that can develop as a result of disk herniation is often self-limited, a percentage of affected individuals require surgery.[11,12] Surgical intervention may relieve pain faster, but it does not restore disk height or its original load-bearing capacity.[12] In addition, postdiskotomy syndrome, which is characterized by persistent pain after disk surgery, can develop. Although back pain associated with disk degeneration can be treated by spinal fusion, this is not always successful and results in limited flexibility and degenerative changes in adjacent vertebrae.[13] Intraspinal injection of chymopapain has been used; however, a recent study of 51 patients showed that this treatment was not effective.[14] Alternative treatments such as laser treatment are now being investigated.[15] Gene therapy may provide another means for treating this disease, but it is still in the developmental stage.[16-18] Surgical resection of an IVD and insertion of a prosthetic device as an IVD replacement is technically feasible.[19] However, not surprisingly because of the materials used, this approach has met with limited success.[20-24] Animal studies have shown that allograft and autograft spinal transplants are possible.[25-28] Allografts were shown to develop degenerative changes over time perhaps because they had been frozen prior to implantation.[25] To examine ways to eliminate the problems associated with allografts, implants of autogenous spinal segments were tested in the lumbar spine of dogs.[27] After 4 months, the transplanted spinal segments showed loss of disk height. This likely resulted from the different mechanical stresses that the disk from L5 experiences when placed into the L2 region and vice versa, similar to what has been shown to occur in a tendon when it is placed in a different location.[29] In another study using rhesus monkeys, an IVD and adjacent bone was mobilized and then placed back into the same location in the same animal. There was loss of disk height in the first 4 months that may have been related to the surgical trauma.[28] However, disk height increased by 12 months, indicating that the cells were viable and could still synthesize matrix macromolecules after transplantation. This supports the premise that disk replacement is a viable treatment approach justifying additional evaluation.

Autografts are clearly not an option and the use of allografts is limited by availability, possibility of immune rejection, and potential for disease transmission. We are investigating an alternative approach using tissue engineering to generate a functional spinal unit in vitro. This functional spinal unit consists of IVD and cartilage end plates. The first step involves the formation of NP tissue in vitro. Reinsertion of NP, particularly as intact tissue, can slow disk degeneration, suggesting that replacement of the NP alone may be a viable strategy.[30-32]

There are numerous methods of growing NP cells in vitro, but there are limitations to each of the currently available methods. When grown in monolayer, NP cells do not accumulate sufficient extracellular matrix to form a continuous layer of tissue.[33,34] Although cells grown in or on alginate or agarose beads remain spherical and accumulate mainly type II collagen and large proteoglycans, these cells do not form tissue.[35-38] Intervertebral disk cells grown in pellet culture will form tissue, but the extracellular matrix is unevenly distributed, suggesting improperly formed tissue.[39] It is not clear if these cultures were a pure population of NP cells or a mixture of NP and anulus cells. There is a preliminary report showing that NP cells embedded in biocompatible scaffolds, such as polyglycolic acid or calcium alginate, can form tissue when placed in the subcutaneous tissue of nude mice. However, by 10 weeks the proteoglycan content in the tissue was no longer increasing and was only about 50% that of the native NP.[40] Whether or not NP tissue would also form in culture under these conditions is not known. Furthermore, the presence of a scaffold during tissue formation might affect matrix organization. Thus there continues to be a need for in vitro methods to form NP tissue that resembles native tissue. We describe the development of an approach to form NP tissue in vitro. The initial characterization of this tissue is presented and compared with native NP.

Methods

Substrates
Filter inserts (12 mm diameter, Millicell CM, Millipore Corp, Billerica, MA) were precoated with type II collagen (Sigma Aldrich, St. Louis, MO) as described previously.[41] Porous cylindrical substrates (4 mm diameter, 4 mm height) were formed to serve as substrates for cell attachment by sintering calcium polyphosphate powder (CPP) as previously described.[42] The CPP has an added benefit as it can also act as bone substitute material if used as an implant.[42]

Tissue Formation
Sheep lumbar spines (6 to 12 months of age) and bovine distal caudal spines (6 to 9 months of age) were harvested and the ligament tissue surrounding the disk was removed aseptically. The AF and NP were identified and the NP dissected out and placed in Ham's F12 medium. The dissection of NP tissue was confirmed by histologic evaluation of representative fragments. The tissue was sequentially digested by 0.5% protease (Sigma Aldrich, St. Louis, MO) for 1 hour at 37°C, followed by 0.1% collagenase A (Roche, Laval, Quebec, Canada) overnight at 37°C.[43] The cell suspension was then filtered through a sterile mesh and cells were placed on filter inserts (2×10^6 cells, 3.3×10^6/cm^2) or on the upper surface of porous calcium polyphosphate sub-

strates (2×10^6 cells, 1.6×10^5 cells/mm^2). Cells were maintained in Dulbeccos Modified Eagle Medium supplemented with 20% fetal bovine serum and ascorbic acid (100 μg/mL) under standard tissue culture conditions. The medium was changed every 2 to 3 days and fresh ascorbic acid was added with each change. Cultures were harvested at various times up to 6 weeks.

Punch biopsies of native NP tissue (4 mm diameter) and underlying bone were taken from sheep lumbar, fetal bovine lumbar, bovine lumbar (16 months of age), and bovine caudal spines for the biochemical and mechanical comparison studies with the in vitro formed tissues.

Histologic Assessment of Chondrocyte Cultures

After the cultures were harvested, the tissues were fixed in 10% buffered formalin and embedded in paraffin. Sections (5 mm thick) were cut and stained with either hematoxylin and eosin to visualize cells or with toluidine blue to detect the presence of sulphated proteoglycans. The stained sections were examined with the use of light microscopy.

Determination of Dry Weight, Proteoglycan, Collagen, and DNA Contents

The tissue was removed from the substrate or the bone plug (for the native NP) and lyophilyzed overnight. The dry weight was then determined. The tissue was digested with the use of papain (Sigma Aldrich, St. Louis, MO; 40 μg/mL in 20 mM ammonium acetate, 1 mM EDTA, and 2 mM dithiotreitol) for 48 hours at 65°C. The tissue sample was evaluated for DNA, proteoglycan, and collagen content as described previously.[44,45] The DNA content of the tissues was determined from aliquots of the papain digest using the Hoechst 33258 dye binding assay (Polysciences, Warrington, PA) and fluorometry (emission wavelength 365 nm, excitation wavelength 458 nm). The proteoglycan content was determined by measuring the amount of sulphated glycosaminoglycans using the dimethylmethylene blue dye binding assay and spectrophotometry (wavelength 525 nm). The collagen content was determined using the chloramine-T/Ehlrich's reagent dye binding assay and spectrophotometry (wavelength 560 nm) following overnight hydrolysis.

Mechanical Properties

The compressive mechanical properties of the NP tissue (in vitro formed and native tissues) were assessed in uniaxial, unconfined compression while submersed in culture media at 37°C using a Mach-1 mechanical tester (Biosyntech, Laval, PQ).[45] The values for applied force and resulting deformations were collected using LabVIEW 5.0 data acquisition software (National Instruments, Austin, TX) interfaced to a personal computer. Nondestructive mechanical testing was conducted to assess the compressive properties of the tissues using large deformation dynamic testing. Samples were preloaded to 0.5 g, which was defined as the zero strain state. Dynamic deformation was applied by compressing the tissue using a sinusoidal waveform at 0.1 Hz between 0% and 25% strain for a total of 20 cycles. The dynamic stress-strain response was determined by normalizing the final

Table 1 Biochemical Properties of Native NP Tissue From Different Animals

	Sheep Lumbar	Bovine Lumbar	Fetal Bovine Lumbar	Bovine Caudal
DNA (μg/mg dry weight)	1.0 ± 0.1	1.0 ± 0.04	1.7 ± 0.08	1.3 ± 0.1
GAG (μg/mg dry weight)	358.3 ± 1	512.9 ± 15	388.2 ± 7	226.9 ± 11
OH-Pro(μg/mg dry weight)	39.6 ± 0.2	36.8 ± 2.3	37.0 ± 0.9	51.4 ± 2.9
Proteoglycan:Collagen	0.9	1.4	1.1	0.4

Table 2 Effect of Substrate on In Vitro Formed NP Tissue

	Filter Insert	3D Substrate (CPP)
Dry weight (mg)	1.6 ± 0.1	1.3 ± 0.1
DNA (μg/mg dry weight)	3.7 ± 0.1	4.9 ± 0.8
GAG (μg/mg dry weight)	197.3 ± 19	163.6 ± 8

force-deformation curve for the cross-sectional area of the sample and sample thickness. The amount of energy lost during deformation (hysteresis) was defined as the percentage difference between the area under the loading and unloading curves. The sample was unloaded and allowed to equilibrate for 15 minutes before the next test. The sample was then subjected to sequential step compressions of 2.5% strain to a maximum of 25% strain. At each step, the resulting force decay was recorded until equilibrium was reached. Equilibrium was defined as a change in force less than 0.2 g/min. The equilibrium stress was plotted as a function of the applied strain.

Results

Composition of NP from Different Animals

As shown in Table 1, the cellularity of the NP obtained from the different animals (sheep lumbar, fetal bovine lumbar, young adult bovine lumbar, and bovine caudal) is similar, although the fetal spines are significantly more cellular. The collagen content is similar in the lumbar spines of sheep and cows but is approximately 1.4-fold greater in the NP of bovine caudal spines. Proteoglycan content, however, is more variable as the amounts of glycosaminoglycans range from 226.9 ± 11 to 512.9 ± 15 μg/mg dry weight.

Characterization of NP Tissue Formed on Different Substrates

A continuous layer of NP tissue formed on both filter inserts and on the surface of a three-dimensional porous calcium polyphosphate substrate within 2 weeks. By 6 weeks, the tissue on the CPP was more cellular and contained less proteoglycan than the tissue formed on the filter insert (Table 2). This suggested that both types of substrates support tissue formation although the tissues that form show some differences.

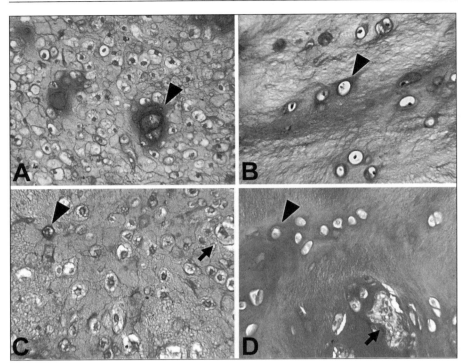

Figure 1 Histologic appearance of in vitro formed and native nucleus pulposus tissue. Photomicrographs showing tissue formed in vitro by sheep lumbar **(A)** or bovine caudal **(C)** nucleus pulposus cells. Native sheep lumbar nucleus pulposus **(B)** and bovine caudal nucleus pulposus **(D)** are shown for comparison. The arrowhead indicates the increased pericellular proteoglycan and the arrow indicates the notochordal cells. Toluidine blue stain, 250× magnification.

Histologic Appearance of the NP Tissue

The appearances of NP tissues formed in vitro were compared with those of the native NP. Histologic evaluation showed that there was a continuous layer of tissue composed of cells surrounded by abundant extracellular matrix by 2 weeks (data not shown). The matrix contained sulphated proteoglycans as demonstrated by Toluidine blue staining (Figure 1). Similar to the in vivo NP, scattered individual cells had greater pericellular staining intensity suggestive of localized enhanced proteoglycan accumulation. Notochordal cells, identified by their larger size and vacuolated cytoplasm, were present in the tissue formed by the bovine caudal cells. No notochordal cells were seen in the tissue formed by sheep lumbar NP cells.

Composition of In Vitro Formed NP

The tissues formed in vitro were more cellular and had less collagen than their respective native NP (Table 3). The proteoglycan content was similar to

Table 3 Biochemical Properties of In Vitro Formed NP Tissue Compared With Native Tissue

	Sheep Lumbar (%)*	Bovine Caudal (%)*
DNA (μg/mg dry weight)	274.2 ± 48.4	250 ± 22.9
GAG (μg/mg dry weight)	45.7 ± 2.3	93.2 ± 2.6
OH-Proline (μg/mg dry weight)	16.1 ± 0.8	24.2 ± 1.4

*Expressed as percent of native tissue

Table 4 Mechanical Properties of NP Tissue*

	In Vitro Formed NP†	Native NP
Equilibrium Stress (kPa)‡	12.1 ± 0.8	9.4 ± 1.3
Hysteresis (%)§	40.4 ± 1.6	38.7 ± 0.1

*Results expressed as mean ± SEM
†Tissue formed by bovine caudal NP cells
‡at 25% strain
§at 0.1 Hz

the native bovine caudal NP but was less than half that of the sheep lumbar NP.

Mechanical Properties of Tissue Formed by Caudal NP Cells

There were no significant differences in the weight-bearing capacity (equilibrium stress) and elasticity (percent hysteresis) compared with the native bovine caudal NP obtained from animals of a similar age (Table 4).

Discussion

Studies of the mechanisms regulating regeneration of NP have been limited in part by the absence of an appropriate in vitro model. In addition, the development of a biological approach to treat disk degeneration has been hindered by an inability to develop engineered tissue with appropriate biomechanical properties to withstand the forces placed on it immediately after implantation. This study describes our investigations to develop methods to form NP tissue in vitro. The results have shown that it is possible to form NP tissue in the absence of a scaffold when the cells are placed on the top of an appropriate porous substrate. Histologic evaluation confirmed that the NP tissue formed in vitro has some features similar to the native NP. When bovine caudal NP cells form the tissue, it has similar compressive properties to the native NP. This was unexpected as the amount of collagen of the NP tissue formed in vitro is approximately one fourth that of the amount of collagen of the native NP. The lack of effect of the relative paucity of collagen on the mechanical properties of the tissue is not well understood. It is possible that the contribution of collagen to weight-bearing properties is minimal and that collagen was present in quantities sufficient to maintain the compressive properties of the tissue.[1] Alternatively, collagen may be more important in resisting shearing forces than compressive forces. Additional studies are needed to investigate the relative contribution of collagen to shearing and compressive forces.

Our observation that the tissue formed in vitro had less collagen than the native NP is similar to results showing low collagen content in other engineered tissue such as cartilage. These results suggest that additional factors may be involved in regulating collagen synthesis and accumulation.[45,46] We have shown that applying intermittent cyclic shear stimulation during cartilage tissue formation will increase the amount of collagen in the tissue if it is applied for 4 weeks.[47] Minimal amounts of stimulation were necessary as the application of loading for 6 minutes every second day was sufficient to have an effect. This raises the possibility that mechanical loading may be important in regulating tissue composition in weight-bearing tissues and that it may be necessary to mechanically stimulate NP cells during tissue formation to attain collagen levels similar to the in vivo tissue.

Using this approach of growing cells at high density on the surface of a porous substrate, it is possible to generate tissue with and without notochordal cells by varying the source of the cells. We observed that the tissue formed by sheep lumbar NP cells had no notochordal cells whereas the bovine caudal cells generated tissue with notochord cells. This is of potential importance as there are data that suggest that NP degeneration may be due to the loss of these cells, which are known to synthesize proteoglycans, with aging.[9] In keeping with this observation, dogs that maintain notochordal cells throughout life do not develop disk degeneration or develop it at a later age.[48,49] The tissue formed by bovine caudal cells had proteoglycan content similar to that of native tissue and a significantly greater collagen content than the tissue formed by sheep cells, which lack notochordal cells. This in vitro culture system should facilitate investigation of the contribution of notochordal cells to maintenance of NP tissue and disk degeneration.

Biochemical evaluation of NP from different sources demonstrated that NP from bovine and sheep lumbar spines had similar cellularity and collagen content. The age of the animal seemed to affect cellularity as the fetal bovine tissue was more cellular than that of NP tissue from young adults. Proteoglycan content appeared to be more variable and thus the proteoglycan to collagen ratio varied from 0.4 to 1.4. This suggests that there is heterogeneity among NP tissue obtained from different animals and that evaluations of tissue formed in vitro should include a comparison to the in vivo tissue from which the cells were obtained.

Summary

In summary, NP tissue with appropriate compressive mechanical properties can be formed in vitro. Future efforts should be directed toward identifying the in vitro conditions necessary to form a functional IVD consisting of both NP and AF. This is a critical step in the development of a tissue engineered treatment for degenerative disk disease. Another important step is the identification of appropriate cells, such as mesenchymal stem cells, that can be used to form disk tissues. It is hoped that these studies will ultimately lead to development of procedures to restore normal disk height and function by repairing or replacing the degenerated intervertebral disk.

Acknowledgments

This work was supported by The Arthritis Society of Canada. CS was supported by a NSERC fellowship. We thank Isabelle Schell for secretarial assistance.

References

1. Bogduk N: The inter-body joints and the intervertebral disks, in Bogduk N (ed): *Clinical Anatomy of the Lumbar Spine and Sacrum*. Churchill Livingstone, New York, NY, 1997, pp 13-31.

2. Eyre DR: Biochemistry of the intervertebral disk. *Int Rev Conn Tiss Res* 1979;8:227-291.

3. Simon SR (ed): *Orthopedic Basic Science*. Rosemont, IL, American Academy of Orthopaedic Surgeons, 1994, pp 558-568.

4. Buckwalter JA: Aging and degeneration of the human intervertebral disk. *Spine* 1995;20(11):1307-1314.

5. Antoniou J, Steffen T, Nelson F, et al: The human lumbar intervertebral disk: Evidence for changes in the biosynthesis and denaturation of the extracellular matrix with growth maturation, aging and degeneration. *J Clin Invest* 1996;98(4):996-1103.

6. Miller JAA, Schmatz BS, Schultz AB: Lumbar disk degeneration: Correlation with age, sex and spine level in 600 autopsy specimens. *Spine* 1988;13(2):173-178.

7. Rudert M, Tillman M: Lymph and blood supply of the human intervertebral disk. *Acta Orthop Scand* 1993;64:37-43.

8. Hadjipavlou AG, Simmons JW, Pope MH, Necessary JT, Goel VK: Pathomechanics and clinical relevance of disk degeneration and annular tear: A point-of-view review. *Am J Orthop* 1999;28:561-571.

9. Aguiar DJ, Johnson SL, Oegema TR Jr: Notochordal cells interact with nucleus pulposus cells: Regulation of proteoglycan synthesis. *Exp Cell Res* 1999;246:129-137.

10. Kawaguchi Y, Osada R, Kanamori M, et al: Association between an aggrecan gene polymorphism and lumbar disk degeneration. *Spine* 1999;24(23):2456-2460.

11. Kraemer J: Natural course and prognosis of intervertebral disk diseases. *Spine* 1994;20(6):635-639.

12. Borenstein DG: Epidemiology, etiology, diagnostic evaluation, and treatment of low back pain. *Curr Opin Rheumatol* 1999;11:151-157.

13. Javedan SP, Dickman CA: Cause of adjacent-segment disease after spinal fusion. *Lancet* 1999;354(9178):530-531.

14. Leivseth G, Salvesen R, Hemminghytt S, Brinckmann P, Frobin W: Do human lumbar disks reconstitute after chemonucleolysis? A 7-year follow-up study. *Spine* 1999;24(4):342-347.

15. Choy DS: Percutaneous laser disk compression: 12 years experience with 752 procedure in 518 patients. *J Clin Laser Med Surg* 1998;16(6):325-331.

16. Nishida K, Kang JD, Suh JK, Robbins PD, Evans CH, Gilbertson Lars G: Adenovirus-mediated gene transfer to nucleus pulposus cells. *Spine* 1998;23(22):2437-2443.

17. Nishida K, Kang JD, Gilbertson LG, et al: Modulation of the biologic activity of the rabbit intervertebral disk by gene therapy: An in vivo study of adenovirus-mediated transfer of the human transforming growth factor b1 encoding gene. *Spine* 1999;24(23):2419-2425.

18. Evans CH, Robbins PD: Genetically augmented tissue engineering of the musculoskeletal system. *Clin Orthop* 1999;367:S410-S418.

19. Zigler JE, Anderson PA, Bridwell K, Vaccaro A: Specialty update: What's new in spine surgery. *J Bone Joint Surg Am* 2001;83(8):1285-1292.

20. Hou TS, Tu KY, Xu YK, Li ZB, Cai AH, Wang HC: Lumbar intervertebral disk prosthesis: An experimental study. *Chin Med J* 1991;104(5):381-386.

21. Enker P, Steffee A, Mcmillin C, Keppler L, Biscup R, Miller S: Artificial disk replacement: Preliminary report with a 3-year minimum follow-up. *Spine* 1993;18(8):1061-1070.

22. Bao QB, McCullen GM, Higham PA, Dumbleton JH, Yuan HA: The artificial disk: Theory, design and materials. *Biomaterials* 1996;17:1157-1167.

23. Kostiuk JP: Intervertebral disk replacement experimental study. *Clin Orthop* 1997;337:27-41.

24. Zeegers WS, Bohnen LMLJ, Laaper M, Verhaegen MJA: Artificial disk replacement with the modular type SB Charite III: 2-year results in 50 prospectively studied patients. *Eur Spine J* 1999;8:210-217.

25. Olson EJ, Hanley EN Jr, Rudert MJ, Baratz ME: Vertebral column allografts for the treatment of segmental spine defects: An experimental investigation in dogs. *Spine* 1991;16(9):1081-1088.

26. Diwan AD, Parvataneni HK, Khan SN, Sandhu HS, Girardi FP, Cammisa FP Jr: Current concepts in intervertebral disk restoration. *Orthop Clin N Am* 2000;31(3):453-464.

27. Frick SL, Hanley EN Jr, Meyer RA Jr, Ramp WK, Chapman TM: Lumbar intervertebral disk transfer: A canine study. *Spine* 1994;15(16):1826-1835.

28. Luk KDK, Ruan DK, Chow DHK, Leong JCY: Intervertebral disk autografting in a bipedal animal model. *Clin Orthop* 1997;337:13-26.

29. Malaviya P, Butler DL, Boivin GP, et al: An in vivo model for load-modulated remodeling in the rabbit flexor tendon. *J Orthop Res* 1999;18:116-125.

30. Okuma M, Mochida J, Nishimura K, Sakabe K, Seiki K: Reinsertion of stimulated nucleus pulposus cells retards intervertebral disk degeneration: An in *vitro* and in *vivo* experimental study. *J Orthop Res* 2000;18:988-997.

31. Nishimura K, Mochida J: Percutaneous reinsertion of the nucleus pulposus: An experimental study. *Spine* 1998;23(14):1531-1539.

32. Gruber HE, Johnson TL, Leslie K, et al: Autologous intervertebral disk cell implantation: A model using psammomys obesus, the sand rat. *Spine* 2002;27(13):1626-1633.

33. Ichimura K, Tsuji H, Matsui H, Makiyama N: Cell culture of the intervertebral disk of rats: Factors influencing culture, proteoglycan, collagen and deoxyribonucleic acid synthesis. *J Spinal Disord* 1991;4(4):428-436.

34. Poiraudeau S, Monteiro I, Anract P, Blanchard O, Revel M, Corvol MT: Phenotypic characteristics of rabbit intervertebral disk cells: Comparison with cartilage cells from the same animals. *Spine* 1999;24(9):837-844.

35. Chelberg M, Banks GM, Geiger DF, Oegema TR: Identification of heterogeneous cell populations in normal human intervertebral disk. *J Anat* 1995;86:43-53.

36. Chiba K, Andersson GBJ, Masuda K, Thonar EJ: Metabolism of the extracellular matrix formed by intervertebral disk cells cultured in alginate. *Spine* 1997;22(24):2885-2893.

37. Maldonado BA, Oegema TR: Initial characterization of the metabolism of intervertebral disk cells encapsulated in microspheres. *J Orthop Res* 1992;10:677-690.

38. Gruber HE, Stasky AA, Hanley EN Jr: Characterization and phenotypic stability of human disk cells in vitro. *Matrix Biol* 1997;16:285-288.

39. Lee JY, Hall R, Pelinkovic D, et al: New use of a three-dimensional pellet culture system for human intervertebral disk cells: Initial characterization and potential for tissue engineering. *Spine* 2001;24(21):2316-2322.

40. Kusior LJ, Vacanti CA, Bayley JC, Bonassar LJ: Tissue engineering of nucleus pulposus in nude mice. *Trans Orthop Res Soc* 1999;45:807.

41. Boyle J, Bo L, Cruz TF, Kandel RA: Characterization of proteoglycan accumulation during formation of cartilaginous tissue formed in vitro. *Tissue Eng* 1999;5(1):25-34.

42. Pilliar RM, Filiaggi MJ, Wells JD, Grynpas MD, Kandel RA: Porous calcium polyphosphate scaffolds for bone substitute applications: In vitro characterization. *Biomaterials* 2001;22:963-972.

43. Sun Y, Hurtig M, Pilliar RM, Grynpas M, Kandel RA: Characterization of nucleus pulposus-like tissue formed in vitro. *J Orthop Res* 2001;19:1078-1084.

44. Yu H, Grynpas M, Kandel RA: Composition of cartilaginous tissue with mineralized and non-mineralized zones formed in vitro. *Biomaterials* 1997;18(21):1425-1431.

45. Waldman SD, Grynpas MD, Pilliar RM, Kandel RA: Characterization of cartilaginous tissue formed on calcium polyphosphate substrates in vitro. *J Biomed Mat Res* 2002;62(93):323-330.

46. Vunjak-Novakovic G, Matrin I, Obradovic B, et al: Bioreactor cultivation conditions modulate the composition and mechanical properties of tissue-engineered cartilage. *J Orthop Res* 1999;17:130-138.

47. Waldman SD, Spiteri CG, Grynpas MD, Pilliar RM, Kandel RA: Long-term intermittent shear deformation improves the quality of cartilagenous tissue formed in vitro. *J Orthop Res* 2003;24:590-596.

48. Cole TC, Burkhardt D, Frost L, Gosh P: The proteoglycans of the canine intervertebral disk. *Biochim Biophys Acta* 1985;829:127-138.

49. Butler WF: Comparative anatomy and development of the mammalian disk, in Ghosh P (ed): *The Biology of the Intervertebral Disk*. Boca Raton, FL, CRC Press, 1988, vol 1, p 99.

Chapter 29

Tissue Engineering in Muscle: Current Challenges and Directions

Robert G. Dennis, PhD

Abstract

This review describes the use of skeletal muscle tissue engineering to gener-ate functional muscle tissues from isolated myogenic cells. Though all tissue functions arise from fundamental cellular mechanisms, the organization of tissues and organs confers function that is not possible to achieve with indi-vidual cells or masses of unorganized cells in a scaffold. Thus, it is impera-tive to have a working definition of muscle function and understand how the structure of muscle contributes to the emergence of that function. In vitro tis-sue engineering can be classified as either scaffold-based or self-organizing. A major challenge of engineering muscle tissue is the identification of suit-able tissue interfaces to allow the application of external cues (such as growth factors) to guide tissue development and to allow the controlled gen-eration of mechanical power.

Introduction

Skeletal muscle accounts for nearly half of the total mass of the average adult human and is unique in its ability to actively modify its mechanical properties within tens of milliseconds to allow humans to react to their envi-ronment. Tissue engineering of skeletal muscle could be broadly defined to include any alteration to or enhancement of the musculature of a living organism. This definition, though interesting, would not be specific enough to be useful, as it would include the agricultural use of steroids to rapidly increase the total lean body mass of livestock, the use of resistance training by athletes to induce hypertrophy, and surgical procedures including trans-plants and flaps in which preexisting skeletal muscle is modified and used in clinically relevant procedures (including graciloplasty, cardiomyoplasty, and musculoskeletal reconstructive surgery). Though all of these approach-es to the modification and use of skeletal muscle are of interest, this chapter will address skeletal muscle tissue engineering to generate functional mus-cle tissues from isolated myogenic cells in vitro.

Successful tissue engineering must include a focus on the organization of large numbers of cells into higher-order structures that confer function. These structures may be known as tissues or organs depending on the level of anatomic complexity and structural integration. Though all tissue functions arise from fundamental cellular mechanisms, the organization of tissues and organs confers function that is not possible to achieve with individual cells or masses of unorganized cells in a scaffold. A pile of bricks does not provide the functionality of a house, nor does a crate full of car parts function as an automobile. To understand tissue engineered skeletal muscle, we must have a clear working definition of muscle function and understand how the structure of muscle contributes to the emergence of that function.

Muscle Function

Skeletal muscle performs several important functions including production of proteins, generation of heat and involvement in thermal homeostasis, and serving as an energy substrate under certain physiologic situations. The fundamental function of skeletal muscle, however, is the controlled and efficient generation of mechanical force, work, and power. Collectively, these mechanical functions are referred to as the contractility of muscle. It is important to realize the physical difference between force, work, and power when describing muscle function because each describes an increasingly dynamic and theoretically demanding level of functionality. The generation of force is perhaps the most simple mechanical function of muscle. Force can be applied either actively or passively. A solid object in a gravitational field exerts a force on a surface at rest. Compressed springs exert a force against the constraining fixture. Work is the product of force and displacement. The dynamic aspect of work requires much more tissue plasticity than does force generation. This is reflected in the sliding-filament mechanism of the contractile machinery within each muscle fiber. Power is the rate of work generation and can be viewed as the controlled flow of mechanical energy. This generation of controlled power requires highly specialized tissue interfaces. The very high level mechanical function of muscle is reflected in the detailed structure of the three key tissue interfaces of every skeletal muscle, namely the vascular bed, the neuromuscular junction (NMJ), and the myotendinous junction (MTJ). Controllability arises from the built-in sensory organs of skeletal muscle. Muscle spindles (intrafusal fibers), for example, respond to the rate and magnitude of mechanical strain and the Golgi tendon organs act as force transducers embedded within tendons.

From the ability of muscle to actively generate controlled force, work, and power arise several distinct structural functions of skeletal muscle. Muscle tissue can act as a motor, a brake, a spring, and a strut[2] with the proper level of system integration, including sensors and the ability to rapidly apply feedback control to muscle tissue contractions. By generating active power, muscle acts as a motor. Through the absorption and dissipation of mechanical energy, muscle acts as a brake or shock absorber. Similar to a spring, muscle can store mechanical energy, then return a portion of that energy for later use. By transferring mechanical forces from one part of the skeletal system to another, muscle acts as a strut. During dynamic activities such as locomotion, each of these functions arises from the core ability of muscle to generate controlled mechanical power.

It is the ability to generate controlled and efficient mechanical power that is the greatest challenge to engineering skeletal muscle tissue. There are few reports that address the contractility of skeletal muscle function and none have included quantitative assessments of mechanical power.[1,3-6] Thus, it is accurate to assert that skeletal muscle tissue engineering is in its infancy and it may be decades before functional skeletal muscle engineered in vitro will be available in the clinical setting. However, we can define potential applications for engineered skeletal muscle[1] and describe the technologic challenges that must be addressed to realize the full potential of engineered skeletal muscle.[2]

General Approaches to Skeletal Muscle Tissue Engineering In Vitro

In vitro tissue engineering can be classified into the two broad categories of scaffold-based or self-organizing. The first and by far the most common approach to engineering muscle is the scaffold-based approach. This approach is attractive because of the large body of information about scaffold-based tissue engineering for other tissue systems, such as bone and cartilage. However, there are drawbacks to the use of scaffolds for engineered skeletal muscle. Unlike hard tissues, skeletal muscle has a very small percentage (generally less than 5%) of extracellular matrix (ECM). By comparison, tendon and cartilage have approximately 70% and 90% ECM, respectively. Thus, the selection of ECM for skeletal muscle must allow for a mechanically resilient structure with at least 95% void space that can be occupied by cells. Also, muscle is organized into long narrow fibers that result from myocyte migration, alignment, and fusion. Most available scaffold materials inhibit this important process of muscle development, thereby preventing the genesis of functional muscle tissue. During normal activities, muscle can undergo mechanical strains in excess of 10%, far beyond that experienced by any other musculoskeletal tissue. Modern scaffold materials generally do not tolerate the repeated application of such large mechanical strains. Finally, the presence of the scaffold material provides a force shunt that results in a derangement of the transmission of mechanical signals to individual muscle fibers. Interfering with the process of chemomechanical transduction in scaffold-based engineered muscle would result in arrested development. In addition, the force shunting within scaffold-based muscle inhibits the generation of active contractile force and thus interferes with the primary function of muscle as a tissue.

Self-organized skeletal muscle tissue does not use a preexisting scaffold to define the size and shape of the engineered tissue. This approach to tissue engineering can be successfully achieved in vitro for small mesoscopic specimens of up to approximately 0.5 mm in diameter and several centimeters in length. The self-organization of muscle tissue relies on the formation of a coherent monolayer of myogenic cells, including myotubes and fibroblasts, that detach from a substrate to form into a cylinder of contractile muscle tissue.[1] This process has been successfully applied to the self-organization of tendon and ligament, bone, and cardiac muscle in vitro (RG Dennis, Ann Arbor, MI, unpublished data, 2003). Engineered skeletal muscles self-organized in vitro have the disadvantage of being composed of multiple cell types. Thus, certain molecular biology techniques are more complex with self-organized muscle tissue than with monocultures grown on scaffolds. In addition, self organization involves tissue reorganization while under tension in culture. The tensile stresses arise both from the tractile forces generated by fibroblasts in the cell monolayer and the active spontaneous contractions of the myotubes. This tension contributes to the initial separation of the cohesive cell monolayer from the culture substrate. The reorganization of the tissue between the anchor points in culture into a cylindrical structure with long parallel myotubes is also influenced by tension. The resulting structures tend to have minimal surface area. The small cylindrical structures formed lack the typical range of morphology observed in native whole muscle. However, self-organized muscle tissues tend to be at least as mechanically robust as scaffold-based constructs and their contractile function can be readily assessed in the laboratory. In general, self-organizing muscle constructs have a greater percentage of viable cells per unit of cross-sectional area compared with scaffold-based constructs. Thus, self-organizing constructs tend to generate more active contractile force per unit cross section than scaffold-based engineered muscle.

Potential Applications of Engineered Skeletal Muscle

The ability of skeletal muscle to generate controlled mechanical power is only one of the unique aspects of skeletal muscle. Skeletal muscle has a physical scalability and structural plasticity that allows it to be formed into an enormous range of sizes and shapes. The ability to generate increasing amounts of force is accomplished by adding muscle fibers in parallel. An increased range of displacement is accomplished by lengthening muscle fibers and thereby adding sarcomeres in series within the fiber. Layering of parallel muscle fibers allows the formation of a large range of anatomic structures, including structures that approximate tubes, rods, sheets, fans, plates, toroids, and hollow spheres. The presence of sensory elements, such as muscle spindles and Golgi tendon organs, allows skeletal muscle to rapidly adapt and respond to stimuli. The mechanical efficiency and power density of muscle as a mechanical actuator is generally equal or superior to any modern synthetic technology such as electric motors, piezo-electric actuators, hydraulics, and pneumatics.[7] Thus, it is reasonable to consider a spectrum of applications for engineered skeletal muscle that reaches well beyond its use for the surgical correction of injury or deformity. Functional in vitro engineered skeletal muscle will eventually find application in bioreactors for drug testing and drug discovery, the verification of the efficacy of gene therapy for the correction of congenital muscular disease, the construction of tissue-based implantable devices such as pumps and sphincters, tissue-based robotic actuators and biomimetic devices, hybrid tissue-synthetic prosthetic devices that use both living and nonliving components, and ultimately, tissue engineered animal proteins as food.

Technological Challenges for Engineered Muscle

A major challenge to the development of engineered skeletal muscle tissues is the development of suitable tissue interfaces. Ideally, interfaces would be able to respond to external cues to guide development and allow the controlled generation of mechanical power. This area of research in skeletal muscle tissue engineering has essentially been unexplored until recently. Each of the key tissue interfaces for skeletal muscle presents specific technical challenges, but all are required before it will be possible to realize any level of success with engineered skeletal muscle. The three key tissue interfaces of skeletal muscle are the vascular bed to ensure adequate blood supply to support metabolism, the NMJ to provide the appropriate innervation, and the MTJ to facilitate force transmission

The importance of the vascular interface is readily appreciated considering the metabolic profile of skeletal muscle. Skeletal muscle is so metabolically active that it is impossible to achieve viable tissue thicknesses greater than approximately 200 μm in avascular specimens in culture. In general, the diameter of functional engineered skeletal muscle is limited to approximately 400 μm. The generation of a vascular bed within the engineered muscle is an obstacle that has not been overcome at this time. Once this has been achieved, it will be necessary to develop perfusion systems that can interface with the vascular bed.

The phenotype of skeletal muscle is controlled in large measure by contractile activity and innervation.[8-12] Thus, the importance of the NMJ to muscle function cannot be overstated. Though there have been some reports of nerve-muscle coculture systems,[13-16] there has not been a report of in vitro synaptogenesis with subsequent impact on the contractility of muscle tissue. In addition to dominating the expression of adult myosin isoforms in skeletal muscle, innervation also results in dramatic changes in the excitability of muscle tissue. Excitability may be quantified by measuring the bulk tissue properties of chronaxie (C_{50}) and rheobase (R_{50}) of a muscle tissue specimen, either from an animal or engineered in vitro.[1,5,17,18] The C_{50} is a measure of the duration of an electrical pulse that is required to elicit a specified contraction and R_{50} is a measure of the required amplitude of the same electrical pulse. As a first order approximation, assuming a muscle specimen to be a simple resistive load on within electrical stimulation circuit, the energy required from a single electrical pulse to stimulate a muscle would be proportional to $C_{50} \times (R_{50})^2$ as derived from Ohm's Law.[19] Thus, an increase in either C_{50} or R_{50} indicates a decrease in excitability. Denervation results in dramatic loss of excitability as shown by an increase of both C_{50} and R_{50} by an order of magnitude or more; this increases the required stimulation pulse energy by a factor of approximately 1,000. Using current tissue culture technology, engineered skeletal muscle develops aneurally and the development of the tissue is arrested at an early developmental phenotype. A major challenge to the engineering of skeletal muscle tissue will be to develop cell culture technologies to allow coculture and chronic stimulation of intact neuro-motor tissue constructs.

Perhaps the most overlooked tissue interface for engineered skeletal muscle is at the tendon (MTJ). The importance of the mechanisms of force transmission from the contractile machinery within cells to the skeletal system has only recently been elucidated. Derangements to this complex mechanical transmission system give rise to very serious disease states, including Duchenne muscular dystrophy. The controlled flow of mechanical energy from muscle is channeled directly through tendons to the skeletal system, then ultimately to the external environment. In addition to providing the correct molecular surface signals, the engineered MTJ must also provide mechanical impedance matching to minimize the concentration of stress at the ends of the contractile section of the muscle tissue.

With the key tissue interfaces in place, it will be necessary to develop a sophisticated bioreactor technology to provide perfusion, chronic electrical stimulation, and the application of mechanical strain to the engineered skeletal muscle constructs to guide the development in vitro. In reality, this is a relatively straightforward engineering challenge that is simply awaiting the emergence of an adequately complete tissue model system. The optimization of the various input parameters from the bioreactor to the muscle specimens will be a complex task, but there is adequate baseline data from developmental biology to provide sound initial estimates for most of these parameters. As always in tissue engineering, the core challenge is at the biologic level and the development of organotypic tissue interfaces will enable technology for the development of truly functional skeletal muscle tissues.

Summary

The controlled and efficient generation of mechanical force, work, and power are fundamental functions of skeletal muscle. In vitro tissue engineering can be classified into the two broad categories of scaffold-based or self-organizing constructs. Scaffold-based tissue engineering is the most common approach to engineering muscle tissue replacements. Engineering of muscle tissue must include a means to organize large numbers of cells into structures that are able to perform the functions of muscle. Indeed, the development of tissue interfaces able to generate controlled and efficient mechanical power represents a major challenge. At this time, skeletal muscle tissue engineering is in its infancy and it may be decades before functional skeletal muscle engineered in vitro will be available in the clinical setting.

References

1. Dennis RG, Kosnik PE, Gilbert ME, Faulkner JA: Excitability and contractility of skeletal muscle engineered from primary cultures and cell lines. *Am J Physiol Cell Physiol* 2001;280:C288-C295.
2. Dickinson MH, Farley CT, Full RJ, Koehl MAR, Kram R, Lehman S: How animals move: An integrative view. *Science* 2000;288:100-106.

3. Powell CA, Smiley BL, Vandenburgh HH: Novel techniques for measuring tension development in organized tissue constructs. *FASEB J* 2000;14:A444.

4. Vandenburgh HH, Swasdison S, Karlisch P: Computer-aided mechanogenesis of skeletal muscle organs from single cells in vitro. *FASEB J* 1991;5:2860-2867.

5. Dennis RG, Kosnik PE: Excitability and isometric contractile properties of mammalian skeletal muscle constructs engineered in vitro. *In Vitro Cell Dev Biol Anim* 2000;36:327-335.

6. Kosnik PE, Faulkner JA, Dennis RG: Functional development of engineered skeletal muscle from adult and neonatal rats. *Tissue Eng* 2001;7:573-584.

7. Hollerbach JM, Hunter IW, Ballantyne J: A comparative analysis of actuator technologies for robotics, in Khatib O, Craig J, Lozano-Perez T (eds): *The Robotics Review*. Cambridge, MA, MIT Press, 1991, pp 301-342.

8. Close R, Hoh JFY: Effects of nerve cross-union on fast-twitch and slow-graded muscle fibres in toad. *J Physiol (Lond)* 1968;198:103.

9. Close R, Hoh JFY: Post-tetanic potentiation of twitch contractions of cross-innervated rat fast and slow muscles. *Nature* 1969;221:179.

10. Hoh JFY, Close R: Effects of nerve cross-union on twitch and slow-graded muscle fibres in toad. *Australian J Experimental Biology and Medical Science* 1967;45:51.

11. Close R: Effects of cross-union of motor nerves to fast and slow skeletal muscles. *Nature* 1965;206:831.

12. Close R: Dynamic properties of fast and slow skeletal muscles of rat after nerve cross-union. *J Physiol (Lond)* 1969;204:331.

13. Ecob M, Whalen RG: The role of the nerve in the expression of myosin heavy-chain isoforms in a nerve muscle-tissue culture system. *J Muscle Res Cell Motil* 1985;6:56.

14. Ecob M: The location of neuromuscular-junctions on regenerating adult-mouse muscle in culture. *J Neurol Sci* 1984;64:175-182.

15. Ecob MS, Butler Browne GS, Whalen RG: The adult fast isozyme of myosin is present in a nerve-muscle tissue-culture system. *Differentiation* 1983;25:84-87.

16. Ecob MS: The application of organotypic nerve cultures to problems in neurology with special reference to their potential use in research into neuromuscular diseases. *J Neurol Sci* 1983;58:1-15.

17. Broadie KS: Development of electrical properities and synaptic transmission at the embryonic neuromuscular junction. *Neuromusc Junc Drosophila* 1999;43:45-67.

18. Dennis RG, Dow DE, Hsueh A, Faulkner JA: Excitability of engineered muscle consructs, denervated and stimulated-denervated muscles of rats, and control skeletal muscles in neonatal, young, adult and old mice and rats. *Biophys J* 2002;82:364A.

19. Dennis RG, Kosnik PE: Mesenchymal cell culture: Instrumentation and methods for evaluating engineered muscle, in Atala A, Lanza R (eds): *Methods in Tissue Engineering*. San Diego, CA, Academic Press, 2002, pp 307-316.

Future Directions

Tissue Engineering of Ligament, Tendon, Meniscus, Intervertebral Disk, and Muscle

Considerable advances have been made during the last few years in addressing the scientific issues focused on the tissue engineering of ligaments, tendons, and meniscus. Less is known about the regeneration of intervertebral disk and muscle. Although investigators have identified the important characteristics of the biomaterials, cells, and bioactive factors involved in the regeneration approaches for each of these tissues, an enormous amount of work still remains to be performed to provide the scientific basis for successful treatment strategies. The structural and biochemical constituents of each tissue must be further defined in order to understand the biomechanical requirements required to withstand the considerable in vivo functional loads. Parameters must be established that drive the development of mechanical tissue engineered structures either in vivo or ex vivo using bioreactor technologies. Critical to the success of any tissue engineered treatment plan is the techniques necessary to ensure the functional integration of the new tissue with the host. Also, the complex factors influencing the remodeling of the new tissue must be understood to enable a biomechanically successful construct. Clearly, progress in one area can be applied to another tissue; however, there are specific issues that are special to each tissue that may slow progress. The following future directions provide the basis for ultimate application of the tissue engineering regeneration of these diverse musculoskeletal structures.

Ligaments and tendon
Characterize the clinical goals of bioengineering ligaments and tendon more clearly
Ligaments and tendons are not as well characterized as other connective tissues. They clearly require better structure-function, and "structure-specific" definition of replacement requirements from a clinical perspective. Similarly, clinical situations in which bioengineered full tendon and ligament "replacements/substitutes" or "acute repair augmentation devices" would benefit patients and be useful to clinicians requires better definition.
• Build a tendon segment to unite and integrate torn tendon ends. There must be a specific mechanism/design to integrate to tendon stumps, myotendinous junction, or bone; adequate strength to prevent failure and provide tensioning capability; ability to act as a scaffold for cell ingrowth; prevent adhesion to extratendinous tissues; have shape control so that it is formed to a tendon-like structure; allow gliding through pulleys and extratendinous tissues; and remodeling capability. It should have a capacity to be vascularized and innervated at the correct locations, and have pressure and tension sensors.

Characterize the requirements for bioengineered ligaments and tendon substitutes

"Minimal requirements" for all acute tendon and ligament repairs and chronic replacements need to be prioritized and defined clearly (clinical function, mechanical functions, and biological/physiological essential requirements). Related to this, a clear set of validated clinical outcome measures, markers of success, and standardized methods of evaluation for each clinical and/or tissue deficiency needs to be defined so that any substitute can be evaluated.

- Develop a material that can act as a cell and/or gene carrier. The material may be combined with growth factors such as PDGF, IGF-I, TGF-β, or BMPs that can activate cells to migrate into the matrix, proliferate, and then make matrix proteins to provide strength. Likewise, the cells should be responsive to cytokines such as IL-1β that can drive remodeling.

Define how success will be evaluated

Models in which to validate all bioengineered structure-specific constructs identified in all of the aims above need to be confirmed through a process of comparison with actual human structure-specific requirements. Appropriate animal models are required to test the cells and constructs for each tendon repair case such as supraspinatus, flexor digitorum pollicus, or Achilles tendon, since their anatomy, function, and nutritional requirements are very different. It will be impossible to pursue clinical trials without such model validation of safety and efficacy.

Use current "gold standards" (normal tissues and autografts) as controls for experiments and to pursue most promising approaches as measured against those standards

Using the principles defined above, promising work on bioengineered tendon and ligament substitutes needs to be pursued aggressively, using normal controls as "primary gold standards" for comparison and commonly available techniques of repairs or graft replacements as "secondary gold standards." As a subset of this work, the important principles of optimizing bioengineered substitutes (which cells +/- which matrix +/- which artificial scaffold, if any) in vitro, and in vivo, need to be defined. This would include defining the key relationships of biological and biomechanical factors that can create normal functions in vitro, but most importantly, obtaining and maintaining those properties over time, in vivo, needs to be a primary focus of repairing or replacing these tissues.

Evaluate complementary approaches to solve clinical problems in more than one way

The three basic approaches that have been pursued in the past need to be pursued collectively. First, the notion of either modifying or "engineering scars" (which represent natural "neotissue"), either in situ or ex vivo with reimplantation to achieve results that are closer to normal tissue quality, represents one feasible approach. A second is to engineer a full replacement or strut with cells/matrix/scaffolds as noted above. A third is to use engineering approaches to prevent whole tissue grafts (cell-seeded or otherwise modified) and bioacceptable allografts or xenografts from deteriorating in vivo.

Meniscus
Expand the knowledge base of meniscus development, structure, and cell biology
Little is known regarding meniscus development. A further understanding of which molecules contribute to the formation of the meniscus may be important in regenerative schema. It is not well understood how the three-dimensional structure of the meniscus can influence the tissue regenerative environment. Crucial to this is the characterization of number, type, and metabolism of cells that populate the various meniscus regions.

Develop differential strategies specific for tissue engineering of meniscal tears versus segmental replacement
There remains a need to address the two types of clinical treatment modalities. Each modality has distinctive requirements regarding the need for material properties. Treatments for meniscal tears do not necessarily require the structural properties that are clearly a prerequisite for segmental repair. Delivery of growth factors or gene therapy may be effective in meniscal tears but may not be adequate for production of extensive tissue needed for more extensive defects.

Optimization of delivery systems in vivo
There is no well characterized or accepted scaffold or matrix material that has been shown to be optimal for tissue engineered approaches to the meniscus. The optimal delivery system should be one that promotes cellular integration in tears but does not inhibit the repair response. Additionally, systems that employ arthroscopic approaches using gels are potentially important for clinical acceptance.

Investigate the potential of alternate cell sources for meniscus repair strategies
Meniscal cells are described as fibroblastic in character. There are other cell compartments that are more easily accessible that may provide cells that can be equally effective if placed in a meniscal repair environment. An example might be the use of dermal-derived fibroblasts obtained by simple biopsy. Alternatively, stem cells derived from marrow, fat, or muscle may be a source of regenerative cells.

Explore perioperative approaches for meniscus regeneration
Methods for tissue engineering repair in the constraints of the operating room are highly desirable. These include matrices which contain bioactive molecules either precoated or obtained from the patient during the procedure. Techniques which can harvest cells from various compartments could be advantageous in the operating room. The delivery of appropriate growth factors or transgenes that could stimulate integrative repair could be a perioperative procedure.

Define noninvasive clinical outcome measures for success in meniscal regeneration

Developing imaging technologies that can visualize soft-tissue repair in vivo are a requirement for assessing the clinical outcome of meniscus repair. There is a need for scoring or measurement systems specific for meniscus repair.

Muscle

To generate healthy, adult phenotype skeletal muscle in culture, the principle challenges will be to provide suitable tissue-tissue interfaces to allow appropriate signal transduction to and from the muscle tissue, then to determine the sequence of cues that will be necessary to recapitulate the developmental processes of muscle. Specific efforts must therefore be directed toward the following goals:

Develop a suitable in vitro model system for the muscle-tendon junction (MTJ)

All current engineered muscle constructs eventually fail at the interface between the tissue and the supporting structures at each end of the muscle construct when the muscle specimens are subjected to repeated active contractions. This is the result of a mechanical impedance mismatch at this interface, as well as a lack of the specific myotendinous structures that are required for the safe transmission of mechanical force, work, and power from the contractile machinery within each muscle fiber to the skeletal system. The development of a suitable MTJ is therefore critical for the generation of functional whole muscles in vitro.

Determine the optimal mechanical stimulation protocol, involving both active force generation and passive length changes, to induce muscle growth and the maintenance of healthy adult phenotype

The important of normal activity for muscular development and the maintenance of muscle mass and contractility is well accepted. However, the baseline requirements are less well understood. It will be necessary to determine optimal active and passive mechanical stimulation protocols to guide the development of muscle in vitro, and to maintain the desired tissue mass and phenotype.

Develop a suitable in vitro model system for the neuromuscular junction (NMJ)

Since the early days of neuromotor cross union studies in the 1960s, it has become clearly apparent that it is the activity of the motor nerve that drives muscle phenotype (fast versus slow fibers), and not the other way around. The plasticity of muscle phenotype when subjected to changing stimulation patterns has opened tremendous opportunities for the reengineering of skeletal muscle for surgical transposition and use in graciloplasty and cardiomyoplasty. However, in the absence of the correct neural signals, muscle will quickly degenerate, rapidly losing both mass and contractility. Even if a muscle organ is to be developed in an aneural in vitro environment, it will in any event be critical to have in place the necessary NMJ architecture to allow innervation for muscles that are surgically implanted. To maintain the muscle mass and function after implantation, reinnervation must be rapid and complete. This will be greatly facilitated by the preexistence of NMJ structures, engineered into the muscle during in vitro development.

Determine the relative importance of activity versus innervation for the guidance of muscle development and the control of muscle phenotypic outcome

Recent research has demonstrated that adult muscle phenotype can be maintained almost indefinitely in the absence of innervation, if an "optimized" stimulation protocol is identified and employed. This does not necessarily mean, however, that adult phenotype can be achieved in the absence of innervation. The relative importance of aneural contractile activity versus innervation and nerve-induced muscular contraction must be determined for muscle organs during and throughout development.

Develop a co-culture system that will lead to angiogenesis in the engineered muscle tissue

Without perfusion, engineered skeletal muscle is limited to a maximum viable radius of approximately 200 µm. Though specimens of this size are certainly useful for basic in vitro research, the development of full-sized skeletal muscle specimens will be required in order for tissue engineered muscle to have a significant impact on clinical practice. Even for facial reconstruction, the cross sections are several millimeters, at the least. The availability of a technology to both provide a vascular system for muscle and to perfuse the vascular bed will revolutionize muscle tissue engineering.

Develop a bioreactor technology to control and monitor the electromechanical interventions and perfusion of engineered muscles, and to provide feedback-based control, based on measured physiologic function during in vitro development, to promote the desired phenotypic outcome of the engineered muscle

Though fundamentally technological in nature, bioreactor systems to emulate the mechanical environment during development and normal daily activity will be essential to the development of model systems into the basic mechanisms of muscle organogenesis. In the absence of the correct mechanical signals, muscle will rapidly degenerate, even when all other systemic and nonmechanical cues are in place. Consider, for example, muscle wasting during disuse when a limb is in a cast, or when the forces of gravity are removed, even for short periods of time. It is pure folly to consider the development of this critical technology as merely "incidental" to basic muscle research. It is in fact a core and limiting requirement in current and future in vitro muscle research.

Develop in vitro methods involving controlled electromechanical stimulation to elucidate the poorly understood mechanisms of muscle tissue organogenesis, specifically in terms of the signals that control the overall structures and anatomy of whole muscle and the myotendinous transition
It is difficult or impossible to accurately measure the mechanical stress state within muscle during early development, though certainly these stresses play a crucial role in the genesis of the wide range of muscle architectures observed, even within individual organisms. A single myogenic stem cell could in principle give rise to a whole muscle organ to any arbitrary architecture, with widely divergent thickness, overall shape, myotendinous structure, fiber length distribution, and pennation angle. These anatomic features are critical to the differing functional demands placed on muscle organs in each anatomic position within an organism. An understanding of the anatomic organogenesis of muscle is critical if functional muscle tissue replacements are ever to be realized for any application, clinical or otherwise.

Intervertebral disk
Investigate the mechanisms regulating intervertebral disk degeneration
Studies have show that intervertebral disk (IVD) degeneration is very common and more than 90% of individuals over age 50 years will show degenerative changes. These are superimposed on aging alterations. A better understanding of the mechanisms that regulate the development of aging and degenerative changes is critical to designing appropriate individualized treatment strategies. This will allow design of tissues and/or scaffolds that could modulate these processes.

Determine why disk degeneration is so common and why symptoms develop in only a proportion of individuals
Although most people show changes associated with IVD degeneration, herniation of the nucleus pulposus and back pain will not occur in all of these individuals. Are there biologic features, biochemical changes, or genetic changes that predispose individuals to development of disk disease and can these be used to design "smart" IVD tissue?

Identify the in vitro conditions that will allow formation of an annulus fibrosus
Currently it is possible to form nucleus pulposus tissue in vitro. The annulus fibrosus, the outer component of the IVD, is a highly organized collagenous structure. To date, there has been no report of tissue engineering of this tissue.

Develop a spinal articulation unit composed of IVD, cartilage end plate, and underlying bone
Surgical replacement of the nucleus pulposus might be technically difficult. Also by the time the nucleus pulposus herniates, there are changes in the annulus fibrosus that might predispose to reherniation if only the nucleus pulposus is replaced. It might be more feasible to replace the entire disk. This in vitro formed structure could also serve as a model, the study of

which will allow better understanding of how macromolecular composition and organization contribute to forming functional tissue.

Develop an animal model of disk degeneration

Currently the models used commonly to induce disk degeneration are rat tail and rabbit spine. It will be necessary to develop a large animal model that more closely mimics the human condition in order to be able to evaluate the efficacy of tissue engineered tissues.

Section Five

Methodologies Used in the Construction of Engineered Tissue

Oliver Betz, PhD
Lawrence J. Bonassar, PhD
Scott P. Bruder, MD, PhD
Christina Cosman, MS
Bradford L. Currier, MD
Gary C. du Moulin, PhD
Cynthia A. Entstrasser, BS
Christopher H. Evans, PhD
Steven C. Ghivizzani, PhD
Jeffrey M. Gimble, MD, PhD
Steven A. Goldstein, PhD
Alan J. Grodzinsky, ScD
Farshid Guilak, PhD

Esmaiel Jabbari, PhD
Grace L. Kielpinski, BS
John D. Kisiday, PhD
Lichun Lu, PhD
Antonios G. Mikos, PhD
Glyn Palmer, PhD
Arnulf Pascher, MD
Andre Steinert, MD
Dave Toman, PhD
Gordana Vunjak-Novakovic, PhD
Michael J. Yaszemski, MD, PhD
Shuguang Zhang, PhD

Chapter 30

Quality Control in Producing Cells for Tissue Engineering

Gary C. du Moulin, PhD
Cynthia A. Entstrasser, BS
Grace L. Kielpinski, BS

Abstract

The development and commercialization of tissue engineered products must include the design and implementation of robust quality control programs. Two elements essential for the development of systems of adequate quality are quality assurance (QA) and quality control. Essential elements of QA programs include control over the manufacturing process, manufacturing personnel, production facility, raw materials used in the product, and equipment used to produce the product. For tissue engineered products, not all variability can be controlled because of the inherent variation in the living cells or tissues used as the starting material. Quality control refers to those elements of the quality system that require technical analytic capability. Many standardized analytical methods are not yet available for tissue engineering based therapies. Thus these assays must be developed and validated to assess purity, potency, identity, sterility, yield, and viability. With successful application of quality control systems, tissue engineered products are expected to become a viable therapeutic option for a variety of orthopaedic indications.

Introduction

The use of living human cells to facilitate the repair of defects in structural tissues is a rapidly emerging treatment option. Through a greater understanding of cell biology, it is possible to cultivate, propagate, and manipulate many kinds of living human cells into therapeutically useful modalities. Although these technologies continue to be developed and validated through clinical trials, several products have already been commercialized and have become successful adjuncts to orthopaedic surgery practice. To have clearance for marketing from the United States Food and Drug Administration (FDA), companies or institutions developing these products needed to develop robust quality control programs to demonstrate safety and efficacy. A review of successfully completed development programs can provide insight for other tissue engineering programs.

Autologous Cultured Chondrocytes

Articular cartilage loses the ability to regenerate over time. Several clinical procedures have been used to treat the symptoms of articular cartilage defects. Lavage and débridement offer temporary relief of pain, but they fail to yield long-term results in many patients. The same is true for drilling, microfracture, and abrasion arthroplasty, which typically cause the development of fibrocartilage. Fibrocartilage lacks the specific mechanical properties of hyaline cartilage that normally covers articulating surfaces. Autologous chondrocyte implantation has been shown to result in formation of some hyaline cartilage and to provide relief of pain for many patients. The application of tissue engineering to repair articular defects of the knee was developed in Sweden.[1] This technology involved the processing and culturing of chondrocytes from an autologous cartilage biopsy into an injectable cell suspension. Use of autologous cells circumvented the complications of allograft rejection and viral transmission by returning the patient's own cells for transplantation. In 1997, the FDA approved Carticel (Genzyme Biosurgery, Cambridge, MA) as a biologic product. Since then, approximately 8,000 patients have received autologous chondrocytes.

Autologous tissue engineering programs present several challenges.[2] Every patient seeking treatment requires customized cell processing. A processing failure can lead to a patient not receiving treatment. Special care has to be taken to prevent the possibility of cross-contamination of patient cells and to maintain the microbial integrity of each processing lot. Traditional batch processing techniques are difficult to apply to autologous tissue engineering, making economies of scale problematic. Autologous cell processing on a clinically relevant scale can be approached by applying well-accepted manufacturing principles and practices to customized patient therapy. Cell processing is conducted using standard operating procedures conducted under good manufacturing practices in which all processing events are thoroughly documented in batch production records. Quality systems have been developed to ensure that a safe, reproducible product can be returned to the patient. Two elements essential for the development of systems of adequate quality are QA and quality control.

Quality Assurance Program[3]

Tissue engineered products must be produced with a high level of consistency. Stringent controls must be established for key manufacturing criteria. QA programs ensure that all aspects of the manufacturing process remain within established parameters. Essential elements of QA programs include control over the manufacturing process, manufacturing personnel, production facility, raw materials used in the product, and equipment used to produce the product. Variability inherent in all of these elements must be determined and minimized through an effective compliance development program. For tissue engineered products, not all variability can be controlled because of the inherent variability in the living cells or tissues used as the starting material. Thus QA programs must attempt to reconcile variability to the extent possible. For example, patient biopsies collected in various clinical surroundings under differing conditions may result in differences in biopsy size, storage condition, and site of harvest. Because each patient presents a unique set of clinical circumstances, these and other variables are outside the control of the cell processing facility but can significantly affect growth and performance of the living cell construct.

Maintaining patient identity is vital and requires meticulous documentation of the identity of the patient, physician, hospital of origin, and the site and time of biopsy. A unique lot number must be assigned to identify the biopsy material and its associated file. Steps must be taken to properly identify each sample from biopsy through final product delivery. Each patient lot receives the full complement of in process and final release testing. This information is documented and verified in the individual lot history file. Rigorous cleaning, sterilization, and environmental control programs preserve the aseptic nature of the processing environment. Inspection, testing, and release of raw materials as well as maintenance and calibration of equipment mitigate the potential for significant process variability. Biopsies transported to the cell processing facility and the final cell product returned to the patient must be packaged and shipped in specially designed containers that provide a controlled, protective, environment for living cells. Containers must maintain the integrity of cells within a defined temperature range over a period of time to allow transit to and from points of origin. Cell transit and processing activities are governed by the known limited shelf-life of the living cells. Personnel training is particularly important because ex vivo cell processing is typically not amenable to automation. Therefore, many procedures in tissue engineering must be performed manually by highly skilled technicians. Finally, QA systems must be able to be completed under rigid time constraints to optimize viability and cell function, and to minimize delays in product release.

QA programs for tissue engineering production operations are designed to ensure that the final product meets the highest standards of safety, effectiveness, and reliability. Functionally, five major activities comprise QA programs:

1. A document control group responsible for the control of all product related specifications, standard operating procedures, and other product related records.
2. A compliance group that develops, audits, and maintains systems to ensure, measure, and deliver quality tissue engineered products.
3. A patient operations group that focuses on the biopsy accessioning procedures, document issuance and review, and safe release and shipment of the final product. Given the potentially variable nature of cell processing operations, a QA system should promptly recognize and address incidents of nonconformance with established procedures. A nonconformance reporting system must be implemented to control and reconcile inherent variability and out of specification conditions resulting from cell culture processes. Mechanisms for corrective and preventative action must be provided.

4. A compliance development function to process new regulatory information and guidance with the objective of identifying compliance needs, data trends, and effective ways of achieving continuous improvement.
5. A calibration-metrology group that ensures that equipment and facility controllers are in specification and operate as designed.[4]

Quality Control Program[5]

Quality control refers specifically to those elements of the quality system that require technical analytic capability. Many standardized analytical methods are not yet available for tissue engineering based therapies. Thus these assays must be developed and validated to assess purity, potency, identity, sterility, yield, and viability. These methods are also used to monitor the performance of materials and components, and identify limitations of cell processing procedures. Quality control assays must provide quantifiable data for analysis and identification of trends so that the quality system is capable of continuously directing improvement. Examples of important assays used in the evaluation of tissue engineered products are listed in the following sections.

Sterility Testing

Several sterility tests are conducted throughout the cell culturing process. Testing points are identified at critical cell processing steps to monitor aseptic processing. Sterility tests may also be conducted to assess the microbial load of the biopsy following collection. This is an important indicator of the effectiveness of training of the surgeon and staff who collected and handled the biopsy specimen. Sterility testing protocols must comply with the United States Pharmacopoeia (USP) guidance. All sterility tests are maintained and observed for 14 days.

Endotoxin Testing

All patient lots prepared for release are tested for the presence of endotoxin. The *Limulus* amebocyte lysate (LAL) assay has been developed to measure endotoxin in biologic products, drugs, and clinical specimens.

Cell Viability

Cell viability is determined by mixing a sample of cell suspension with trypan blue. Trypan blue is absorbed by nonviable cells but is excluded from viable cells. Viability is expressed as a percentage of viable cells to total cells viewed microscopically.

Quality control responsibilities may also include environmental monitoring that involves establishing and maintaining, with scheduled monitoring, cleanroom classifications at appropriate baseline levels. Guidelines for aseptic processing specify that the work environment be class 100 for critical processing operations. Microbiologic control of the clean room environment is maintained by close surveillance of air, water, surfaces, and personnel.

Routine monitoring of the environment with established alert levels provides early warning to the QA department. Thus the QA department is aware of the adequacy of cleaning and sanitization procedures before action levels are exceeded. In addition, products are monitored for microbiologic control during crucial manufacturing steps.

The use of media fills in autologous cell processing is an extremely useful QC measure because it demonstrates the effectiveness of microbiologic controls, environmental controls, personnel training, and predetermined interventions. This measure builds confidence into the aseptic processing capabilities of the manufacturing line. Because of the unique nature of tissue engineered products in which maximum lot size is equal to one unit, statistical analysis may be inappropriate for determining the success or failure of media fills. Therefore, acceptance criteria are based on multiple runs with no failures. The scope and duration of media fills must encompass all critical cell culture processing steps with additional emphasis on final harvesting and assembly procedures.

The manufacture of tissue engineered products requires the use of reagents and components that may introduce variability to the manufacturing process. This variability must be overcome through demonstrated controls of the characteristics of raw materials to ensure a consistent product. Inherent variability should be limited solely to biopsy material and not to raw materials used in the cell culture process. Consistency and conformance to specifications for all raw materials must be demonstrated. Through inspection, testing, and auditing programs, the quality control department monitors material performance and vendor compliance to established specifications. Critical materials used in the cell culture process, such as enzymes, are functionally tested before use in patient cell culture lots with the use of either a well-characterized reference human cell line or freshly harvested animal tissues. A materials review board or similar organization reviews trends observed in data collected through these monitoring programs. Adjustments to current specifications and classification schemes can be made when needed. Vendor selection and qualification are also determined through these programs.

In addition to the guidance available through the FDA, the USP has published an informational chapter[6] that should be consulted when developing manufacturing and quality systems for tissue engineering programs devoted to the musculoskeletal clinical practice.

Future Directions

Enhancements to Safe and Effective Cell Culture Technology

As tissue engineered product development programs and the overall field of regenerative medicine matures, enhancements that will improve cell culture technology will undoubtedly be implemented. Areas under review are centered on programs to improve the yield of cells from biopsy materials, effectively using growth factors, and the development of closed cell culture systems that minimize contamination. In some cases, methods used in the blood banking industry can be applied. Closed systems that use bags and tubing can minimize contamination if appropriate sterile connection devices are used. Use of plastic molding technology to produce devices that combine enlarged surface areas with multiple levels called "cell factories" can facilitate the efficient cultivation of cells.

Materials and Component Especially Designed for Human Use

Media and reagents required for optimized cell growth will be improved through the development of defined media formulations that minimize or eliminate the use of animal-based components. High quality, genetically produced cytokines and growth factors are being produced for potential use in the manufacture of tissue engineered products. Biomaterials and plastics for implantation must be manufactured in a manner that minimizes risk of particulate or endotoxin contamination.

Improved Quality Control Testing Platforms

Because of the limited shelf-life of tissue engineered products, new methods to quickly and appropriately determine their unique characteristics are needed. Technology platforms that are available today include polymerase chain reaction, flow cytometry, and rapid diagnostic methods useful in the detection of sterility failures. These and other technologies must be validated and shown to be superior or equivalent to existing techniques or compendial methods.

Quality Assurance Improvements

The manufacture of therapeutics for individual patients requires that QA programs be enhanced to improve safety of these unique products. The loss of lot or patient segregation must be prevented by judicious use of bar coding systems that ensure the correct identification of patients. For large scale manufacturing programs, the use of electronic batch records, manufacturing execution systems, and electronic labeling systems all have a major role in the ultimate success of these novel therapeutic modalities.

Improved and Less Invasive Cell Implantation Protocols

As tissue engineered products move through the developmental cycle, it is critical that orthopaedic surgeons who will use these products have an opportunity to provide input as to the product design and features. This input is needed to help improve the application of the product through minimally invasive implantation. The use of minimally invasive techniques is expected to be associated with improved rehabilitation times and more rapid return to an improved quality of life. In some clinical situations, arthroscopic delivery of tissue engineered products to enhance efficacy and reduce postoperative morbidity may be appropriate.

Summary

The successful manufacture of tissue engineered products for clinical use involves several challenges. Quality systems, including QA and quality control, must be developed and implemented to ensure that tissue engineered products are safe and effective. A particular challenge in the manufacture of individualized tissue engineered products is the inherent variability in the living cells or tissues used as the starting material. QA programs are needed to control each step of the tissue engineering process including biopsy acquisition, tissue processing, and mode of delivery to the patient. Maintaining the identity of samples and procedures to ensure the viability of cells throughout transit and processing are particularly important. QA systems must be performed under rigid time constraints to optimize viability and cell function. Challenges in quality control include the development and validation of assays to assess purity, potency, identity, sterility, yield, and viability of tissue engineered products. With ongoing studies of tissue engineering techniques, it is anticipated that quality systems will be optimized to support the use of these products in the clinical setting.

References

1. Brittberg M, Lindahl A, Nilsson A, et al: Treatment of deep cartilage defects in the knee with autologous chondrocyte transplantation. *N Engl J Med* 1994;331:889-895.
2. Schaeffer S, Stone B, Wolfrum JM, du Moulin GC: *Ex vivo* autologous cell therapies, in Sofer P, Zabriskie DW (eds): *Biopharmaceutical Process Development.* New York, NY, Marcel Dekker, 2000, pp 361-372.
3. Mayhew TA, Williams GR, Senica MA, Kuniholm G, du Moulin GC: Validation of a quality assurance program for autologous cultured chondrocyte implantation. *Tissue Eng* 1998;4:325-334.
4. Ostrovski S, du Moulin GC: Achieving compliance in validation and metrology for cell and gene therapy. *Biopharm* 1995;8:20-28.
5. Kielpinski G, Kehinde O, Kaplan BM, et al: Quality control for ex vivo cell therapy. *Biopharm* 1997;10:34-40.
6. Cell: Gene Therapy and Tissue Engineering Expert Committee: Cell and Gene Therapy Products, in *United States Pharmacopeia-National Formulary 26*, Rockville, MD, US Pharmacopeial Convention, 2003, pp 2257-2285.

Chapter 31

Functional Tissue Engineering of Cartilage: Scaffolds and Bioreactors

Gordana Vunjak-Novakovic, PhD

Abstract

Lost or damaged cartilage has minimal capacity for self repair and should ideally be replaced by an engineered graft that can help reestablish normal structure, cell signaling, and biomechanical function on a long-term basis. The clinical utility of engineered cartilage will likely depend on its capacity to provide regeneration, rather than repair. The ability to undergo orderly remodeling and predictably restore site-specific functional properties are features of engineered cartilage that are important for successful clinical use. Cells, biomaterial scaffolds, and biochemical and mechanical regulatory factors have been used in a variety of ways to generate cartilaginous tissues, either in vitro using bioreactors or in vivo after implantation. One approach involves the integrated use of: (1) chondrocytes or their precursors that can be selected, expanded, and transfected as required, (2) scaffolds that can serve as structural and logistic templates for tissue development and that biodegrade at a controlled rate, and (3) bioreactors that provide conditions necessary for the cells to assemble functional tissue matrix. This paradigm involves the in vitro generation of immature but functional tissue constructs that have the capacity to integrate, remodel, and mature in vivo following implantation. Scaffolds and bioreactors are discussed within the general context of functional tissue engineering of cartilage.

Cells, Scaffolds, and Bioreactors

Functional restoration of articular cartilage remains a challenge, and none of the existing treatment options gives a predicable and consistently good outcome.[1] Tissue engineering can potentially provide long-term restoration of lost or damaged cartilage. One approach to tissue engineering of functional cartilage involves the in vitro cultivation of cartilaginous constructs that have capacity to further develop following in vivo implantation. In this method, the constructs would integrate firmly and completely to the adjacent bone and cartilage and, ultimately, establish site- and scale-specific structural and mechanical properties. It is thought that these goals can be met if the engineered graft can mediate matrix remodeling in a manner similar to that of immature tissue. There are several issues to consider in the implementation of tissue engineered cartilage repair based on in vitro growth of immature, but functional, constructs and their subsequent remodeling and maturation in vivo.

In vitro engineering of cartilage involves the use of chondrogenic cells, biomaterial scaffolds, and bioreactors. Chondrogenic cells that might be used include chondrocytes or mesenchymal stem cells that are selected and/or expanded as required. These cells may also be transfected to express the genes of interest and seeded at high initial density. Biomaterial scaffolds are, ideally, highly porous structures that biodegrade at a known rate. Bioreactors are specialized environments and some have been designed to promote chondrogenesis. These components provide a system that can serve as a physiologically relevant model for controlled in vitro studies of tissue development and function. The study of engineered grafts with structural and functional properties similar to those of immature articular cartilage is of particular interest.

Cells used to engineer cartilage have varied with respect to donor age (embryonic, neonatal, immature, or adult), differentiation state, and method of preparation, including selection, expansion, and gene transfer. The cell type used is expected to affect scaffold structure, rate of degradation, and requirements for culture such as the need for medium supplements. The impact of the cell type on in vivo function of engineered cartilage, such as potential for integration, must also be considered.

Scaffolds are generally designed to provide a structural and logistic template for cell attachment and tissue development. Several scaffolds have been evaluated. These have differed with respect to composition, geometry, structure, mechanical properties, and rates of degradation.[2] Materials studied include collagen, hyaluronic acid, and synthetic polymers. Gels, fibrous meshes, and sponges have been used. Structures with differences in porosity, pore distribution, and connectivity have been tested. These materials also differed in mechanical properties, such as compressive stiffness and elasticity. Two representative, well characterized biomaterial scaffolds that were successfully used for cartilage tissue engineering are a fibrous polyglycolic acid mesh (Figure 1, *A*) and a benzylated hyaluronan sponge (Figure 1, *B*).

Tissue engineering bioreactors are designed to: (1) establish spatially uniform concentrations of attached cells within scaffolds of an appropriate size, (2) control conditions such as temperature, pH, osmolality, levels of oxygen, nutrients, metabolites, and regulatory molecules in culture medium, (3) facilitate mass transfer between the cells and the culture environment, and (4) provide physiologically relevant physical signals, such as interstitial fluid flow, shear, pressure, and mechanical compression.[3] Bioreactors commonly used for cartilage tissue engineering include spinner flasks,[4] (Figure 1, *C*), rotating vessels,[5] (Figure 1, *D*), and perfused chambers with flow of medium around[6] or through the constructs[7,8] (Figures 1, *E* and *F*).

Modulation of Chondrogenesis

Cells, biomaterial scaffolds, biochemical, and physical regulatory signals can be used in a variety of ways to engineer cartilage, in vitro and in vivo.

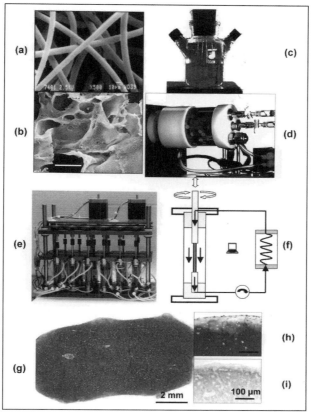

Figure 1 Tissue engineering of cartilage using scaffolds and bioreactors. Chondrogenic cells are seeded on scaffolds (**A** and **B**) and cultured in bioreactors (**C** through **F**) under conditions that promote functional assembly of the tissue matrix (**G** through **I**). Representative scaffolds include: a, fibrous polygycolic acid mesh and b, Hyaluronan sponge. Representative bioreactors include c, flasks with constructs fixed in place and cultured either statically or with magnetic stirring, d, rotating vessels with constructs freely suspended in dynamic laminar flow e, perfused vessels with mechanical stimulation in situ, and f, interstitial flow of culture medium. Six-week construct: (**G** and **H**) safranin O, (**I**) type II collagen. *(A and B are reproduced with permission from Pei M, Solchaga LA, Seidel J, et al: Bioreactors mediate the effectiveness of tissue engineering scaffolds. FASEB J 2002;16:1691-1694; C and D Copyright Sam Ogden; E is reproduced with permission from Altman GH, Stark P, Lu HH, et al: Advanced bioreactor with controlled application of multidimensional strain for tissue engineering. J Biomech Eng 2002;124:742-749; G is reproduced with permission from Freed LE, Hollander AP, Martin I, Barry JR, Langer R, Vunjak-Novakovic G: Chondrogenesis in a cell-polymer-bioreactor system. Exp Cell Res 1998;240:58-65; H and I are reproduced with permission from Martin I, Vunjak-Novakovic G, Yang J, Langer R, Freed LE: Mammalian chondrocytes expanded in the presence of fibroblast growth factor-2 maintain the ability to differentiate and regenerate three-dimensional cartilaginous tissue. Exp Cell Res 1999;253:681-688.)*

Recapitulating some aspects of the in vivo environment of tissue development may stimulate the cells to regenerate functional tissue structures. Creating an environment to regenerate tissues generally involves the presence of reparative cells, facilitated transport of biochemical substances, and the use of physical regulatory signals.

The presence and density of cells that are active in biosynthesis are likely to determine the capacity of an engineered graft for continued development, integration, and long-term survival following implantation in vivo. Cell source and cell density also largely determine the mechanical properties of engineered cartilage at the time of implantation, and thus affect ease of handling of the graft. Immature cartilage contains a high concentration of progenitor cells and a high cell-to-matrix ratio.[9] In contrast, mature cartilage has a low cell concentration and a limited capacity to heal. High initial cell densities in engineered constructs were critical for rapid accumulation and functional assembly of extracellular matrix components in differentiated chondrocytes[2,10] and for the induction of chondrogenesis in bone marrow-derived precursor cells.[11] The integration of engineered cartilage with native cartilage[12] and bone[13] also depended on the presence of cells capable of the progressive formation of cartilaginous tissue. Many surgical techniques for cartilage repair, including the Carticel procedure using autologous cultured chondrocytes (Genzyme Biosurgery, Cambridge, MA),[14] are designed to bring in new cells or facilitate access to progenitor cells. These techniques demonstrate that intrinsic cartilage repair can be enhanced by introducing cells and factors that are normally present in developing cartilage but not in mature tissue.

The structure of a scaffold determines the transport of nutrients, metabolites, and regulatory molecules to and from the cells. The mechanical properties of the scaffold determine the mechanotransduction at the cellular and tissue levels and thus affects its suitability for a particular application. The patterns of chondrogenesis are quite different for cells embedded in gels and those seeded onto meshes or sponges. Cells in gels are associated with formation of cell clusters that accumulated matrix but remained separated with matrix-free gel even after 6 weeks of culture.[15] Cells seeded onto meshes or sponges exhibited initiation of chondrogenesis in the high cell density region at the construct surface, with progressive development of a continuous cartilaginous matrix over their entire cross-sections during 6 weeks in culture.[10]

Tissue engineering would greatly benefit from the generation of new bioinductive scaffolds capable of delivering multiple growth factors,[16] interacting with fluids in bioreactors[17,18] and releasing growth factors in response to mechanical loading.[19] This suggests the exciting possibility that a biomaterial scaffold can provide control points in directing cell responses both by its structural features and by covalent coupling of regulatory molecules.

The hydrodynamic factors of a cell culture can modulate chondrogenesis in at least two ways: (1) by associated effects on mass transport of biochemical factors between the developing tissue and culture medium (eg, oxygen, nutrients, growth factors), and (2) by direct physical stimulation of the cells (eg, shear, pressure). In vivo, mass transfer within articular cartilage involves diffusion in conjunction with fluid flow that accompanies tissue loading and unloading. In vitro, mass transfer has been shown to determine the size and composition of engineered constructs and native cartilage explants.[20] The composition, morphology, and mechanical properties of engineered cartilage grown in mechanically active environments were generally better than those formed in static environments, presumably because of enhanced mass transport to and from the cells.[3] Chondrogenesis was markedly improved in constructs cultured in the dynamic laminar flow of rotating bioreactors[10,21] and in constructs perfused with culture medium at physiologic interstitial flow velocities.[7,22,23] In native cartilage, the same effects are achieved by loading-induced interstitial flow of fluid. Physical stimuli conveyed by the flow of culture medium, although different in nature and intensity from physical signals associated with joint loading, can thus be used to promote in vitro synthesis and functional assembly of cartilaginous matrix.

Functional Properties of Engineered Cartilage

After 4 to 6 weeks of culture, constructs engineered using bovine chondrocytes were typically 5 to 6 mm thick and uniformly cartilaginous (Figures 1, *G* and *I*). Constructs contained 75% as much glycosaminoglycan (GAG) and 40% as much total collagen (hydroxyproline) per unit wet weight as normal cartilage. These constructs had normal tissue morphology except for the lack of zonal organization and cell columnarization.[10,21] The collagen network had normal fibril density and diameter, and normal fractions of collagen types II, IX, and X. However, the number of pyridinium cross-links per collagen molecule was only one third that of normal tissue.[24] The zero-strain compressive modulus was in the range from 0.1 to 0.5 MPa, depending on the type of the scaffold and if specific growth factors were used to supplement the culture medium or were transfected to the cells.[17,21,25,26] For comparison, the modulus was 270 kPa for the articular surface[27] and 710 or 950 kPa for the deep zone of bovine cartilage.[5,27]

The compositions and mechanical properties of constructs were generally in the range of those for fetal cartilage. These results suggest that bioreactors can yield engineered constructs resembling immature rather than mature native cartilage, even after prolonged culture. It is likely that the functional deficiencies of cartilage engineered in vitro are attributable to the absence of specific biochemical and physical factors that are normally present in vivo. The application of mechanical loading that was physiologic in both the type and magnitude of strain enhanced the in vitro assembly of functional tissue matrix.[15] A similar result was reported for human ligaments engineered using mesenchymal stem cells.[28] Following implantation, engineered cartilage remodeled into physiologically stiff cartilage and new subchondral bone.[13,29] Notably, columnar cells and a tidemark at an appropriate depth were observed in engineered cartilage following in vivo, but not in vitro, exposure to physiologic loading.[13]

The optimal duration of in vitro cultivation, or even the need for in vitro cultivation, has not been determined. Prolonged (7-month) cultivation in a bioreactor resulted in normal fraction of GAG, aggregate modulus, and

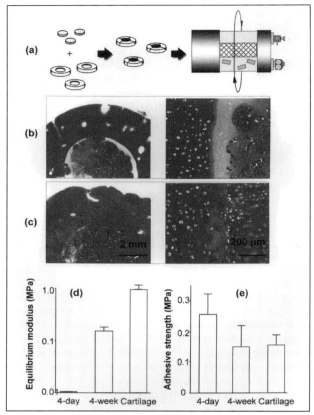

Figure 2 Bioreactor studies of integration. **A,** Model system: engineered constructs (5 mm diameter × 2 mm thick disks) were sutured into native cartilage (10/5 mm diameter × 2 mm thick rings) and the resulting composites were cultured in bioreactors for 1 to 8 weeks. **B** and **C,** Integration interface: 2 weeks of culture (**B**), 4 weeks of culture (**C**). Stain: safranin O/fast green; disk is in the center and on the right. **D** and **E,** Functional evaluation: Confined-compression equilibrium modulus of the engineered cartilage (**D**), and adhesive strength of the integration interface (**E**).

hydraulic permeability, but the collagen fraction and dynamic stiffness remained subnormal.[5] Controlled bioreactor studies demonstrated a trade-off between the stiffness of engineered cartilage and its integration potential.[12] Disk-shaped constructs or cartilage explants were sutured into cartilage rings made of intact or trypsin-treated cartilage and cultured for 1 to 8 weeks. Evaluations of structure and function included measurements of compressive stiffness of the central disk and adhesive strength of the integration interface (Figure 2, *A*). Immature constructs integrated better than either more mature constructs or cartilage explants. The integration of immature constructs involved cell proliferation and the progressive formation of cartilaginous tissue (Figures 2, *B* and *C*). In contrast, the integration of more

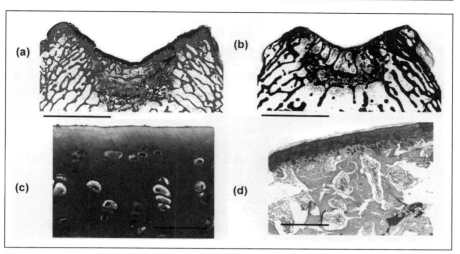

Figure 3 Osteochondral repair in rabbit. Composites made of engineered cartilage and osteoconductive support were press-fit into 7 mm × 5 mm × 5 mm osteochondral defects in rabbit knee joints. 6-month explants: **A,** Alcian blue, **B,** von Kossa, **C,** Toluidine blue, and **D,** type II collagen. Scale bars: (**A** and **B**) 5 mm, (**C**) 50 μm, (**D**) 200 μm. *(Reproduced with permission from Schaefer D, Martin I, Jundt G, et al: Tissue engineered composites for the repair of large osteochondral defects.* Arthritis Rheum *2002;46:2524-2534.)*

mature constructs or native cartilage involved only the secretion of extracellular matrix components. Compressive stiffness improved from immature to mature constructs and cartilage explants (Figure 2, *D*). However, the adhesive stiffness at the disk-ring interface was markedly greater for immature constructs than for either mature constructs or cartilage explants (Figure 2, *E*). Ideally, the duration of in vitro cultivation should ensure that constructs have both adequate stiffness and capacity for integration.

Large osteochondral defects in adult rabbits were repaired using composites based on in vitro engineered cartilage.[13] Articular chondrocytes were expanded, seeded on polyglycolic acid scaffolds in bioreactors, cultured for 4 to 6 weeks, sutured to an osteoconductive support, and implanted. Histologic, biochemical, and biomechanical (indentation) evaluations were performed. After 6 months, defects implanted with engineered constructs were repaired with cartilage that had normal thickness and characteristic architectural features, including the tidemark and columnar arrangement of chondrocytes. Repair tissue also had new subchondral bone (Figures 3, *A* through *D*) and Young's moduli comparable to that of native articular cartilage. Integration with subchondral bone was excellent, but integration with cartilage was inconsistent. Thus functional engineered cartilage can provide a mechanically stable tissue-like template that remodeled in vivo into osteochondral tissue with physiologically thick and stiff cartilage.

Summary and Research Needs

The primary functions of orthopaedic tissues are biomechanical and the main goal of all orthopaedic tissue engineering is the restoration of normal biomechanical function. Ideally, lost or damaged cartilage should be replaced by an engineered graft that can reestablish normal structure, composition, cell signaling, and biomechanical function of the damaged or diseased tissue. In light of this paradigm, the clinical utility of tissue engineering will likely depend on the capacity of engineered grafts to replicate the site-specific properties of the native tissue in defects of different sizes. The engineered graphs should also provide regeneration, rather than repair, and undergo orderly remodeling in response to environmental factors.[30] These goals can be met by using new cells capable of chondrogenesis and facilitating transport of nutrients, metabolites, and regulatory signals.[31] The approaches currently in use or being evaluated involve filling the defect with cells alone[14] or with a cell seeded scaffold[32] with the implantation of functional tissue constructs engineered in vitro.[13]

In general, engineered tissues have histologic and biochemical characteristics similar to those of native tissue but have failed to achieve normal mechanical properties. The current school of thought is that the restoration of native tissue structure is indispensable for a good long-term clinical outcome. The transition from immature to mature cartilage is believed to be necessary for the tissue to assume its full physiologic function.[33] The architecture of the repair tissue (eg, columnar cells, tidemark at an appropriate depth), integration with adjacent cartilage and bone, and the ability to withstand physiologic loading are perhaps more important indicators of successful repair than the reestablishment of exact biochemical properties.[30]

The success criteria, animal models, and evaluation methods for tissue engineered cartilage repair still need to be established. An evolving discipline called functional tissue engineering is focused on the role of biomechanical factors in tissue regeneration, with an overall goal to develop rational design principles that can be used to guide orthopaedic tissue engineering.[33] Some of the key principles include definition of the functional properties of native tissues for normal and pathologic situations at various developmental stages. Other important goals include the selection and prioritization of the requirements for engineered constructs and tissue regeneration, and the establishment of standards to evaluate engineered constructs and tissue repair.[33]

The current research needs include: (1) quantitative studies of tissue development (eg, mechanotransduction in native and engineered tissues, mathematical models of tissue development, structure-function correlations, coordinated in vitro and in vivo studies), (2) new technologies (eg, minimally invasive testing/monitoring methods, custom-designed scaffolds with incorporated genes and regulatory molecules, bioreactors with mechanical stimulation in situ, improved methods for autologous cell sourcing), and (3) criteria for functional tissue engineering.

References

1. Buckwalter JA, Mankin HJ: Articular cartilage repair and transplantation. *Arthritis Rheum* 1998;41:1331-1342.

2. Freed LE, Vunjak-Novakovic G: Tissue engineering of cartilage, in Bronzino JD (ed): *The Biomedical Engineering Handbook,* ed 2. Boca Raton, FL, CRC Press, 2000, pp 124-1-124-26.

3. Freed LE, Vunjak-Novakovic G: Tissue engineering bioreactors, in Lanza RP, Langer R, Vacanti J (eds): *Principles of Tissue Engineering,* ed 2. San Diego, CA, Academic Press, 2000, pp 143-156.

4. Vunjak-Novakovic G, Freed LE, Biron RJ, Langer R: Effects of mixing on the composition and morphology of tissue-engineered cartilage. *AIChE J* 1996;42:850-860.

5. Freed LE, Langer R, Martin I, Pellis N, Vunjak-Novakovic G: Tissue engineering of cartilage in space. *Proc Natl Acad Sci USA* 1997;94:13885-13890.

6. Mizuno S, Allemann F, Glowacki J: Effects of medium perfusion on matrix production by bovine chondrocytes in three-dimensional collagen sponges. *J Biomed Mater Res* 2001;56:368-375.

7. Pazzano D, Mercier KA, Moran JM, et al: Comparison of chondrogensis in static and perfused bioreactor culture. *Biotechnol Prog* 2000;16:893-896.

8. Davisson T, Sah RL, Ratcliffe A: Perfusion increases cell content and matrix synthesis in chondrocyte three-dimensional cultures. *Tissue Eng* 2002;8:807-816.

9. Jackson DW, Simon TM: Tissue engineering principles in orthopaedic surgery. *Clin Orthop* 1999;367:S31-S45.

10. Freed LE, Hollander AP, Martin I, Barry JR, Langer R, Vunjak-Novakovic G: Chondrogenesis in a cell-polymer-bioreactor system. *Exp Cell Res* 1998;240:58-65.

11. Butnariu-Ephrat M, Robinson D, Mendes DG, Halperin N, Nevo Z: Resurfacing of goat articular cartilage from chondrocytes derived from bone marrow. *Clin Orthop* 1996;330:234-243.

12. Obradovic B, Martin I, Padera RF, Treppo S, Freed LE, Vunjak-Novakovic G: Integration of engineered cartilage. *J Orthop Res* 2001;19:1089-1097.

13. Schaefer D, Martin I, Jundt G, et al: Tissue engineered composites for the repair of large osteochondral defects. *Arthritis Rheum* 2002;46:2524-2534.

14. Brittberg M, Lindahl A, Nilsson A, Ohlsson C, Isaksson O, Peterson L: Treatment of deep cartilage defects in the knee with autologous chondrocyte transplantation. *N Engl J Med* 1994;331:889-895.

15. Buschmann MD, Gluzband YA, Grodzinsky AJ, Kimura JH, Hunziker EB: Chondrocytes in agarose culture synthesize a mechanically functional extracellular matrix. *J Orthop Res* 1992;10:745-758.

16. Richardson TP, Peters MC, Ennett AB, Mooney DJ: Polymeric system for dual growth factor delivery. *Nat Biotechnol* 2001;19:1029-1034.

17. Pei M, Solchaga LA, Seidel J, et al: Bioreactors mediate the effectiveness of tissue engineering scaffolds. *FASEB J* 2002;16:1691-1694.

18. Gooch KJ, Blunk T, Courter DL, et al: IGF-I and mechanical environment interact to modulate engineered cartilage development. *Biochem Bioph Res Co* 2001;286:909-915.

19. Lee KY, Peters MC, Anderson KW, Mooney DJ: Controlled growth factor release from synthetic extracellular matrices. *Nature* 2000;408:998-1000.

20. Vunjak-Novakovic G, Obradovic B, Madry H, Altman G, Kaplan D: Bioreactors for orthopaedic tissue engineering, in Caplan AI, Goldberg V (eds): *Orthopaedic Tissue Engineering: Basic Science and Practice.* New York, NY, Marcel Dekker Inc, in press.

21. Vunjak-Novakovic G, Martin I, Obradovic B, et al: Bioreactor cultivation conditions modulate the composition and mechanical properties of tissue engineered cartilage. *J Orthop Res* 1999;17:130-138.

22. Sittinger M, Bujia J, Minuth WW, Hammer C, Burmester GR: Engineering of cartilage tissue using bioresorbable polymer carriers in perfusion culture. *Biomaterials* 1994;15:451-456.

23. Dunkelman NS, Zimber MP, Lebaron RG, Pavelec R, Kwan M, Purchio AF: Cartilage production by rabbit articular chondrocytes on polyglycolic acid scaffolds in a closed bioreactor system. *Biotechnol Bioeng* 1995;46:299-305.

24. Riesle J, Hollander AP, Langer R, Freed LE, Vunjak-Novakovic G: Collagen in tissue-engineered cartilage: Types, structure and crosslinks. *J Cell Biochem* 1998;71:313-327.

25. Pei M, Seidel J, Vunjak-Novakovic G, Freed LE: Growth factors for sequential cellular de- and re-differentiation in tissue engineering. *Biochem Bioph Res Co* 2002;294:149-154.

26. Madry H, Padera R, Seidel J, et al: Gene transfer of a human insulin-like growth factor I cDNA enhances tissue engineering of cartilage. *Hum Gene Ther* 2002;13:1621-1630.

27. Chen AC, Bae WC, Schinagl RM, Sah RL: Depth- and strain-dependent mechanical and electromechanical properties of full-thickness bovine articular cartilage in confined compression. *J Biomech* 2001;34:1-12.

28. Altman G, Horan R, Martin I, et al: Cell differentiation by mechanical stress. *FASEB J* 2002;16:270-272.

29. Wakitani S, Goto T, Pineda SJ, et al: Mesenchymal cell-based repair of large, full-thickness defects of articular cartilage. *J Bone Joint Surg Am* 1994;76:579-592.

30. Vunjak-Novakovic G, Goldstein SA: Biomechanical principles of cartilage and bone tissue engineering, in Mow VC, Huiskes R (eds): *Basic Orthopaedic Biomechanics and Mechanobiology,* ed 3. Lippincott-Williams and Wilkins, in press.

31. Newman AP: Articular cartilage repair. *Am J Sports Med* 1998;26:309-324.

32. Nixon AJ, Fortier LA, Williams J, Mohammed H: Enhanced repair of extensive articular defects by insulin-like growth factor-I-laden fibrin composites. *J Orthop Res* 1999;17:475-487.

33. Guilak F, Butler DL, Goldstein SA: Functional tissue engineering: The role of biomechanics in articular cartilage repair. *Clin Orthop* 2001;391:S295-S305.

Chapter 32

Injectable Polymers and Hydrogels for Orthopaedic and Dental Applications

Esmaiel Jabbari, PhD
Lichun Lu, PhD
Bradford L. Currier, MD
Antonios G. Mikos, PhD
Michael J. Yaszemski, MD, PhD

Abstract

The use of injectable polymers and hydrogels that cross-link and harden in situ represent a treatment option for bone defects. This approach is minimally invasive with less tissue dissection than surgical treatments of similar orthopaedic and dental applications. Important factors to consider for in situ polymerization are the biocompatibility of the initiator system and the rise in temperature that occurs during cross-linking. An ideal synthetic injectable substitute should be osteoconductive and allow for neovascularization, so that cells from the surrounding tissue can migrate into and proliferate on the injected scaffold. Although preliminary results with naturally derived polymeric materials are promising, the low mechanical properties and limited availability of these compounds has prompted investigations of the use of synthetic biodegradable and injectable polymers and hydrogels. The copolymerization of fumaric acid and propylene glycol leads to the formation of polypropylene fumarate (PPF). In PPF, the unsaturated carbon-carbon double bond of the fumarate group is used in cross-linking of the polymer. Advantages of PPF are that it is injectable and hardens in situ by a free radical cross-linking reaction of the unsaturated carbon-carbon fumarate double bond in response to a suitable initiator and accelerator. In an attempt to introduce cross-linking groups in copolymers with thermoreversible properties, block copolymers of PPF and polyethylene glycol (PEG) have been synthesized for orthopaedic and dental applications. We developed an alternate PEG derivative, oligopolyethylene glycol fumarate (OPF), as an injectable, in situ cross-linkable, and biodegradable hydrogel for dental applications.

Introduction

Tissue engineering strategies for orthopaedic and dental applications involve the use of highly porous and interconnected scaffolds with acceptable mechanical properties to facilitate the adhesion, spreading, migration, proliferation, and differentiation of appropriate cell types. The desired cell type depends on whether the tissue under consideration is bone, cartilage, tendon, ligament, or meniscus.

Clinical needs are the driving force for the development of injectable polymeric systems in orthopaedic and dental applications. The use of injectable systems may help to minimize tissue dissection and retraction in certain clinical situations. Injectable polymers and hydrogels could be valuable in numerous orthopaedic surgery applications. These include treatment of distal radius fractures, vertebroplasty, kyphoplasty, minimally invasive posterior spinal fusion, unicameral bone cysts, thoracoscopic anterior interbody grafting after disk excision in scoliosis, focal cartilage defects, and alveolar ridge augmentation.[1] Polymers or hydrogels used clinically can be thermoplastics or thermosets. Thermoplastics maintain the ability to soften upon heating. Thermosets cross-link during curing and no longer have this softening ability when heated. The polymers can be processed into preformed scaffolds or they can be injectable.[2]

Autografts, allografts, and synthetic materials are currently being used in a variety of orthopaedic surgery applications. Autograft tissue is in limited supply, and morbidity at the harvest site is a concern. Allografts have a small but potentially serious risk of disease transfer and immunogenic response from the host tissue. Stress shielding and particulate wear are potential concerns with the use of nondegradable polymers such as polymethyl methacrylate (PMMA), for filling defects or as a bone cement in orthopaedic practice.[3] Synthetic biodegradable and biocompatible polymers and hydrogels are an attractive option for orthopaedic and dental applications because of unlimited supply, minimum risk of disease transfer, and the potential to design the mechanical properties for each application by modifying the processing of the polymer. In addition, the rate of polymer degradation can be designed to coincide with the rate of tissue regeneration for a given application.

One approach to bone replacement involves the use of prefabricated scaffolds for cell transplantation. Highly porous and interconnected scaffolds have been fabricated to promote three-dimensional tissue growth, nutrient diffusion, matrix production, and vascularization. These prefabricated scaffolds present a large surface area for cell growth and its porosity reduces the barriers to diffusion of essential compounds.

An alternate approach involves the use of injectable and in situ cross-linkable biodegradable polymers or hydrogels. Rigid scaffolds must be implanted surgically, but the use of injectable scaffolds that harden in situ can reduce the invasiveness of the implantation procedure. Moreover, injectable scaffolds are more suitable for treating irregularly shaped defects in applications such as vertebroplasty, unicameral bone cysts, and focal cartilage defects.

An ideal synthetic injectable polymer or hydrogel for orthopaedic and dental applications should be biocompatible, have desired mechanical properties, and degrade in a controlled manner concurrent with tissue regeneration. In addition, it should degrade into biocompatible molecules, have a suitable working time and rheological properties for injection via syringe, and harden in situ to attain the desired mechanical properties. The resulting scaffold must have the ability to become highly porous and interconnected. An ideal synthetic injectable substitute must be osteoconductive and allow for neovascularization so that cells from the surrounding tissue can migrate to and proliferate on it. Important factors for in situ polymerization include the biocompatibility of the initiator system and the temperature rise during cross-linking.

Naturally Occurring Polymers as Injectable Scaffolds and Hydrogels

Naturally based injectable hydrogels include chitosan, which exhibits a reversible liquid-gel transition around pH 7, alginate, which forms a gel ionically when reacted with di- or tri-valent counter ions, and hyaluronic acid. Hyaluronic acid hydrogels form a thick gel by shear-thinning upon removal of the shearing force.[4] Natural injectable polymers for hard tissue applications are based mainly on hydroxyapaptite.[5] Although preliminary results with naturally derived polymeric materials are promising, their poor mechanical properties and limited availability have prompted evaluations of the use of synthetic biodegradable and injectable polymers and hydrogels.

Synthetic Injectable Polymers

Synthetic biodegradable polymers used in orthopaedic practice and dentistry include poly-α-hydroxy esters, other polyesters including PPF, poly-ε-caprolactone, polyphosphazenes, polyanhydrides, polyacrylamides, and functionalized polyethylene oxides.

Poly-α-hydroxy esters, specifically polylactic acid, polyglycolic acid, and their copolymers (PLGA) have a long history of use as synthetic biodegradable surgical sutures. They are also used for plates and screws for fracture fixation devices in low-load bearing applications, and as artificial scaffolds for cell transplantation. Some PLGA copolymers are biocompatible and marketed for certain clinical indications. In general, PLGA polymers undergo bulk degradation over a period of weeks to years leading to a sudden decrease of mechanical properties without a change in the overall dimensions of the polymer. Rigid scaffolds or foams of PLGA copolymers can be made by casting from an organic solvent, such as methylene chloride, and used as a bone graft or as a plug for osteochondral defects.[6] Because organic solvents are used for casting, it is difficult to use PLGA polymers in injectable applications in the clinical setting.

Polyphosphazenes (PPHOS) are a unique class of polymers that contain alternating nitrogen and phosphorous atoms in their backbone structure. The polymer backbone does not contain carbon atoms. These polymers are versatile in that side groups can be added after synthesis. For example, changing the amount of a hydrolytically unstable side group can alter the rate of degradation. For several blends of PPHOS/PLGA, a near zero-order degradation rate is observed. Poly-ethyl glycinato-methoxyphenoxy-phosphazene is a good substrate for osteoblast-like cell attachment and growth. Functional side groups have been added both throughout the three-dimensional structure of polyphosphazines as well as on the polymer surface.[7]

Polyanhydrides are another important class of biodegradable polymers with well defined degradation characteristics that can be controlled by varying the ratio of hydrophobic to hydrophilic monomers such as 1,3-bis-p-carboxyphenoxy-propane and sebacic acid.[8] Polyanhydrides degrade by surface erosion, unlike polyesters, which degrade uniformly throughout their three-dimensional structure, leading to a sudden increase in acid production in the late stages of degradation. The use of polyanhydrides in orthopaedic and dental applications has been limited by their poor mechanical strength. Photoinitiators in the visible wavelength range have been investigated for in situ cross-linking of polyanhydrides.

Polypropylene Fumarate

Anhydride monomers can be copolymerized with aliphatic diols in an attempt to facilitate polymer processing, maintain good compressive properties, and to improve solubility of highly crystalline polyanhydrides, such as polyfumaric acid. The copolymerization of fumaric acid and propylene glycol leads to the formation of PPF in which the unsaturated carbon-carbon double bond of the fumarate group is used in cross-linking of the polymer.

One application of this novel polymer is as a biodegradable scaffold for treatment of bone defects. After cross-linking, PPF is mechanically suitable for use as a temporary trabecular bone replacement.[9] The antibiotics gentamicin and vancomycin have been incorporated in a PPF-based composite without any loss of mechanical properties.[10] Scaffolds of PPF can be synthesized by transesterification of diethyl fumarate and propylene glycol or by condensation of propylene glycol and fumaric acid. With these methods, polymer chains with average molecular weights ranging from 700 to 5,000 Daltons (Da) can be produced. The chains of greatest molecular weight are attained by transesterification of the oligomer.[11]

We have investigated the degradation behavior, biocompatibility, and compressive properties of a PPF and β-tricalcium phosphate (PPF/β-TCP) composite.[12] When the PPF/β-TCP composite was implanted subcutaneously for up to 12 weeks, a mature fibrous capsule encased the implants.[13] The osteoconductivity of the composite scaffold was assessed in a rat segmental defect model. Significant bone ingrowth was evident for both prepolymerized and in situ polymerized samples 5 weeks after implantation.[14] The compressive strength and compressive modulus of the PPF/β-TCP composite range from 2-10 MPa and 60-250 MPa, respectively. Without β-TCP, cross-linked PPF maintains mechanical properties similar to those of trabecular bone through 7 weeks of degradation. An initial increase in strength and modulus was observed with in vitro degradation of PPF/β-TCP composites during a 12-week period. When implanted subcutaneously in rats, the rates of degradation of the composites in vitro followed similar trends, with initial increases in both strength and modulus.

An advantage of PPF is that it is injectable. It also hardens in situ by a free radical cross-linking reaction of the fumarate unsaturated carbon-carbon double bond in the presence of a suitable initiator and accelerator. After injection, the in situ cross-linking reaction of PPF can be initiated in a manner identical to the free-radical polymerization of PMMA bone cement that is used clinically. Components of this injectable and in situ cross-linkable system as a scaffold for bone defects include N-vinyl pyrrolidinone (NVP) as the cross-linker, benzoyl peroxide as the initiator, and N,N-dimethyl-p-toluidine as the accelerator. This initiation system has been successfully used for cross-linking PPF with a temperature rise during the cross-linking reaction that has not exceeded 45°C.[15]

Incorporation of the NVP monomer into the cross-linked network is very important to consider because NVP monomer, like methyl methacrylate monomer, can be toxic. Incorporation depends on the polymer/cross-linker ratio and is independent of the accelerator concentration.[16] For equal weights of PPF and VP, 50% of the cross-linker was incorporated into the cross-linked network. A disadvantage of NVP is that this cross-linker is not degradable, which affects the degradation profile of the PPF/NVP injectable system. Therefore, use of degradable cross-linkers and lower concentrations of cross-linker, but maintaining a high degree of cross-linking, would improve the degradation profile of a PPF based system for bone defects in load-bearing applications.

Synthetic Injectable Hydrogels

Injectable hydrogel systems have been developed for filling cartilage defects and as a carrier for transplanted cells. Promising results have been obtained with PEG that has been modified with methacrylate end groups. This polymer can be injected and then cross-linked with ultraviolet light to form a hydrogel in situ. Chondrocytes embedded in these hydrogels survived the cross-linking process, proliferated, and produced cartilage-like extracellular matrix.[17] Specimens removed from athymic mice at 2, 4, and 7 weeks after implantation showed cartilage formation with increasing glycosoaminoglycan (GAG) and collagen content.[18] Constructs up to 8 mm thick have been produced using the methacrylate functionalized PEG. The chondrocytes embedded in the constructs remained viable at the center of these polymeric scaffolds at least for the first 2 weeks of in vitro culture. Over 4 to 6 weeks of in vitro culture, similar constructs showed production of GAG and type II collagen throughout the scaffold.[19] However, no evidence of degradation has been reported over the course of several weeks in vitro or in vivo for methacrylate functionalized PEG, which could limit the extent of new cartilage tissue formation in these gels.

OPF

An alternate PEG derivative, OPF, has been developed as an injectable, in situ cross-linkable, and biodegradable hydrogel for dental applications.[20,21] Synthesized by condensation polymerization of PEG with fumaryl chloride,

OPF is an oligomer of PEG with fumarate moieties. The fumarate groups in this macromer allow for cross-linking in situ as well as degradation via hydrolysis.[22] Like injectable PPF, a chemical initiation system consisting of ammonium persulfate and ascorbic acid is used to form hydrogels without the need for ultraviolet light.[23] The attachment of marrow stromal cells on an OPF hydrogel has been investigated with a model cell adhesion specific peptide.[24] The model peptide Gly-Arg-Gly-Asp (GRGD) was incorporated into the OPF hydrogel after being coupled to acrylated PEG of molecular weight 3,400 Da.[25]

Hydrogels with a wide variety of physical properties can be synthesized by altering the PEG chain length of OPF, the cross-link density, or the initial peptide concentration. As the peptide concentration increased, the attachment of marrow stromal cells to OPF hydrogels with PEG molecular weights of 930 and 2,860 Da increased. However, the number of attached marrow stromal cells to an OPF hydrogel of PEG molecular weight 6,090 Da remained constant regardless of the peptide density. The length of the PEG chain in OPF also influenced the degree of cell attachment. For example, when 1 mmol peptide/g of OPF hydrogel was incorporated into the OPF, the degree of cell attachment relative to initial seeding density was $93.9 \pm 5.9\%$, $64.7 \pm 8.2\%$, and $9.3 \pm 6.6\%$ for OPF with PEG chains of 930 Da, 2,860 Da, and 6,090 Da, respectively. On the other hand, the cross-linking density of the OPF hydrogel did not significantly affect cell attachment. The interaction was sequence specific because attachment of marrow stromal cells to the GRGD modified hydrogel was competitively inhibited when cells were incubated in the presence of soluble GRGD before cell seeding. These results indicate that altering the peptide concentration can modulate cell attachment to the OPF hydrogel.

The OPF oligomer has also been cross-linked with N,N'-methylene bisacrylamide to fabricate injectable scaffolds that cross-link in situ as a cell carrier for mesenchymal stem cells (MSC).[26] This system is potentially useful for the treatment of osteochondral defects. A novel combination of redox initiators consisting of ammonium pesulfate and N,N,N',N'-tetramethylethylenediamine was used in this system to obtain neutral pH values. In this injectable system, MSC were successfully seeded. The encapsulated MSC cultured in complete osteogenic media exhibited alkaline phosphatase activity and an increasing amount of mineralized matrix for up to 21 days.

Thermoreversible Amphiphilic Copolymers

Copolymers consisting of PEG and poly-α-hyroxy esters possess lower critical solution temperatures (LCST) and sol-gel transition temperatures that are dependent on PEG block length.[27] These copolymers undergo gelation in water upon reaching their LCST. Another class of biodegradable copolymers, polyorganophosphazenes, that incorporate methoxy-polyethylene glycol (mPEG), also have a LCST.[28] The LCST of polyorganophosphazene-based copolymers decreases with increasing sodium chloride concentration. These thermoreversible copolymers do not undergo additional chemical cross-linking.

In an attempt to introduce cross-linking groups in copolymers with thermoreversible properties, block copolymers of PPF and mPEG have been synthesized in our laboratory.[29] Copolymers with mPEG molecular weights of 570 and 800 Da have LCST of 40°C and 55°C, respectively, at a concentration range of 5% to 25% (by weight) in phosphate buffered saline. Salt concentration of 10% by weight lowered the LCST of these copolymers by 20°C to 30°C depending on the molecular weight of mPEG. Marrow-derived osteoblasts were cultured on the PPF-b-mPEG thermoreversible hydrogels modified in bulk with a covalently linked Arg-Gly-Asp-Ser (RGDS) peptide in the concentration range of 10 to 1,000 nmol/mL.[30] A concentration of 100 nmol/mL RGDS was sufficient to promote adhesion of marrow-derived osteoblasts. Moreover, marrow-stromal osteoblast migration was highest at 100 nmol/mL linked RGDS concentration.

The surface concentration of a ligand such as RGDS relative to its bulk concentration in the hydrogel was quantified with an enzyme linked immunosorbent assay, using mouse monoclonal anti-biotin antibody, horseradish peroxidase-conjugated antimouse immunoglobulin G, and a chemiluminescent substrate.[31] Results indicate that the active ligand concentration on the surface of the hydrogel depends on hydrogel composition and the relative magnitude of the PEG spacer arm of the ligand compared with the PEG block length of the copolymer. The results suggest that steric hindrance caused by mobile PEG chains of the copolymer contribute to the decreased ligand surface concentration.

In situ cross-linkable biodegradable macroporous scaffolds based on PPF-b-mPEG thermoreversible hydrogels were fabricated by coupled free radical and pore formation reactions.[32,33] Cross-linking was initiated by redox reaction of ammonium persulfate and ascorbic acid. Pores were formed by the reaction between ascorbic acid and sodium bicarbonate with the evolution of carbon dioxide. Hydrogels with porosity up to 84% by volume were obtained with this method. Depending on the concentration of the initiator, median pore size ranged from 50 to 200 mm, which is suitable as an injectable scaffold for dental applications.

The MSC seeded with a collagen gel onto RGD-modified macroporous hydrogels showed a significant increase in cell numbers after 28 days of culture, and significant calcium deposition was apparent after 28 days. Confocal microscopy revealed that MSC suspended in a collagen gel and cultured on RGD-modified hydrogels adhered to the surface of the hydrogel. However, MSC suspended in a collagen gel and cultured on unmodified hydrogels were located within the pores of and not in direct contact with the hydrogel surface.[31] These results demonstrate that biomimetic in situ cross-linkable hydrogels facilitate the adhesion of MSC onto the polymer. In addition, they support the differentiation of MSC to the osteoblastic phenotype in the presence of osteogenic culture media.

Summary

Bone defects may be treated with the use of injectable polymers and hydrogels that cross-link and harden in situ. An advantage of the injectable approach will be its minimally invasive nature for those clinical scenarios where it is a viable treatment option. Although both natural and synthetic polymers can be used, the poor mechanical properties and limited availability of natural compounds has prompted evaluations of the use of synthetic polymers and hydrogels. An ideal synthetic injectable substitute should allow for neovascularization so that cells from the surrounding tissue can migrate into and proliferate on the injected scaffold. The biocompatibility of the initiator system and the rise in temperature that occurs during in situ cross-linking are important considerations. A PEG derivative, OPF, is an injectable, in situ cross-linkable, and biodegradable hydrogel for dental applications. Hydrogels with a wide variety of physical properties can be synthesized by altering the PEG chain length of OPF, the cross-link density, or the initial peptide concentration. Initial results have been promising, although much additional work is needed to bring these technologies to the clinical setting.

References

1. Yaszemski MJ, Oldham JB, Lu L, Currier BL: Clinical needs for bone tissue engineering technology, in Davies JE (ed): *Bone Engineering*. Toronto, Canada, 2000, pp 541-547.

2. Hutmacher DW: Scaffolds in tissue engineering bone and cartilage. *Biomaterials* 2000;21:2529-2543.

3. Bostrom RD, Mikos AG: Tissue engineering of bone, in Atala A, Mooney D (eds): *Synthetic Biodegradable Polymer Scaffolds*. Boston, MA, Birkhauser, 1997, pp 215-234.

4. Gutowska A, Jeong B, Jasionowski M: Injectable gels for tissue engineering. *Anat Rec* 2001;263:342-349.

5. Temenoff JS, Mikos AG: Injectable biodegradable materials for orthopedic tissue engineering. *Biomaterials* 2000;21:2405-2412.

6. Thomson RC, Shung AK, Yaszemski MJ, Mikos AG: Polymer scaffold processing, in Lanza RP, Langer R (eds): *Principles of Tissue Engineering*. Austin, TX, Lands, 2000, pp 251-262.

7. Allcock HR: The synthesis of functional polyphosphazenes and their surfaces. *Applied Organomettalic Chemistry* 1998;12:659-666.

8. Domb AJ, Langer R: Solid-state and solution stability of poly(anhydrides) and poly(esters). *Macromolecules* 1989;22:2117-2122.

9. Gresser JD, Trantolo DJ, Nagaoka H, et al: Bone cement Part 1, in Wise DL, Trantolo DJ, Altobelli DE, Yaszemski MJ, Gresser JD (eds): *Human Biomaterials Applications*. Totowa, Humana Press, 1996, pp 169-185.

10. Gerhart TN, Roux RD, Horowitz G, Miller RL, Hanff P, Hayes WC: Antibiotic release from an experimental biodegradable bone cement. *J Orthop Res* 1988;6:585-592.

11. Peter SJ, Suggs LJ, Yaszemski MJ, Engel PS, Mikos AG: Synthesis of poly(propylene fumarate) by acylation of propylene glycol in the presence of a proton scavenger. *J Biomater Sci Polym* 1999;10:363-373.

12. Peter SJ, Nolley JA, Widmar MW, et al: In vitro degradation of a poly(propylene fumarte)/b-tricalcium phosphate composite orthopaedic scaffold. *Tissue Eng* 1997;3:207-215.

13. Peter SJ, Miller ST, Zhu G, Yasko AW, Mikos AG: In vivo degradation of a poly(propylene fumarate)/b-tricalcium phosphate injectable composite scaffold. *J Biomed Mater Res* 1998;41:1-7.

14. Yaszemski MJ, PayneRG, Hayes WC, Langer R, Aufdemorte TB, Mikos AG: The ingrowth of new bone tissue and initial mechanical properties of a degrading polymeric composite scaffold. *Tissue Eng* 1995;1:41-52.

15. Peter SJ, Kim P, Yasko AW, Yaszemski MJ, Mikos AG: Crosslinking characteristics of an injectable poly(propylene fumarate)/b-tricalcium phosphate paste and mechanical properties of the crosslinked composite for use as a biodegradable bone cement. *J Biomed Mater Res* 1999;44:314-321.

16. Gresser JD, Hsu S, Nagaoka H, et al: Analysis of a vinyl pyrrolidone/poly(propylene fumarate) resorbable bone cement. *J Biomed Mater Res* 1995;29:1241-1247.

17. Bryant SJ, Anseth KS: Hydrogel properties influence ECM production by chondrocytes photoencapsulated in poly(ethylene glycol) hydrogels. *J Biomed Mater Res* 2001;59:63-72.

18. Elisseeff J, Anseth KS, Simms D, McIntosh W, Randolph M, Langer R: Transdermal photopolymerization for minimally invasive implantation. *Proc Natl Acad Sci USA* 1999;96:3104-3107.

19. Bryant SJ, Anseth KS: The effect of scaffold thickness on tissue engineered cartilage in photocrosslinked poly(ethylene oxide) hydrogels. *Biomaterials* 2001;22:619-626.

20. Jo S, Engel PS, Mikos AG: Synthesis of poly(ethylene glycol)-tethered poly(propylene fumarate) and its modification with GRGD peptide. *Polymer* 2000;41:7595-7604.

21. Suggs LJ, Payne R, Yaszemski MJ, Alemany L, Mikos AG: Synthesis and characterization of a block copolymer consisting of poly(propylene fumarate) and poly(ethylene glycol). *Macromolecules* 1997;30:4318-4323.

22. Jo S, Shin H, Shung AK, Fisher JP, Mikos AG: Synthesis and characterization of oligo(poly(ethylene glycol) fumarate) macromer. *Macromolecules* 2001;34:2839-2844.

23. Temenoff JS, Athanasiou KA, LeBaron RG, Mikos AG: Effect of poly(ethylene glycol) molecular weight on tensile and swelling properties of oligo(poly(ethylene glycol) fumarate) hydrogels for cartilage tissue engineering. *J Biomed Mater Res* 2001;59:429-437.

24. Shin H, Jo S, Mikos AG: Modulation of marrow stromal osteoblast adhesion on biomimetic oligo[poly(ethylene glycol) fumarate] hydrogels modified with Arg-Gly-Asp peptides and a poly(ethyleneglycol) spacer. *J Biomed Mater Res* 2002;61:169-179.

25. Jo S, Shin H, Mikos AG: Modification of oligo(poly(ethylene glycol) fumarate) macromer with a GRGD peptide for the preparation of functionalized polymer networks. *Biomacromolecules* 2001;2:255-261.

26. Jabbari E, Behravesh E, Mikos AG: Development of a biodegradable redox initiated oligo(poly(ethylene glycol) fumarate) based hydrogel as an injectable in situ crosslinkable cell carrier. Proceedings of the Annual meeting of American Institute of Chemical Engineers, November 3-8, Indianapolis, IN, USA, 2002.

27. Jeong B, Bae YH, Kim SW: In situ gelation of PEG-PLGA-PEG triblock copolymer aqueous solutions and degradation thereof. *J Biomed Mater Res* 2000;50:171-177.

28. Lee SB, Song SC, Jin JI, Sohn YS: A new class of biodegradable thermosensitive polymers. 2: Hydrolytic properties and salt effect on the lower critical solution temperature of poly(organophosphazenes) with methoxypoly(ethylene glycol) and amino acid esters as side groups. *Macromoleulces* 1999;32:7820-7827.

29. Behravseh E, Shung AK, Jo S, Mikos AG: Synthesis and characterization of triblock copolymers of methoxy poly(ethylene glycol) and poly(propylene fumarate). *Biomacromolecules* 2002;3:153-158.

30. Behravesh E, Zygourakis K, Mikos AG: Adhesion and migration of marrow-derived osteoblasts on injectable in situ crosslinkable poly(propylene fumarate-co-ethylene glycol) based hydrogels with a covalently linked RGDS peptide. *J Biomed Mater Res* 2003;65A:260-270.

31. Behravesh E, PhD: Thesis. Rice University, 2002.

32. Behravesh E, Jo S, Zygourakis K, Mikos AG: Synthesis of in situ crosslinkable macroporous biodegradable poly(propylene fumarate-co-ethylene glycol) hydrogels. *Biomacromolecules* 2002;3:374-381.

33. Behravesh E, Timmer MD, Lemoine JJ, Liebschner MAK, Mikos AG: Evaluation of the in vitro degradation of macroporous hydrogels using gravimetry, confined compression testing, and microcomputed tomography. *Biomacromolecules* 2002;3:1263-1270.

Chapter 33

Collagen Uses in Current and Developing Orthopaedic Therapies

Dave Toman, PhD

Abstract

Purified collagen has been widely used in medical applications for several years and purified type I collagen is the most commonly used type of collagen. Although bovine collagen is readily available, human collagen is preferred as it carries a reduced risk of hypersensitivity reactions. In orthopaedic practice, collagen is used as a scaffold for cell infiltration and repair, and as a delivery device for growth factors, cells, and DNA or viral vectors. Products of interest for orthopaedic applications include collagen sponges, pastes, fibrillar matrices, and gels. Combining other components with collagen offers the potential for matrices with enhanced properties or that interact with factors to provide optimal effect at the surgical site. A collagen matrix can thus contribute to a milieu that provides key signals for cell differentiation. Several products using collagen are currently in various stages of development. It is anticipated that collagen-based products that become available during the next 5 years will have an important impact on the therapeutic options in orthopaedic practice.

Introduction

The advances in biology over the past 20 years have greatly enhanced understanding of connective tissue composition, cell-cell and cell-matrix interaction and signaling, tissue and organ development, and tissue response and repair.[1] Multiple biologic activators and mediators with the potential to aid in the repair or augmentation of several tissues have been described. As these factors have been isolated and tested in therapeutic applications in preclinical models, it became evident that an appropriate matrix is a key element in the success of new treatments. Ideally, a matrix should retain the biologic activators at the site of action for a specified duration and provide a milieu for cell attraction, growth, and differentiation. Although previously the focus has been on the biologic factors, the carrier matrix may play an equal or more important role in determining the ultimate success of tissue repair.

For complex repairs of bone, cartilage, or tendon, the orthopaedic surgeon has relied on the use of additional material obtained either from autografts or allografts. The use of autografts requires that patients have a suitable site for harvest of sufficient tissue. Harvest of autografts represents an additional surgical procedure and is associated with patient morbidity that can be significant. Allograft tissue products may be variable in performance because of storage conditions and treatment to ensure removal of infectious agents. Instead of autografts or allografts, pure and highly characterized products that have been rigorously tested in clinical trials could be used.

Collagen is an abundant natural protein found throughout the body. Purified collagen has been widely used in medical applications for several years. Thus this biomaterial has a long history of biocompatibility, safety, and low immunogenicity in humans.[2] Although collagen is best known for its properties of structural support, it also plays a key role in cell signaling and binding of growth factors, cytokines, and other structural proteins. There are more than 20 different types of collagen,[3] but purified type I collagen is the predominant collagen used in medicine. In orthopaedic applications, collagen is used as a scaffold for cell infiltration and repair, and as a delivery device for growth factors, cells, and DNA or viral vectors. Collagen is used in several applications in orthopaedic practice, but some uses of collagen that are under development follow principles that might be applied to orthopaedic medicine and other areas of medicine.

Role of Collagen in Current and Future Orthopaedic Applications

Collagen provides a scaffold for cell infiltration and remodeling of tissue. Collagraft (Zimmer, Warsaw, IN, and NeuColl, Campbell, CA), a collagen matrix combined with hydroxyapatite and tricalcium phosphate, demonstrates osteoconductive and osteoinductive properties.[4] Clinical studies of Collagraft show that it elicits repair equivalent to that of autologous bone for acute long bone fractures, osseous defects, and bone voids when combined with bone marrow and rigid fixation (Figure 1). Healos, a sponge-like material (DePuy Spine, Raynham, MA), is composed of cross-linked collagen fibers coated with hydroxyapatite in an open pore matrix. Healos has osteoconductive properties and is marketed in the United States for filling bony voids or gaps of the skeletal system. Both of these products can be easily cut and shaped using a surgical scalpel. Currently, Healos is in clinical trials with the BAK Interbody Fusion System (Zimmer, Warsaw, IN) for spinal fusion. Collagen Meniscus Implant (Regen Biologics, Franklin Lakes, NJ), a resorbable collagen scaffold for the repair of a meniscus defect, has been tested in a multicenter clinical trial. It is expected that a New Drug Application will be filed with the Food and Drug Administration (FDA) in late 2004. A highly processed porcine tissue patch (Restore, DePuy, Raynham, MA), composed of greater than 90% collagen, has recently been approved for use in rotator cuff repair, and its use in the repair of medial collateral ligaments is being explored.[5]

A collagen scaffold can be used as a drug delivery system. In 2002, the FDA granted approval for marketing of InFuse Bone Graft/LT Cage Tapered Fusion Device (Medtronic Sofamor Danek, Minneapolis, MN) containing

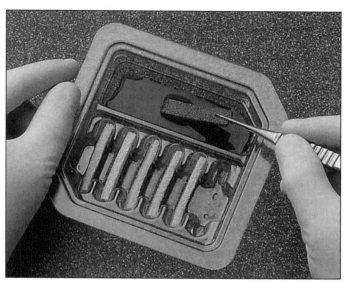

Figure 1 Collagraft is rehydrated for 1 to 3 minutes, then typically coated with autologous bone marrow prior to implantation. The strips may be used as is or molded into the desired shape to fill the bony defect as completely as possible. (*Courtesy of NeuColl.*)

human recombinant bone morphogenetic protein-2 (rhBMP-2) delivered in an Absorbable Collagen Sponge (ACS) (Integra LifeSciences, Plainsboro, NJ) to enhance bone growth. This approval for marketing was based on clinical studies that showed identical fusion rates of the product and autologous bone graft[6] (Figure 2). The collagen sponge holds the rhBMP-2 at the surgical site and provides a scaffold for bone fusion. The rhBMP-2/ACS has been tested for treatment of open tibia shaft fractures stabilized by intramedullary nails with results demonstrating accelerated fracture healing and reduction in the frequency of secondary interventions. A panel review of the FDA supported rhBMP-2/ACS (Wyeth, Madison, NJ) for this indication. The rhBMP-2/ACS product has been approved and is available in Europe (InductOs). Another recombinant osteoinductive protein, rhOP-1 (also called BMP-7, Stryker Biotech, Hopkinton, MA), has been on the market since 2001. This product is delivered to the surgical site in a collagen paste as an alternative to autograft in recalcitrant long bone nonunions in which autograft is unfeasible and alternative treatments have failed. An experimental product known as MP-52 growth factor (also called BMP-14 or rhGDF-5) (DePuy, Raynham, MA), delivered in a Healos matrix, is undergoing evaluation in clinical trials in Europe. Transforming growth factor β-3 (TGF-β3) delivered in a fibrillar collagen matrix to keep the TGF-β3 localized inhibited cranial suture closure in young rats compared with animals used as controls.[7] Anti-TGF-β2 antibody delivered in a collagen gel caused

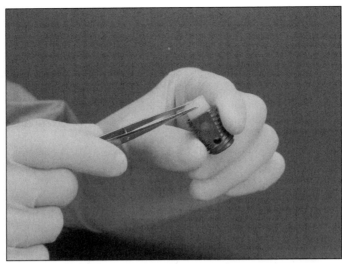

Figure 2 To use InFuse Bone Graft, surgeons reconstitute the rhBMP-2 powder with supplied sterile water and then apply it to collagen sponges. The sponges are inserted inside each of two LT Cage Lumbar Tapered Fusion Devices, which are then implanted between the vertebrae. (*Courtesy of Medtronic, Inc.*)

a reduction in the percent bridging of posterior frontal sutures compared with controls in a calvarial organ culture.[8] These methods, including collagen sponges, pastes, fibrillar matrices, and gels, should be applicable to the delivery of other growth factors to specific tissue sites, including bone and cartilage sites. Growth factors of interest include other members of the TGF-β family such as other BMPs, insulin-like growth factor-1 (IGF-1), IGF-2, inhibins, and growth and differentiation factors[9] such as low density lipoprotein-related protein-5.[10]

Other components can be combined with collagen to generate novel matrices with enhanced properties or that interact with factors to provide optimal effect at the surgical site. In preclinical studies of rhBMP-2 for posterolateral fusion, the collagen sponge required reinforcement with hydroxyapatite and tricalcium phosphate to create a compression-resistant matrix similar in composition to Collagraft.[6] When TGF-β2 was bound to fibrillar type I collagen through a difunctional polyethylene glycol linkage, a stronger and longer fibroblastic response was observed after subcutaneous implantation in rats compared with an admixed formulation of TGF-β2 and collagen.[11] Fibroblast growth factor-2 (FGF-2) loaded into a collagen-heparin sulfate matrix showed a threefold higher binding capacity of FGF-2 to the matrix and a more sustained release in vitro and enhanced angiogenesis in a rat model compared with a collagen matrix without heparin sulfate.[12] A novel matrix, combining collagen with DNA, was used to deliver BMP and induce bone formation in rats.[13]

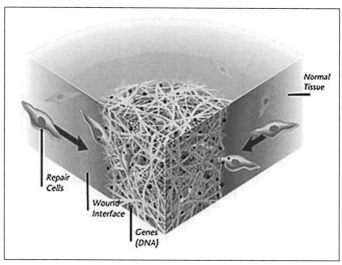

Figure 3 A Gene Activated Matrix (GAM), consisting of DNA and a structural matrix carrier, at a wound site. Cells surrounding the GAM proliferate and migrate into the GAM and thus are targeted for gene transfer and expression. While in the matrix, transfected cells act as local in situ bioreactors, producing DNA-encoded proteins that stimulate tissue repair. (*Courtesy of Selective Genetics.*)

Collagen can encapsulate and deliver cells to the surgical site. Trials of transplanted chondrocytes in a collagen matrix to repair an osteochondral defect in the knee joint showed elimination of knee locking and reduced pain and swelling in 35 patients after 1 year. These patients had a hyaline-like smooth and firm surface at second look arthroscopy.[14] A honeycomb structure of collagen, using either type I, II, or III collagen, can be used to generate a scaffold for cell culture for future studies for bone and cartilage repair.[15]

A collagen matrix can contribute to a milieu that provides key signals for cell differentiation. Mesenchymal stem cells (MSC) grown in a collagen matrix with the addition of BMP-4 transform into a chondrocyte phenotype (Osiris Therapeutics, Baltimore, MD).[16] When seeded in a collagen gel, MSC generated a composite that achieved 50% to 60% of the stiffness and strength of normal tendons after 4 weeks in a rabbit model of an Achilles tendon defect (Osiris Therapeutics, Baltimore, MD).[17] In an exciting study, muscle cells embedded in a collagen gel showed promise in the treatment of full-thickness articular cartilage defects in a rabbit model.[18] This result may indicate that cells isolated from a heterologous tissue site and placed in an appropriately engineered matrix may provide an alternative source of cells that are difficult to obtain.

Collagen can be used as a matrix to deliver plasmid DNA or viral vectors for gene therapy applications. Genes are delivered in a collagen matrix in applications for repair of tendon, ligament, bone, muscle, and other tissues (Selective Genetics, San Diego, CA)[19] (Figure 3). Collagen provides the scaffold for cell ingrowth. The cells become transfected with the gene and then express the encoded protein factor for a longer duration than with delivery of DNA alone. Preclinical trials of the treatment of tibial fractures by delivery of a gene for human parathyroid hormone (PTH) in a collagen sponge showed filling of a 1-cm and 1.6-cm defect with new bone in 8 to 18 weeks in beagles.[20] Delivery of gene(s) for PTH alone or in combination with those for FGF-2 in a collagen matrix are being tested for lumbar fusion as well as genes for platelet-derived growth factor and FGF-2 for more rapid healing of tendons. A separate study showed that the use of collagen as a delivery vehicle prolonged tissue expression of naked DNA and adenovirus to 40 days and 65 days, respectively, in an animal model.[21] Subcutaneous administration of collagen sponges delivering reporter genes complexed to polymers have shown longer persistence of gene expression in rats compared with administration of DNA-collagen sponges.[22] Preliminary in vitro experiments demonstrated that chondrocytes seeded into a collagen-glycosaminoglycan matrix containing plasmid DNA can be transfected with a reporter gene.[23]

Collagen Sources

Most medical-grade collagens originate from bovine, porcine, or equine sources, with bovine hide and tendons as the primary sources. Human collagen is available from several companies, but quantities are generally limited. Commercially available medical-grade collagen is produced under strictly controlled manufacturing environments and the raw starting material must be from verified sources. Some products have additional safeguards such as isolation from closed, certified herds in the United States and the use of processing treatments that inactivate infectious agents, including prions.[24]

An estimated 2% to 3% of the population has evidence of a localized hypersensitivity to bovine collagen based on use in aesthetic applications,[25] but human collagen has been shown not to elicit a hypersensitivity reaction in these patients (D Toman, PhD, unpublished data). Several companies are developing additional sources of collagen, focusing on human collagen isolated from more controlled sources. Inamed Corporation (Santa Barbara, CA) has initiated marketing of human collagen products isolated from cultured foreskin fibroblasts. Two other companies are developing recombinant sources of human collagen. Cohesion Technologies (Palo Alto, CA) has developed human type I collagen in both the milk of closed herd-transgenic cattle and in yeast.[26] Fibrogen (South San Francisco, CA) has developed recombinant type I, type II, and type III collagens in yeast.[27] Although no significant adverse events have occurred with the use of bovine collagen in orthopaedic applications, the use of human collagen will help reduce potential risks of hypersensitivity. Consideration of hypersensitivity reactions is becoming more important as increasingly sophisticated tissue engineering applications are used in orthopaedic practice.

Summary and Conclusions

Collagen has proved to be valuable in the use of several approved medical devices. Several other products based on collagen are being evaluated in late-stage preclinical and/or clinical trials. Orthopaedic surgeons should witness the approval of multiple products to aid in bone, cartilage, and tendon healing within the next 5 years. The development of medical products for protein, cell, and DNA delivery requires interdependent processes to formulate the active therapeutic agent and the carrier matrix. It is anticipated that collagen will play an important role as a component of these carrier matrices.

References

1. Streuli C: Extracellular matrix remodelling and cellular differentiation. *Curr Opin Cell Biol.* 1999;11:634-640.

2. Keefe J, Wauk L, Chu S, DeLustro F: Clinical use of injectable bovine collagen: A decade of experience. *Clin Mater* 1992;9:155-162.

3. Gelse K, Poschl E, Aigner T: Collagen. *Adv Drug Delivery Rev* 2003;55:1531-1546.

4. Alvis M, Block JE, Lalor P, et al: Osteoinduction by a collagen mineral composite combined with isologous bone marrow in a subcutaneous rat model. *Orthopedics* 2003;26:77-80.

5. Musahl V, Abramowitch SD, Gilbert TW, et al: The use of porcine small intestinal submucosa to enhance the healing of the medial collateral ligament: A functional tissue engineering study in rabbits. *J Orthop Res* 2004;22:214-220.

6. McKay B, Sandhu HS: Use of recombinant human bone morphogenic protein-2 in spinal fusion applications. *Spine* 2002;27:566-585.

7. Opperman LA, Moursi AM, Sayne JR, Wintergerst AM: Transforming growth factor-beta 3 (TGF-b3) in a collagen gel delays fusion of the rat posterior interfrontal suture in vivo. *Anat Rec* 2002;267:120-130.

8. Moursi AM, Winnard PL, Fryer D, Mooney MP: Delivery of TGF-b2-perturbing antibody in a collagen vehicle inhibits cranial suture fusion in calvarial organ culture. *Cleft Palate Craniofac J* 2003;40:225-232.

9. Yoon ST, Boden SD: Osteoinductive molecules in orthopaedics: Basic science and preclinical studies. *Clin Orthop* 2002;395:33-43.

10. Patel MS, Karsenty G: Regulation of bone formation and vision by LRP5. *N Engl J Med* 2002;346:1572-1574.

11. Bentz H, Schroeder JA, Estridge TD: Improved local delivery of TGF-b2 by binding to injectabel fibrillar collagen via difunctional polyethylene glycol. *J Biomed Mater Res* 1998;39:539-548.

12. Pieper JS, van Wachem PB, van Luyn MJA, Brouwer LA, Veerkamp JH, van Kuppevelt TH: Loading of collagen-heparin sulfate matrices with bFGF promotes angiogenesis and tissue generation in rats. *J Biomed Mater Res* 2002;62:185-194.

13. Murata M, Arisue M, Sato D, Sasaki T, Shibata T, Kuboki Y: Bone induction in subcutaneous tissue in rats by a newly developed DNA-coated atelocollagen and bone morphogenic protein. *Br J Oral Maxillofac Surg* 2002;40:131-135.

14. Ochi M, Uchio Y, Tobita M, Kuriwaka M: Current concepts in tissue engineering technique for repair of cartilage defect. *Artif Organs* 2001;25:172-179.

15. Itoh H, Aso Y, Furuse M, Noishiki Y, Miyata TA: Honeycomb collagen carrier for cell culture as a tissue engineered scaffold. *Artif Organs* 2001;25:213-217.

16. US patent # 5,908,784. In Vitro Chondrogenic Induction of Human Mesenchymal Stem Cells, issued 1999.

17. Butler DL, Awad HA: Perspectives on cell and collagen composites for tendon repair. *Clin Orthop* 1999;367:S324-S332.

18. Adachi N, Sato K, Usas A, et al: Muscle derived, cell based ex vivo gene therapy for treatment of full-thickness articular cartilage defects. *J Rheumatol* 2002;29:1920-1930.

19. Bonadio J: Tissue engineering via local gene delivery. *J Mol Med* 2000;78:303-311.

20. Bonadio J, Smiley E, Path P, Goldstein S: Localized, direct plasmid gene delivery *in vivo*: Prolonged therapy results in reproducible tissue regeneration. *Nat Med* 1999;5:753-759.

21. Ochiya T, Nagahara S, Sano A, Itoh H, Terada M: Biomaterials for gene delivery: Atelocollagen-mediated controlled release of molecular medicines. *Curr Gene Ther* 2001;1:31-52.

22. Scherer F, Schillinger U, Putz U, Stemberger A, Plank C: Nonviral vector loaded collagen sponges for sustained gene delivery in vitro and in vivo. *J Gene Med* 2002;4:634-643.

23. Samuel RE, Lee CR, Chivizzani SC, Evans CH, Yannas IV, Olsen BR: Spector M: Delivery of plasmid DNA to articular chondrocytes via novel collagen-glycosaminoglycan matrices. *Hum Gene Ther* 2002;13:791-802.

24. US patent # 5,616,689. Method of Controlling Structure Stability of Collagen Fibers Produced from Solutions or Dispersions Treated with *Sodium Hydroxide* for Infectious Agent Deactivation, issued 1997.

25. DeLustro F, Smith ST, Sundsmo J, Salem G, Kincaid S, Ellingsworth L: Reaction to injectable collagen: Results in animal models and clinical use. *Plast Reconstr Surg* 1987;79:581.

26. Olsen DR, Leigh SD, Chang R, et al: Production of human type I collagen in yeast reveals unexpected new insights into the molecular assembly of collagen trimers. *J Biol Chem* 2001;276:24038-24043.

27. Myllyharju J, Nokelainen M, Vuorela A, Kivirikko KI: Expression of recombinant human type I-III collagens in the yeast pichia pastoris. *Biochem Soc Trans* 2000;28:353-357.

Chapter 34

Self-Assembling Peptide Hydrogel Scaffolds for Cartilage Tissue Engineering

John D. Kisiday, PhD
Shuguang Zhang, PhD
Christina Cosman, MS
Sanaz Saatchi, MS
Alan J. Grodzinsky, ScD

Abstract

The development and design of scaffolds for use in tissue engineering applications requires careful consideration. We devised a novel self-assembling peptide hydrogel scaffold for cartilage repair and developed a method to encapsulate chondrocytes within the peptide hydrogel. The ability to achieve appropriate composition and ultrastructure of cartilage-like extracellular matrix using three-dimensional encapsulated cells can be influenced by the application of mechanical loading during culture and choice of medium. It was observed that specific dynamic compressive loading regimes increased chondrocyte biosynthesis and increased retention of newly synthesized molecules compared with free-swelling controls. The insulin, transferrin, and selenium medium supplement (ITS) with 0.2% fetal bovine serum (FBS) was effective for stimulation of both primary chondrocyte division and chondrocytic extracellular matrix synthesis.

Introduction

Emerging medical technologies for effective and lasting repair of articular cartilage include the delivery of cells or cell-seeded scaffolds to a defect site to initiate de novo tissue regeneration. A major challenge in choosing an appropriate scaffold for cartilage repair is the identification of a material that can simultaneously stimulate high rates of cell division and high rates of cell synthesis of phenotypically specific extracellular matrix (ECM) macromolecules until repair evolves into steady state tissue maintenance. We devised a novel self-assembling peptide hydrogel scaffold for cartilage repair and developed a method to encapsulate chondrocytes within the peptide hydrogel.[1,2] Certain peptides are able to self-assemble into stable hydrogels at low (0.1% to 1%) peptide concentrations. In addition, the self-assembling nature of the peptide hydrogel and the flexibility of molecular design may offer advantages in controlling scaffold degradation, cell attachment, and the delivery of tethered stimulatory growth factors to the microenvironment of encapsulated cells. We hypothesize that a self-assembling peptide hydrogel would provide an appropriate environment for the retention of the chondrocyte phenotype and for the synthesis of a mechanically functional cartilage ECM. In addition, dynamic mechanical compression applied during in vitro culture is postulated to stimulate increased deposition of ECM and further improve biomechanical properties.

Self-Assembling Peptide Scaffolds for Tissue Engineering

A new class of peptide-based biomaterials has been actively pursued as a molecular engineered scaffold for tissue repair. Certain peptides are able to self-assemble into stable hydrogels at low (0.1% to 1%) peptide concentrations.[3-5] Such self-assembling peptides are characterized by amino acids sequences of alternating hydrophobic and hydrophilic side groups. Sequences of charged amino acid residues can have alternating positive and negative charges. Self-assembling peptides form stable β-sheet structures when dissolved in deionized water. Exposure to electrolyte solution initiates β-sheet assembly into interweaving nanofibers.[3-5] Such self-assembly occurs rapidly when the ionic strength of the peptide solution exceeds a certain threshold or when the pH is such that the net charge of the peptide molecules is near zero.[6] Intermediate steps of self-assembly have been investigated by observing relatively slow nanofiber formation and subsequent network assembly in deionized water, without triggering rapid self-assembly by the addition of electrolytes.[7] The self-assembling peptide hydrogel has unique features suitable for a tissue engineering scaffold. The nanofiber structure is almost three orders of magnitude smaller than that of most polymer microfibers. This peptide hydrogel presents a unique polymer structure with which cells may interact. The peptide sequences may be designed for specific cell-matrix interactions that influence cell differentiation and tissue formation.[8] The synthetic nature of the peptide minimizes the risk of delivery of pathogens that may be associated with materials derived from animals or humans.[8] Although more studies are needed, there are some evaluations of the immunogenicity of peptide. The injection of two peptides (EAKA$_4$ and RADA$_4$) into leg muscles of Fisher 344 rats resulted in no detectable toxic reaction after 9 days and 5 weeks, respectively.[5]

Casting and Culture of Chondrocyte Seeded Peptide Gels

The peptide KLD-12 was dissolved in 295 mM sucrose solution to maintain physiologic osmotic pressure.[1] Chondrocytes isolated from the femoropatellar grooves of newborn bovine calves were resuspended in sucrose solution. The peptide solution was added to the chondrocytes to obtain a final peptide concentration of 0.4 to 0.5% (w/v) and a cell density of 15×10^6 cells/mL. The suspension was injected into a stainless steel casting frame, which was

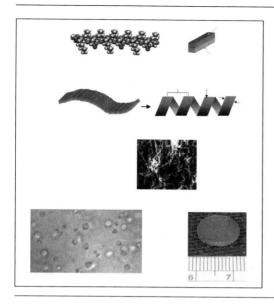

Figure 1 Depiction of self-assembly of KLD-12 peptide, which can take place in the presence of cells.[1,2]

then placed in PBS to initiate self-assembly of the peptide gel into a slab structure. After 25 min, the seeded peptide gel was transferred to a Petri dish for long-term culture (details shown in Figure 1).

Development of Cartilage-like Tissue in Unloaded Gel

Constructs were cultured in Dulbecco's Minimum Essential Media (DMEM) + 10% FBS or 1% ITS + 0.2% FBS. During 4 weeks of culture in vitro, chondrocytes retained their morphology and developed a cartilage-like, mechanically functional ECM that was rich in proteoglycans and type II collagen. These findings indicated that a stable chondrocyte phenotype was maintained.[1] Time dependent accumulation of this ECM was paralleled by increases in equilibrium modulus and dynamic stiffness up to about 25% of the values observed for native calf tissue. The content of viable chondrocytes within the peptide gel during the first 7 days of culture increased at a rate that was 2.5-fold greater than that in a parallel chondrocyte-seeded agarose gel, which is a well-defined reference chondrocyte culture system. By 4 weeks of culture, 93% of the newly synthesized 35S-labeled macromolecules and 77% of the newly synthesized 3H-labeled macromolecules were retained in the scaffold, and approximately 70% of the 3H-proline was incorporated into collagen. These values are similar to those observed in native calf cartilage explants.[9]

Effects of Loading During in vitro Growth

Additional studies identified dynamic compressive loading regimes that increased chondrocyte biosynthesis and increased retention of newly syn-

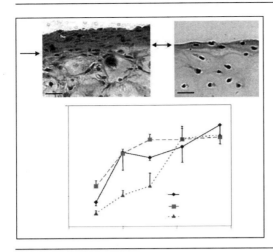

Figure 2 Influence of DMEM supplemented with 10% FBS versus ITS supplemented with 0.2% FBS on (A,B) outgrowth of dedifferentiated cells, and (C) glycosaminoglycans synthesis.

thesized molecules compared with free-swelling controls. Alternate day loading for four periods of 0.75 hours on/5.25 hours off at 1 Hz caused a significant increase in proteoglycan synthesis, total glycosaminoglycan accumulation, and static and dynamic mechanical stiffness. These results demonstrate the potential of dynamic compression to differentially accelerate accumulation of extracellular matrix components in cartilage tissue engineered constructs during in vitro culture.[10]

Choice of Medium

ITS medium with 0.2% FBS was effective for stimulation of both primary chondrocyte division and chondrocytic extracellular matrix synthesis[11] (Figure 2). In addition, surface dedifferentiation was approximately proportional to serum concentrations, with significantly less surface dedifferentiation, proliferation, and fibrous tissue deposition in ITS plus 0.2% FBS medium relative to cultures in 10% FBS (Figure 2). Therefore, this ITS media is advantageous over medium with greater concentrations of serum in that fibrous tissue formation was reduced while maintaining high levels of biosynthesis. This study also illustrated that minimal concentrations of serum are sufficient to enhance chondrocyte biosynthesis and cell division in ITS medium. Specific components of serum may be combined with ITS or other media to further enhance biosynthesis. Thus a defined medium based on ITS with added components of serum may support optimal chondrocyte activities for both cartilage tissue engineering and basic research science applications.

Summary and Conclusions

Taken together, these results demonstrate the potential of a self-assembling peptide hydrogel as a scaffold for the synthesis and accumulation of a cartilage-like ECM in a three-dimensional cell culture for cartilage tissue repair. The peptide KLD-12 used in this study represents one of a class of specially designed self-assembling peptides made through molecular engineering that can be tailored to the specific cell and tissue application of interest. In general, the ability to achieve appropriate composition and ultrastructure of cartilage-like extracellular matrix using three-dimensional encapsulated cells may be critically dependent on application of mechanical loading during culture and choice of medium.

Acknowledgments

This work was supported by NIH Grant AR33236, the DuPont-MIT Alliance, and the Cambridge MIT Institute.

References

1. Kisiday J, Jin M, Kurz B, et al: Self-assembling peptide hydrogel fosters chondrocyte extracellular matrix production and cell division: Implications for cartilage tissue repair. *PNAS* 2002;99:9996-10001.

2. Kisiday JD, Jin M, Kurz B, et al: Cartilage tissue engineering using a new self-assembling peptide gel scaffold, in Hascall VC, Kuettner KE (eds): *The Many Faces of Osteoarthritis*. Birkhauser Verlag, Basel, 2002, pp 423-428.

3. Zhang S, Holmes T, Lockshin C, Rich A: Spontaneous assembly of a self-complementary oligopeptide to form a stable macroscopic membrane. *Proc Natl Acad Sci USA* 1993;90:3334-3338..

4. Zhang S, Holmes TC, DiPersio CM, Hynes RO, Su X, Rich A: Self-complementary oligopeptide matrices support mammalian cell attachment. *Biomaterials* 1995;16:1385-1393.

5. Holmes TC, de Lacalle S, Su X, Liu G, Rich A, Zhang S: Extensive neurite outgrowth and active synapse formation on self-assembling peptide scaffolds. *Proc Natl Acad Sci USA* 2000;97:6728-6833.

6. Caplan MR, Moore PN, Zhang S, Kamm RD, Lauffenburger DA: Self-assembly of a beta-sheet protein governed by relief of electrostatic repulsion relative to van der Waals attraction. *Biomacromolecules* 2000;1:627-631.

7. Marini DM, Hwang W, Lauffenburger DA, Zhang S, Kamm RD: *Nanoletters* 2002;2:259-295.

8. Holmes TC: Novel peptide-based biomaterial scaffolds for tissue engineering. *Trends Biotechnol* 2002;20:16-21.

9. Sah RL, Kim YJ, Doong JY, Grodzinsky AJ, Plaas AHK, Sandy JD: Biosynthetic response of cartilage explants to dynamic compression. *J Orthop Res* 1989;7:619-636.

10. Kisiday JD, Jin M, Grodzinsky AJ: Effects of dynamic compressive loading duty cycle on in vitro conditioning of chondrocyte-seeded peptide and agarose scaffolds. *Trans Orthop Res Soc* 2002.

11. Kisiday JD, Semino C, Zhang S, Grodzinsky AJ: Evaluation of ITS-supplemented medium for application to 3-D encapsulated chondrocyte culture. *Trans Orthop Res Soc* 2002.

Chapter 35

Design and Optimization of Scaffolds for Cartilage Tissue Engineering

Lawrence J. Bonassar, PhD

Abstract

Scaffolds are an important component of tissue engineered products to repair cartilage. There are many requirements of scaffolds that will be implanted as they must provide support for cells and help to regulate the in vivo environment of the tissue engineered material. Many different types of materials have been evaluated for use in scaffolds, including fibers, foams or sponges, hydrogels, and combinations of materials. Currently, there is no single material or approach to scaffold design that meets all the requirements for a cartilage tissue engineering implant. It is anticipated that innovations in scaffold design and development will address these unmet needs. Technologic advances in scaffold processing are also important to enable scale-up of manufacturing processes.

Introduction

The use of biomaterials as scaffolds for cells is thought to be critical in cartilage tissue engineering. Scaffolds may play several key roles in generation of new tissue. These roles include providing initial mechanical support for cells, particularly in loaded environments, providing organizational cues for assembly of extracellular matrix, delivery of proteins or genetic material to regulate cell behavior, and regulating the biologic and chemical environment of implants.

Given the range of tasks that scaffolds are required to perform, it is not surprising that a wide variety of materials have been used for cartilage tissue engineering. To date, no less than 24 materials or combinations of materials have been used for delivery of chondrocytes or stem cells in attempts to regenerate cartilage (Table 1). These scaffolds fall into one of four categories: (1) fibers, (2) foams or sponges, (3) hydrogels, and (4) composite combinations of other types. Each of these classes of scaffolds has innate advantages and disadvantages attributed to the processes used to manufacture these materials and the resultant physical structure and properties. Understanding these innate constraints will allow the suitability of a scaffold for a particular application to be evaluated. The limitations of any given scaffold design can then be addressed.

Table 1 List of Polymers and Types of Scaffolds Used in Cartilage Tissue Engineering Applications

Fibers	Foams/Sponges	Gels	Composites
PGA[1]	PLA[7]	Alginate[10]	PLA/PGA[20]
PLA[3]	PLGA[8]	Agarose[1]	PEO/PLGA[2]
PLGA[3]	PEU[22]	Chitosan[11]	PLA/Alginate[19]
	Collagen[23]	Collagen[14]	PCL/Fibrin Glue[9]
		Fibrin Glue[12]	Collagen-GAG[24]
		PEO[25]	PGA/Alginate[26]
		PEO-co-PPO[15]	
		PEG[3]	
		PLGA-co-PEG[16]	
		PniPAm-co-Aac[17]	
		KDL peptide[18]	

PGA = polyglycolic acid, PLA = polylactic acid, PLGA = polylactic acid-co-glycolic acid,
PEU = polyesterurethane, PEO = polyethylene oxide, PPO = polyproylene oxide, PEG = polyethylene glycol,
PCL = polycaprolactone, PNiPAm-co-Aac: poly(N-isopropylacrylamide)-co-acrylic acid,
KDL = lysine-aspartic acid-leucine, GAG = glycosaminoglycan

In evaluating and comparing individual scaffolds or types of scaffolds, it is important to identify design criteria relevant to the required function of the materials (Table 2). For the purpose of this discussion, scaffolds will be evaluated by the following criteria: (1) the mechanical properties of the scaffold in relation to the properties of the tissue it is designed to replace and the in vivo mechanical environment, (2) the extent to which the scaffold is sufficiently porous to allow diffusion of nutrients to and waste products away from cells, (3) the degradation rate of the material, which should be slow enough to provide initial support for cells and fast enough to not hinder accumulation of extracellular matrix, (4) the ability of cells to adhere to or be localized within the scaffold, (5) the biologic compatibility of the material with the cells seeded within, (6) the in vivo environment of the implanted scaffold, (7) the ease with which an implant can be placed in vivo, (8) the safety profile of the scaffold, and (9) the cost of the material and the fabrication process. The relative strengths and weaknesses of each type of scaffolds will be discussed in the context of these criteria.

Fibrous Scaffolds

Many of the first applications of cartilage tissue engineering used scaffolds made from fibers of polyesters such as polyglyolic acid (PGA).[1,2] Other chemically similar polymers such as polylactic acid (PLA)[3] and polylactic acid-co-glycolic acid (PLGA)[3] have also been used in cartilage tissue engineering in the form of woven or nonwoven meshes. These scaffolds are desirable because the materials from which they are fabricated have a long history of safe use in many medical applications and they are currently manufactured in quantities sufficient for commercial manufacture. Another

Table 2 List of Advantages and Disadvantages of Various Types of Polymer Scaffolds

	Fibers	Foams	Gels	Composites
Mechanical properties	$--$ to $+$	$+$ to $++$	$--$ to $-$	$+$ to $++$
Porosity/diffusivity	$++$	$--$ to $++$	$++$	$--$ to $++$
Degradation rate	$++$	$++$	$--$ to $++$	$-$ to $++$
Adhesion/seeding	$--$ to $+$	$--$ to $-$	$++$	$-$ to $++$
Cytotoxicity	$-$ to $+$	$-$ to $+$	$+$	$-$ to $++$
Biocompatibility	$-$ to $+$	$+$ to $++$	$+$ to $++$	$-$ to $++$
Ease of delivery	$+$	$+$	$++$	$-$ to $++$
Regulatory	$++$	$++$	$--$ to $++$	$--$ to $-$
Cost	$+$	$+$	$--$ to $++$	$-$

Each type is rated from least suitable $(--)$ to most suitable $(++)$ for eight independent design considerations

advantage is that these scaffolds are biodegradable at controlled rates. These scaffolds are also highly porous and have characteristics favorable for transport of nutrients and other compounds.

The main limitations of fibrous scaffolds are their generally poor mechanical properties that are typically two or three orders of magnitude lower than those of cartilage. Furthermore, although cells adhere to these materials, obtaining efficient seeding rates requires significant time and the use of bioreactors.[4] There is also concern that the end products of scaffold degradation can lower local pH in vivo,[5] leading to inflammatory reactions in the vicinity of implants.

Foams and Sponges

The same polymers used in fibrous scaffolds are also used to make polymer foams or sponges for tissue engineering applications. Thus these foams and sponges share many of the advantages and disadvantages of the fibrous scaffolds. A variety of techniques are available for sponge fabrication, including freeze drying,[6] particulate leaching,[7] critical point foaming,[8] and three-dimensional printing and solid freeform fabrication techniques.[9] Many of these methods are amenable to scale-up manufacturing for commercialization. Foams tend to be stiff thus providing an alternative to fiber-based systems. However, the increased density of the foams makes cell seeding more difficult and has the potential to hinder transport of nutrients and waste products through the implants. Problems with the removal of degradative products are associated with cell necrosis.

Hydrogels

As the name implies, hydrogels are highly hydrated materials that generate osmotic swelling entropically and/or ionically by polymer chains. This osmotic swelling is restrained by physical or chemical cross-links. Prior to gelation, polymer chains are in solution, which allows for cells to be seeded into the scaffold. After gelation, cells are encapsulated in the polymer, ensuring high seeding densities. Another advantage of hydrogels is their ability to be delivered in a minimally invasive manner, including by injection. This relies on a mechanism by which these polymers can be geled in situ. This can be accomplished through a variety of mechanisms including the introduction of soluble ionic cross-linkers (alginate[10] and chitosan[11]), the use of covalent cross-linking agents (fibrin glue[12]), initiation of network formation using photosensitive cross-linkers (PEO[13]) and induction of phase transitions due to changes in pH (collagen[14]) or temperature (PEO-PPO,[15] PLGA-co-PEG,[16] PniPAAm-co-Aac,[17] and KDL peptides[18]). Typically, cross-linked hydrogels are stiffer and degrade more slowly, while polymers that gel by phase transitions tend to be quite weak and degrade more quickly.

In vivo, hydrogels are resorbed either by degradation of the polymer chains or of the cross-links. Thus the degradation time can be highly variable, from less than a week for uncross-linked PEO-PPO gels, to many months for alginate gels. Both the cost and regulatory challenges of the use of hydrogels are widely variable. Gels based on natural polymers, such as alginate, chitosan, and collagen, are reasonable in cost and have established safety profiles. More recently developed systems such as PniPAAm-co-Aac are extremely costly and there are limited data regarding their safety. Many of the thermally gelling systems are block copolymers that couple hydrophilic and hydrophobic polymers (eg, PEO-co-PPO and PEG-co-PLGA) and additional safety data may be required even if the blocks from which they are composed are well characterized.

Composites

Given that there is no clear choice of scaffold material or system that is favorable in all design aspects, one strategy for optimizing scaffold properties is to fabricate composite scaffolds that retain the advantages of the individual materials from which they are composed. A prime example of this strategy is the "strutted hydrogel" designed to mechanically reinforce a weak hydrogel, typically with a foam or sponge.[13,19] This can also be used to try to enhance the seeding of foams by encapsulating cells in the gel, which eliminates the need for cells to adhere to the foam.[9] Using a similar approach, other studies have attempted to stiffen fibrous scaffolds using solvent deposition to mechanically cross-link the fibers using a second polymer.[20]

Although these strategies are often successful, there are potential drawbacks to this approach. In particular, the development of scaffolds based on a composite of materials can be more expensive and time consuming than development of an individual material. As with copolymer systems, the combination of multiple materials with known safety profiles will likely require additional studies to document safety of the composite. Finally, although the mechanical or cell localization properties are often enhanced in a composite scaffold, cytotoxicity and biocompatibility are often worse as both materials may elicit potentially unwanted effects.

Optimizing Scaffold Design

There is no single material or approach to scaffold design that currently meets all the necessary criteria for a cartilage tissue engineering implant. This unmet need continues to drive the development of new materials. However, an alternative approach is to chemically modify existing materials to achieve the desired properties. Furthermore, if a particular method of processing does not result in a scaffold with the required properties, it may be possible to change the processing techniques to achieve the desired results, rather than change the starting material used.

Alterations in scaffold chemistry can have significant effects on a variety of properties, including regulation of cell behavior. In studies of PLA/PGA composites, increasing the PGA content of scaffolds increased the adhesion and proliferation rates of chondrocytes seeded onto implants.[20] In hydrogels composed of PEG-co-PLA copolymers, varying the ratio of PEG to PLA regulated the spatial distribution of macromolecules in the extracellular matrix.[16] In this system, collagen was confined to the pericellular regions of the implants in samples with high PEG content and more uniform collagen distributions were found in samples with high PLA content.

These properties are not only linked to the chemical identity of the material, but the way in which it is processed. Mechanical properties of scaffolds, for example, can be controlled by alterations in fiber diameter in the case of fibrous scaffolds, porosity and pore structure for foams, and polymer concentration and cross-linking for hydrogels. In the case of alginate, increasing the concentration of cross-linker (Ca^{2+} or Ba^{2+}) produces stiffer scaffolds. However, the cross-linking agent itself also regulates scaffold architecture, with Ca^{2+} yielding a gel with small pores and thin walls, while Ba^{2+} produces thick-walled pores (Figure 1).

Innovations in scaffold processing are also important to develop methods that will facilitate scale-up for commercial manufacture. In the case of alginate, gel formation is typically initiated by the addition of highly soluble calcium salts, such as $CaCl_2$, which cross-link the polymer quickly. Using less soluble salts, such as $CaSO_4$, increases the cross-linking time and enables the injection of a seeded gel into a mold.[21] The use of this type of system along with traditional injection molding technology has the potential to produce large quantities of implants at a high level of reproducibility.

Summary

Significant progress has been made in the development of scaffolds for use in tissue engineering applications. More than 20 materials or combinations of materials have been evaluated for the delivery of chondrocytes or stem cells in attempts to regenerate cartilage. Currently, there is no one optimal material or method of producing a scaffold for in vivo implantation. There are several considerations for the development of new scaffolds, including the mechanical properties of the scaffold in relation to the properties of the tissue it is designed to replace, the porosity of the scaffold, the degradation

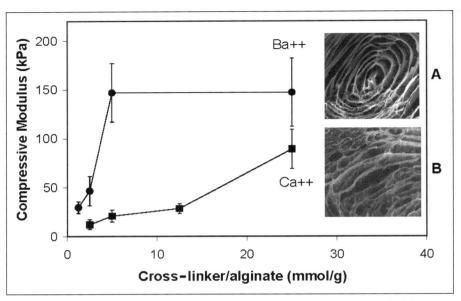

Figure 1 Effect of cross-linker concentration on compressive modulus of alginate hydrogels for samples cross-linked with $CaCl_2$ or $BaCl_2$. A, Scanning electron micrograph of $BaCl_2$ cross-linked gels. B, Scanning electron micrograph of $CaCl_2$ cross-linked gels.

rate of the material, and the biologic compatibility of the material with the cells to be seeded. The safety profile of the scaffold and the cost to manufacture the material should also be considered.

References

1. Vacanti CA, Langer R, Schloo B, Vacanti JP: Synthetic polymers seeded with chondrocytes provide a template for new cartilage formation. *Plast Reconstr Surg* 1991;88:753-759.

2. Freed LE, Marquis JC, Nohria A, Emmanual J, Mikos AG, Langer R: Neocartilage formation in vitro and in vivo using cells cultured on synthetic biodegradable polymers. *J Biomed Mater Res* 1993;27:11-23.

3. Rotter N, Sittinger M, Hammer C, Bujia J, Kastenbauer E: Transplantation of in vitro cultured cartilage materials: Characterization of matrix synthesis. *Laryngorhinootologie* 1997;76:241-247.

4. Vunjak-Novakovic G, Obradovic B, Martin I, Bursac PM, Langer R, Freed LE: Dynamic cell seeding of polymer scaffolds for cartilage tissue engineering. *Biotechnol Prog* 1998;14:193-202.

5. Pazzano D, Mercier KA, Moran JM, et al: Comparison of chondrogensis in static and perfused bioreactor culture. *Biotechnol Prog* 2000;16:893-896.

6. Zhang R, Ma PX: Poly(alpha-hydroxyl acids)/hydroxyapatite porous composites for bone-tissue engineering. *J Biomed Mater Res* 1999;44:446-455.

7. Wake MC, Gupta PK, Mikos AG: Fabrication of pliable biodegradable polymer foams to engineer soft tissues. *Cell Transplant* 1996;5:465-673.

8. Harris LD, Kim BS, Mooney DJ: Open pore biodegradable matrices formed with gas foaming. *J Biomed Mater Res* 1998;42:396-402.

9. Huang Q, Goh JC, Hutmacher DW, Lee EH: In vivo mesenchymal cell recruitment by a scaffold loaded with transforming growth factor beta1 and the potential for in situ chondrogenesis. *Tissue Eng* 2002;8:469-482.

10. Paige KT, Cima LG, Yaremchuk MJ, Vacanti JP, Vacanti CA: Injectable cartilage. *Plast Reconstr Surg* 1995;96:1390-1398.

11. Lahiji A, Sohrabi A, Hungerford DS, Frondoza CG: Chitosan supports the expression of extracellular matrix proteins in human osteoblasts and chondrocytes. *J Biomed Mater Res* 2000;51:586-595.

12. Passaretti D, Silverman RP, Huang W, et al: Cultured chondrocytes produce injectable tissue-engineered cartilage in hydrogel polymer. *Tissue Eng* 2001;7:805-815.

13. Bryant SJ, Nuttelman CR, Anseth KS: The effects of crosslinking density on cartilage formation in photocrosslinkable hydrogels. *Biomed Sci Instrum* 1999;35:309-314.

14. Hunter CJ, Imler SM, Malaviya P, Nerem RM, Levenston ME: Mechanical compression alters gene expression and extracellular matrix synthesis by chondrocytes cultured in collagen I gels. *Biomaterials* 2002;23:1249-1259.

15. Cao Y, Rodriguez A, Vacanti M, Ibarra C, Arevalo C, Vacanti CA: Comparative study of the use of poly(glycolic acid), calcium alginate and pluronics in the engineering of autologous porcine cartilage. *J Biomater Sci Polym Ed* 1998;9:475-487.

16. Bryant SJ, Anseth KS: Controlling the spatial distribution of ECM components in degradable PEG hydrogels for tissue engineering cartilage. *J Biomed Mater Res* 2003;64:70-79.

17. An YH, Webb D, Gutowska A, Mironov VA, Friedman RJ: Regaining chondrocyte phenotype in thermosensitive gel culture. *Anat Rec* 2001;263:336-341.

18. Kisiday J, Jin M, Kurz B, et al: Self-assembling peptide hydrogel fosters chondrocyte extracellular matrix production and cell division. *Proc Natl Acad Sci USA* 2002;99:9996-10001.

19. Caterson EJ, Li WJ, Nesti LJ, Albert T, Danielson K, Tuan RS: Polymer/alginate amalgam for cartilage-tissue engineering. *Ann N Y Acad Sci* 2002;961:134-138.

20. Moran JM, Pazzano D, Bonassar LJ. Characterization of polylactic acid-polyglycolic acid composites for cartilage tissue engineering. *Tissue Eng* 2003;9:63-70.

21. Chang SC, Rowley JA, Tobias G, et al: Injection molding of chondrocyte/alginate constructs in the shape of facial implants. *J Biomed Mater Res* 2001;55:503-511.

22. Saad B, Moro M, Tun-Kyi A, et al: Chondrocyte-biocompatibility of DegraPol-foam: In vitro evaluations. *J Biomater Sci Polym Ed* 1999;10:1107-1119.

23. Fujisato T, Sajiki T, Liu Q, Ikada Y: Effect of basic fibroblast growth factor on cartilage regeneration in chondrocyte-seeded collagen sponge scaffold. *Biomaterials* 1996;17:155-162.

24. Lee CR, Breinan HA, Nehrer S, Spector M: Articular cartilage chondrocytes in type I and type II collagen-GAG matrices exhibit contractile behavior in vitro. *Tissue Eng* 2000;6:555-565.

25. Bryant SJ, Nuttelman CR, Anseth KS: The effects of crosslinking density on cartilage formation in photocrosslinkable hydrogels. *Biomed Sci Instrum* 1999;35:309-314.

26. Mizuno H, Roy AK, Vacanti CA, Kojima K, Ueda M, Bonassar LJ: Tissue engineering of a composite intervertebral disc. *Tissue Eng* 2000;6:666.

Chapter 36

Adipose Derived Adult Stem Cells for Musculoskeletal Tissue Engineering

Farshid Guilak, PhD
Jeffrey M. Gimble, MD, PhD

Abstract

Exogenously introduced cells play an important role in the success of engineered tissue replacements. The development of these replacements has focused on the use of stem cells because of their ability to differentiate into a variety of different cell types. Adult stem cells derived from adipose tissue have characteristics of multipotent adult stem cells, similar to those of bone marrow derived stromal cells, often termed mesenchymal stem cells (MSC). Thus, adipose-derived adult stem cells are promising candidates for the development of functional tissue replacements for musculoskeletal tissues. However, several interrelated basic science and applied questions regarding the safety, efficacy, and cost-effectiveness of the use of these cells must be addressed before their use can be incorporated into clinical practice.

Introduction

Tissue engineering is a rapidly growing field that uses combinations of implanted cells, biomaterial scaffolds, and biologically active molecules to repair or regenerate damaged or diseased tissues. Many diverse and increasingly complex approaches for tissue repair are being developed in this context. The underlying premise of these approaches is that exogenously introduced cells play a necessary role in the success of engineered tissue replacements. Thus, a major consideration is the identification and characterization of appropriate sources of cells for tissue engineering applications. In particular, there has been significant emphasis on the use of progenitor (stem) cells because of their ability to differentiate into a variety of different cell types. The ability to expand these cells in culture also makes them an appealing choice for use in additional applications. Several such cells have been identified in the human body. In this review, we summarize the current knowledge of adult stem cells derived from adipose tissue in the context of their application in tissue engineering. These cells show characteristics of multipotent adult stem cells, similar to those of bone marrow derived stromal cells or MSC, and show significant promise for the development of functional tissue replacements for the musculoskeletal system.

Cell Sources for Tissue Engineering

The ultimate selection of a cell source for tissue engineering may involve several factors. The ability of cells to undergo controlled differentiation (ie, plasticity) and to maintain a differentiated phenotype are essential. Other important considerations include the functional capacity of the cells to synthesize and maintain the appropriate tissue(s) of interest, the availability of the cells with respect to ease of harvest or expandability in culture, and the potential of the cells for allogeneic transplantation. In addition, other factors such as cost-effectiveness and the ability to meet regulatory criteria will have an impact on the overall utility of any approach.

A multitude of cell types have been used for musculoskeletal tissue engineering, including primary cells as well as undifferentiated progenitor cells. Early tissue engineering approaches for musculoskeletal repair focused on the reimplantation of primary cells isolated from the same tissue type that was meant to be repaired or regenerated.[1-3] As the role of environmental factors in controlling cell behavior becomes more fully elucidated, the use of transdifferentiation of primary cells is being explored. In this approach, cells are isolated from a mature tissue to engineer a different tissue.[4,5]

Stem Cells: Definitions and Applications for Tissue Engineering

Significant resources have been applied to the evaluation of the use of stem cells in tissue engineering. A stem cell is defined as a cell that has the ability to divide or self replicate for indefinite periods. Under the appropriate environmental signals, a stem cell can differentiate to different cell types with specialized structures and functions. In this respect, stem cells could provide a valuable source of cells for diverse tissue engineering applications that require an abundant source of progenitor cells.

Several types of stem cells have been defined based on their ability to form different cell types. For example, the fertilized egg is said to be totipotent because it has the potential to generate all the cells and tissues of the embryo. It is currently hypothesized that embryonic stem cells, taken from the blastocyst of the embryo, may provide a source of totipotent cells. Most scientists use the term pluripotent to describe stem cells that can give rise to cells derived from all three embryonic germ layers, namely the mesoderm, endoderm, and ectoderm. The term multipotent is generally used to describe stem cells that can differentiate into two or more cell types. Unipotent describes cells that are capable of differentiating along only one lineage.

Adult Stem Cells

It is now apparent that many adult tissues harbor cells that have the potential to differentiate into multiple cell types. A variety of terms have been used to describe such cells, including MSC, adult stem cells, multipotential adult stem cells, human marrow stromal cells, or mesenchymal progenitors.[6-16] The adult stem cell is defined as an undifferentiated (unspecialized) cell that is found in a differentiated (specialized) tissue. This stem cell can renew itself and become specialized to yield all of the specialized cell types of the tissue from which it originated.[17] The important aspects of this definition are that adult stem cells retain the capacity for self-renewal as well as the potential for differentiation along specific lineages. Due to their accessibility, stem cells from adults have been used extensively in a variety of tissue engineering applications in the musculoskeletal system.

Although no totipotent adult stem cells have been identified, sources of multipotent, and possibly pluripotent adult stem cells have been found in numerous musculoskeletal tissues, including bone marrow,[6,9,12,14] adipose tissue,[18-20] trabecular bone,[11] periosteum,[21,22] synovial tissue,[23,24] muscle,[25-27] and many other tissues. These cells have been shown to serve as critical components for the potential tissue engineered repair of musculoskeletal tissues.[28,29] Although these cells from different tissues share several characteristics, there are significant differences in their potential for proliferation and differentiation, their profile of cell surface markers, and their ease of harvest. Thus these cell types are vastly different with regard to the utility for specific tissue engineering applications.

Adipose Tissue as a Source of Adult Stem Cells

This chapter primarily focuses on the use of human adipose tissue as a source of adult stem cells. A variety of terms have been used to describe these cells, such as preadipocytes,[30-32] stromal cells,[33] adipose endothelial cells,[34] processed lipoaspirate cells,[20] and adipose derived adult stromal/stem (ADAS) cells.[18,35-39]

The isolation of cells from adipose tissue was first documented in a rodent model[40] and then applied to small explants of human adipose tissue.[41,42] Similar but more refined methods are currently used on liposuction waste tissue to isolate ADAS cells.[18,33] The isolation process involves washing and mincing of the tissue, followed by digestion with collagenase, and a subsequent differential centrifugation step that separates the floating mature adipocytes from a stromal-vascular fraction containing the ADAS cells. The stromal/vascular fraction is then suspended in growth media plated, and following overnight incubation, washed free of any nonadherent cells. Cells may then be expanded in culture, cryopreserved, or subjected to specific differentiation protocols. A major advantage of ADAS cells is the availability of large quantities of human adipose tissue as the discarded waste of liposuction procedures. Routinely, a milliliter of liposuction aspirate yields over 400,000 ADAS cells following expansion.[35]

Characterization of ADAS Cell Surface Markers

Specific cell surface proteins are often used as markers to identify and characterize stem cells. In several studies, isolated ADAS cells have been analyzed using flow cytometry and immunohistochemistry to determine their expression of cell surface proteins[20,33-35,39,43] (Table 1). The profiles of surface marker expression reported by independent research groups is highly

Table 1 Surface Protein Profile of Undifferentiated ADAS Cells[20,33-35,39,43]

Positive Surface Protein Markers		Negative Surface Protein Markers	
Protein Name	CD#	Protein Name	CD#
Tetraspan	CD9	Integrin α	CD11
CALLA	CD10	Integrin β	CD14
Aminopeptidase N	CD13	PECAM	CD16
Integrin β1	CD29	LCA	CD18
Hyaluronate Receptor	CD34	ICAM-3	CD31
Integrin α4	CD44	NCAM	CD45
ICAM-1	CD49d	E-Selectin	CD50
DAF	CD54	Factor VIII related	CD56
Complement protectin	CD55	Ag[†,¶]	CD62
Transferrin receptor	CD59	HLA-DR	CD104
5′ Nucleotidase	CD71	Stro-1[2]	
Thy-1	CD73		
Endoglin	CD90		
VCAM[*]	CD105		
Muc18	CD106		
ALCAM	CD146		
HLA-ABC	CD166		

[*]Negative marker[43]
[†]Positive marker[20,43]
[¶]Positive marker on 86% of cells[34]
ALCAM, activated lymphocyte cell adhesion molecule; CALLA, common acute lymphocytic leukemia antigen; CD, cluster of differentiation; DAF, decay accelerating factor; E selectin, endothelial selectin; HLA, histocompatibility locus antigens; ICAM, intercellular adhesion molecule; LCA, leukocyte common antigen; NCAM, neural cell adhesion molecule; PECAM, platelet endothelial cell adhesion molecule; VCAM, vascular cell adhesion molecule

consistent, but not identical (Table 2). For example, the endothelial associated von Willebrand Factor VIII was not uniformly detected.[20,33,34,43] Although the reason for this apparent discrepancy is not known, possible explanations include differences in cell isolation methods, the length of time the cells were in culture, the type and purity of the collagenase used in the digestion procedure, differences in monoclonal antibodies to detect various epitopes on the same protein, and differences in results of immunohistochemistry and flow cytometry methods. Nevertheless, the reported levels of protein expression are remarkably homogeneous across individual donors.[20,33,34,43]

The profile of the surface protein of ADAS cells is similar to that of adult stem cells isolated from adult bone marrow-derived cells, such as MSC,[12] and muscle-derived stem cells.[27] However, the tissue of origin may influence the specific surface protein profile of the adult stem cells. Although muscle-derived adult stem cells express NCAM (CD56), ADAS cells do not.[27,33] Likewise, in direct comparisons, the protein patterns of ADAS cells and bone marrow-derived adult stem cells are similar but not identical.[43] Furthermore, ADAS cells may express slight differences in the profile of cell surface proteins depending on the adipose depot of origin.[33,39] Additional work will be needed to fully evaluate any tissue-specific features

Table 2 ADAS Cell Differentiation Potential

Cell Lineage	Differentiation Cocktail	References
Adipocyte	Dexamethasone Isobutylmethylxanthine Insulin PPAR-γ Ligand Fetal bovine serum	20, 33, 35, 39, 41, 42, 51, 52
Chondrocyte	Ascorbic Acid Insulin Dexamethasone Transforming growth factor β Fetal bovine serum	18, 20, 36, 39
Hematopoietic Support	Supplemental Cytokines (IL-2, IL-7, IL-15) Fetal bovine serum	55
Myocyte	Horse Serum Dexamethasone	20, 43, 48
Neuronal	Butylated Hydroxyanisole Forskolin Valproic acid insulin	20, 43, 50
Osteoblast	Ascorbic Acid β-Glycerophosphate Dexamethasone 1,25 Dihydroxyvitamin D3	19, 20, 37, 39, 43, 46

IL = Interleukin

of adult stem cells and, more importantly, the implications of these profiles on the functional capabilities of the cell type.

Multilineage Differentiation of ADAS Cells: Musculoskeletal Lineages

Human ADAS cells exhibit significant plasticity. To date, ADAS cells have been induced along several lineage pathways relevant to musculoskeletal repair, including chondrogenic, osteogenic, myogenic, neurogenic, and adipogenic cell types. Clonogenic studies indicate that individual clones are truly multipotent and suggest that ADAS cells are stem cells, rather than a mixed population of unipotent progenitors. The differentiation process can be controlled by altering the physical and biochemical environment of the cells (Table 2). For example, the presence of and time of exposure to exogenous growth factors, hormones, vitamins, and other soluble mediators in combination with the control of cell shape has a profound effect on the phenotype of ADAS cells.

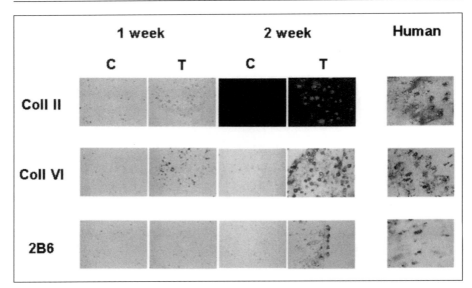

Figure 1 Immunohistochemistry of alginate beads containing stromal cells derived from adipose tissue. Cells were grown in three-dimensional alginate cultures for 2 weeks in control (C) or chondrogenic media (T). Alginate beads were subsequently fixed, sectioned, and labeled for cartilage matrix proteins including collagens type II (row 1) and VI (row 2) and the 2B6 epitope of proteoglycan (row 3). *Adapted with permission from Erickson GR, Gimble JM, Franklin DM, Rice HE, Awad H, Guilak F: Chondrogenic potential of adipose tissue-derived stromal cells in vitro and in vivo.* Biochem Biophys Res Commun *2002;290:763-769.)*

In response to transforming growth factor-β, ascorbate, and dexamethasone, human ADAS cells express biochemical markers characteristic of articular chondrocytes.[18,20,36,39] An important requirement for chondrogenic differentiation appears to be the maintenance of a rounded cell shape in a three-dimensional culture system such as a micromass pellet or as individual cells suspended in gel matrix such as alginate.[18] Within 1 to 2 weeks, the cells express collagen type II, collagen type VI, and proteoglycans such as aggrecan (Figure 1). Following this differentiation process, human ADAS cells retain the chondrocyte phenotype for up to 12 weeks when implanted subcutaneously in vivo in immunodeficient mice.[18] When seeded in hydrogel scaffolds, ADAS cells can form a cartilaginous matrix with some of the functional properties of native articular cartilage.[44,45] Furthermore, recent studies suggest that reduced oxygen tension (5%), as would be present in the normal synovial joint, arrests cell proliferation and enhances the synthesis of matrix molecules by ADAS cells, as compared with ambient oxygen levels (20%).[38]

ADAS cells differentiate into osteoblast-like cells in the presence of ascorbate, β-glycerophosphate, and dexamethasone. In vitro, the cells min-

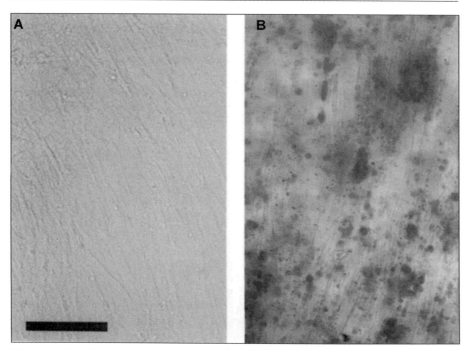

Figure 2 Alizarin red staining of human ADAS cells grown under *(A)* control and *(B)* osteogenic conditions for 3 weeks. Osteogenic conditions induced significant mineralization as compared with control conditions. Scale = 100 μm.

eralize the extracellular matrix and express genes and proteins that are associated with the osteoblast phenotype[19,20,39,43,46,47] (Figure 2). When seeded on porous ceramic biomaterials such as hydroxyapatite and implanted subcutaneously in immunodeficient mice, ADAS cells form new bone within 6 weeks.[37] These findings demonstrate that ADAS cells have osteogenic potential when used in combination with appropriate matrices.

In the presence of horse serum, human ADAS cells express myoD, myogenin, and other transcription factors known to regulate the differentiation of skeletal myocytes.[20,43,48] These cells form myotubules and express myosin light chain kinase in addition to other markers characteristic of the myocyte lineage. Human ADAS cells also express α-smooth muscle actin, indicating that the cells may have the potential to replicate phenotypic characteristics of smooth muscle cells. This potential could be exploited for applications in tissue engineering of the gastrointestinal and urinary tracts. The finding that bone marrow-derived adult stem cells can differentiate into cardiac myocytes[49] suggests that it may be possible to differentiate ADAS cells in a similar manner for the treatment of myocardial ischemic injuries.

In the presence of antioxidants and the absence of serum, human ADAS cells assume a bipolar morphology, characteristic of the neuronal lineages, and express neurogenic markers including nestin, intermediate filament M, and Neu N.[20,50] The induced cells have also been shown to express glial fibrillary acidic protein, consistent with oligodendrocyte differentiation.[50] Although there is evidence that ADAS-derived cells can produce biochemical and morphologic markers characteristic of neurons, the functional characteristics of neurons have not been documented in these cells. Given the clinical need to regenerate neurons, the differentiation of cells into neurons continues to be an area of intense research.

Other Differentiation Pathways of ADAS Cells

In addition to the ability to differentiate into cells of musculoskeletal lineages, ADAS cells are capable of differentiating into other cell types. Several groups have demonstrated the ability of ADAS cells to differentiate into adipocytes in vitro. Standard induction media include insulin, methylisobutylxanthine (a phosphodiesterase inhibitor resulting in elevated cyclic adenosine monophosphate levels), hydrocortisone or dexamethasone, and indomethacin or rosiglitazone (a peroxisome proliferator activated receptor γ ligand) (Table 2). Within 10 days of induction, ADAS cells display increased accumulation of intracellular neutral lipid as detected by Oil Red O or Nile Red staining, increased secretion of the protein leptin, and induction of adipocyte-associated genes such as the fatty acid binding protein aP2[20,35,39,43,51,52] (Figure 3). Undifferentiated and adipocyte-differentiated ADAS cells seeded in various biomaterial scaffolds create new adipose depots when implanted subcutaneously in animal models.[30-32,53,54] These findings suggest that ADAS cells may have potential applications in the formation of new adipose depots in vivo for reconstructive or cosmetic applications. Additional studies will be required before ADAS cells can be used to reconstruct large defects, such as those following lumpectomy procedures in breast cancer patients.

Human ADAS cells express surface adhesive proteins that have been associated with hematopoietic support function in bone marrow stromal cells.[33,55] In addition, ADAS cells express and secrete many hematopoietic cytokines, including macrophage colony stimulating factor, granulocyte-macrophage colony stimulating factor, tumor necrosis factor α, interleukin 6, interleukin 7, interleukin 8, interleukin 11, and stem cell factor.[55] As suggested by these findings, human ADAS cells support the proliferation and differentiation of CD34[+] hematopoietic stem cells isolated from umbilical cord blood in coculture systems.[55] Within 2 weeks, cocultures of these cells contain both undifferentiated stem cells and cells expressing markers of early B-cells, T-cells, natural killer cells, and the various myeloid lineages.

Finally, in other studies, ADAS-like cells have been shown to express the endothelial cell marker, von Willebrand factor.[34] Newly isolated cells were used to seed the surface of synthetic biomaterials used for vascular grafts and were found to prevent thrombus formation in vivo for extended periods.[34,56] These findings suggest a potential role for ADAS cells as endothelial cell substitutes or for other applications in vascular tissue engineering.

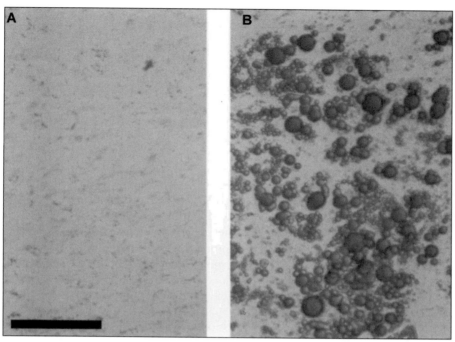

Figure 3 Oil red O staining of cells grown under *(A)* control and *(B)* adipogenic conditions for 2 weeks. Adipogenic conditions induced the formation of lipid vacuoles. Scale = 100 μm.

Challenges and Future Goals in the Application of Stem Cells

Before human ADAS cells or other adult stem cells can be introduced into clinical practice, several issues must be addressed, including interrelated basic science and applied questions regarding the safety, efficacy, and cost-effectiveness of the use of these cells.

Of primary concern are issues of safety, which include factors related to cell harvest, disease transmission, immunogenicity, tumor formation, and long-term function. These issues must often be balanced against financial considerations. For example, allogeneic transplantation of cells may be less expensive than other options, but this approach introduces the possibility of disease transmission from donor to host, or the development of an immune response once implanted. In another example, the use of long-term expansion of cells in culture to provide a low-cost means of generating large numbers of source cells must be weighed against risks associated with the decreased functional potential of cells with age in culture. Long-term expansion in culture also has the potential for introducing mutations or immortalizing cells. The medium used in culture is another example of a safety concern. Most adult stem cells are cultured in the presence of fetal bovine serum. In the United States, cattle herds are certified for having no history of infectious diseases, including bovine spongiform encephelopathy (BSE). However, BSE has been detected in cattle in the European Union and in Japan. Thus the use of fetal bovine serum is viewed as a greater risk in certain countries and alternative culture supplements may be needed.

Of equal importance to the long-term success of stem cell-based tissue engineering is their efficacy in restoring function. Within the context of the musculoskeletal system, most surgical procedures are designed to decrease pain and restore the principally biomechanical functions of bone, cartilage, tendon/ligament, muscle, or other tissues. To address these issues of biomechanical function, a series of guidelines and recommendations for functional tissue engineering have been proposed to aid in the rational design of tissue replacements.[57,58] These guidelines include: (1) improved definitions of functional success for tissue engineering approaches, (2) a more thorough understanding of the in vivo biomechanical requirements and intrinsic properties of native tissues, (3) the development and prioritization of design criteria that meet the biomechanical and metabolic demands of the tissue to be replaced, (4) design and characterization of the biomechanical and biophysical properties of biomaterial scaffolds for tissue engineering, (5) a more thorough understanding of the biophysical environment of cells within engineered constructs, and (6) the use of biophysical stimuli to control cellular differentiation and tissue metabolism in vivo and in vitro. Examples of the applications of these guidelines to a variety of tissues are to be published.[59]

In addition to these biomechanically based guidelines, several basic biologic questions still remain to be addressed that will serve to improve the safety, efficacy, and cost-effectiveness of cell-based therapies. These issues include: (1) improved characterization of mechanisms of growth and differentiation of different sources of adult stem cells, (2) a more thorough understanding of cellular plasticity and fate in the context of tissue engineering, (3) optimization of local and systemic environmental factors, including inflammatory mediators, growth factors, mechanical factors, and oxygen content, (4) characterization of the effects of age, sex, and disease state on the properties of adult stem cells, and (5) identifying the strengths and limitations of animal models in predicting effects in clinical applications.

Intertwined with unanswered questions regarding the safety and efficacy of adult stem cells are issues relating to the manufacture and regulation of cell-based therapies. A critical step in the translation of this technology to the clinic will be the development and regulatory approval of reproducible and cost-effective protocols for the manufacture of large numbers of cells. It is anticipated that regulatory guidelines to establish safety and efficacy will differ greatly depending on the extent of manipulation of cells and engineered tissue replacements ex vivo.

Summary and Conclusions

The application of adult stem cells in tissue engineering shows tremendous promise for the development of products for clinical use. However, despite rapid advances and many early successes, there are few clinical applications of tissue engineering based on adult stem cells. Early work focused on trial-and-error methods and there is a need to balance such an approach with more conservative engineering based approaches for product design and manufacture. Furthermore, other rapidly evolving new technologies may have a significant impact on stem cell-based tissue engineering. It is important to consider the principles of functional tissue engineering in the context of novel growth factors, new biomaterials, gene therapy, and other technologic advances.

Acknowledgments

We wish to acknowledge that our work was supported in part by grants from Artecel Sciences, Inc, the North Carolina Biotechnology Center, the Kenan Institute for Engineering, Technology, and Science, and NIH grants AR50245, AG15768, AR48182, and AR49294. We would like to thank Hani Awad, Geoffrey Erickson, Beverley Fermor, Yuan-Di Halvorsen, Kristen Lott, Henry Rice, Lori Setton, Robert Storms, David Wang, Quinn Wickham, and the R&D staff at Artecel Sciences for their contributions to this work.

References

1. Grande DA, Pitman MI, Peterson L, Menche D, Klein M: The repair of experimentally produced defects in rabbit articular cartilage by autologous chondrocyte transplantation. *J Orthop Res* 1989;7:208-218.

2. Green WT Jr: Articular cartilage repair: Behavior of rabbit chondrocytes during tissue culture and subsequent allografting. *Clin Orthop* 1977;124:237-250.

3. Wakitani S, Kimura T, Hirooka A, et al: Repair of rabbit articular surfaces with allograft chondrocytes embedded in collagen gel. *J Bone Joint Surg Br* 1989;71:74-80.

4. Itay S, Abramovici A, Nevo Z: Use of cultured embryonal chick epiphyseal chondrocytes as grafts for defects in chick articular cartilage. *Clin Orthop* 1987;220:284-303.

5. Kon M: Cartilage formation from perichondrium in a weight-bearing joint: An experimental study. *Eur Surg Res* 1981;13:387-396.

6. Caplan AI, Bruder SP: Mesenchymal stem cells: Building blocks for molecular medicine in the 21st century. *Trends Mol Med* 2001;7:259-264.

7. Caterson EJ, Nesti LJ, Danielson KG, Tuan RS: Human marrow-derived mesenchymal progenitor cells: Isolation, culture expansion, and analysis of differentiation. *Mol Biotechnol* 2002;20:245-256.

8. Colter DC, Sekiya I, Prockop DJ: Identification of a subpopulation of rapidly self-renewing and multipotential adult stem cells in colonies of human marrow stromal cells. *Proc Natl Acad Sci USA* 2001;98:7841-7845.

9. Jiang Y, Jahagirdar BN, Reinhardt RL, et al: Pluripotency of mesenchymal stem cells derived from adult marrow. *Nature* 2002;418:41-49.

10. Nakahara H, Bruder SP, Haynesworth SE, et al: Bone and cartilage formation in diffusion chambers by subcultured cells derived from the periosteum. *Bone* 1990;11:181-188.

11. Noth U, Osyczka AM, Tuli R, Hickok NJ, Danielson KG, Tuan RS: Multilineage mesenchymal differentiation potential of human trabecular bone-derived cells. *J Orthop Res* 2002;20:1060-1069.

12. Pittenger MF, Mackay AM, Beck SC, et al: Multilineage potential of adult human mesenchymal stem cells. *Science* 1999;284:143-147.

13. Prockop DJ: Adult stem cells gradually come of age. *Nat Biotechnol* 2002;20:791-792.

14. Prockop DJ, Sekiya I, Colter DC: Isolation and characterization of rapidly self-renewing stem cells from cultures of human marrow stromal cells. *Cytotherapy* 2001;3:393-396.

15. Verfaillie CM: Adult stem cells: Assessing the case for pluripotency. *Trends Cell Biol* 2002;12:502-508.

16. Yoo JU, Barthel TS, Nishimura K, et al: The chondrogenic potential of human bone-marrow-derived mesenchymal progenitor cells. *J Bone Joint Surg Am* 1998;80:1745-1757.

17. Kirschstein R, Skirboll SR: Stem Cells: Scientific Progress and Future Research Directions, 2001, Department of Health and Human Services.

18. Erickson GR, Gimble JM, Franklin DM, Rice HE, Awad H, Guilak F: Chondrogenic potential of adipose tissue-derived stromal cells in vitro and in vivo. *Biochem Biophys Res Commun* 2002;290:763-769.

19. Halvorsen YC, Wilkison WO, Gimble JM: Adipose-derived stromal cells: Their utility and potential in bone formation. *Int J Obes Relat Metab Disord* 2000;24:S41-S44.

20. Zuk PA, Zhu M, Mizuno H, et al: Multilineage cells from human adipose tissue: implications for cell-based therapies. *Tissue Eng* 2001;7:211-228.

21. Nakase T, Nakahara H, Iwasaki M, et al: Clonal analysis for developmental potential of chick periosteum-derived cells: Agar gel culture system. *Biochem Biophys Res Commun* 1993;195:1422-1428.

22. O'Driscoll SW, Fitzsimmons JS: The role of periosteum in cartilage repair. *Clin Orthop* 2001;391:S190-S207.

23. De Bari C, Dell'Accio F, Tylzanowski P, Luyten FP: Multipotent mesenchymal stem cells from adult human synovial membrane. *Arthritis Rheum* 2001;44:1928-1942.

24. Nishimura K, Solchaga LA, Caplan AI, Yoo JU, Goldberg VM, Johnstone B: Chondroprogenitor cells of synovial tissue. *Arthritis Rheum* 1999;42:2631-2637.

25. Jankowski RJ, Deasy BM, Huard J: Muscle-derived stem cells. *Gene Ther* 2002;9:642-647.

26. Lee JY, Qu-Petersen Z, Cao B, et al: Clonal isolation of muscle-derived cells capable of enhancing muscle regeneration and bone healing. *J Cell Biol* 2000;150:1085-1100.

27. Young HE, Steele TA, Bray RA, et al: Human pluripotent and progenitor cells display cell surface cluster differentiation markers CD10, CD13, CD56, and MHC class-I. *Proc Soc Exp Biol Med* 1999;221:63-71.

28. Bruder SP, Kraus KH, Goldberg VM, Kadiyala S: The effect of implants loaded with autologous mesenchymal stem cells on the healing of canine segmental bone defects. *J Bone Joint Surg Am* 1998;80:985-996.

29. Solchaga LA, Goldberg VM, Caplan AI: Cartilage regeneration using principles of tissue engineering. *Clin Orthop* 2001;391:S161-S170.

30. Halberstadt C, Austin C, Rowley J, et al: A hydrogel material for plastic and reconstructive applications injected into the subcutaneous space of a sheep. *Tissue Eng* 2002;8:309-319.

31. Kral JG, Crandall DL: Development of a human adipocyte synthetic polymer scaffold. *Plast Reconstr Surg* 1999;104:1732-1738.

32. Patrick CW Jr, Zheng B, Johnston C, Reece GP: Long-term implantation of preadipocyte-seeded PLGA scaffolds. *Tissue Eng* 2002;8:283-293.

33. Gronthos S, Franklin DM, Leddy HA, Robey PG, Storms RW, Gimble JM: Surface protein characterization of human adipose tissue-derived stromal cells. *J Cell Physiol* 2001;189:54-63.

34. Williams SK, Wang TF, Castrillo R, Jarrell BE: Liposuction-derived human fat used for vascular graft sodding contains endothelial cells and not mesothelial cells as the major cell type. *J Vasc Surg* 1994;19:916-923.

35. Aust L, Devlin B, duLaney T, et al: Recovery of human adipose derived adult stem cells from liposuction aspirates. Unpublished data 2003.

36. Awad HA, Halvorsen YC, Gimble JM, Guilak F: The effects of TGF-beta-1 and dexamethasone on the growth and chondrogenic differentiation of adipose derived stromal cells. *Tissue Eng* 2003;9:1301-1312.

37. Hicok KC, duLaney TV, Zhou YS, et al: Human adipose derived adult stem cells produce osteoid in vivo. *Tiss Eng* in press.

38. Wang DW, Fermor B, Gimble JM, Guilak F: Effects of hypoxia on the proliferation and chondrogenic potential of human adipose-derived stromal cells. *Trans Orthop Res Soc* 2003;28:875.

39. Wickham MQ, Gimble JM, Erickson GR, Vail TP, Guilak F: Multipotent stromal cells derived from the infrapatellar fat pad of the knee. *Clin Orthop* 2003;412:196-212.

40. Rodbell M: Metabolism of isolated fat cells: Effects of hormone on fat metabolism and lipolysis. *J Biol Chem* 1964;239:375-380.

41. Deslex S, Negrel R, Vannier C, Etienne J, Ailhaud G: Differentiation of human adipocyte precursors in a chemically defined serum-free medium. *Int J Obes* 1987;11:19-27.

42. Hauner H, Entenmann G, Wabitsch M, et al: Promoting effect of glucocorticoids on the differentiation of human adipocyte precursor cells cultured in a chemically defined medium. *J Clin Invest* 1989;84:1663-1670.

43. Zuk PA, Zhu M, Ashjian P, et al: Human adipose tissue is a source of multipotent stem cells. *Mol Biol Cell* 2002;13:4279-4295.

44. Awad HA, Wickham MQ, Leddy HA, Gimble JM, Guilak F: Chondrogenic differentiation of adipose-derived adult stem cells in agarose, alginate, and gelatin scaffolds. *Biomaterials.* Published online, doi:10.1016/j.biomaterials.2003.10.045,2004.

45. Leddy HA, Awad HA, Wickham MQ, Guilak F: Molecular diffusion in articular cartilage and tissue engineered cartilage contructs. *Trans Orthop Res Soc* 2003;28:289.

46. Halvorsen YD, Franklin D, Bond AL, et al: Extracellular matrix mineralization and osteoblast gene expression by human adipose tissue-derived stromal cells. *Tissue Eng* 2001;7:729-741.

47. Huang JI, Beanes SR, Zhu M, Lorenz HP, Hedrick MH, Benhaim P: Rat extramedullary adipose tissue as a source of osteochondrogenic progenitor cells. *Plast Reconstr Surg* 2002;109:1033-1041.

48. Mizuno H, Zuk PA, Zhu M, Lorenz HP, Benhaim P, Hedrick MH: Myogenic differentiation by human processed lipoaspirate cells. *Plast Reconstr Surg* 2002;109:199-209.

49. Toma C, Pittenger MF, Cahill KS, Byrne BJ, Kessler PD: Human mesenchymal stem cells differentiate to a cardiomyocyte phenotype in the adult murine heart. *Circulation* 2002;105:93-98.

50. Safford KM, Hicok KC, Safford SD, et al: Neurogenic differentiation of murine and human adipose-derived stromal cells. *Biochem Biophys Res Commun* 2002;294:371-379.

51. Halvorsen YD, Bond A, Sen A, et al: Thiazolidinediones and glucocorticoids synergistically induce differentiation of human adipose tissue stromal cells: biochemical, cellular, and molecular analysis. *Metabolism* 2001;50:407-413.

52. Sen A, Lea-Currie YR, Sujkowska D, et al: Adipogenic potential of human adipose derived stromal cells from multiple donors is heterogeneous. *J Cell Biochem* 2001;81:312-319.

53. Kimura Y, Ozeki M, Inamoto T, Tabata Y: Time course of de novo adipogenesis in matrigel by gelatin microspheres incorporating basic fibroblast growth factor. *Tissue Eng* 2002;8:603-613.

54. Patrick CW Jr: Tissue engineering strategies for adipose tissue repair. *Anat Rec* 2001;263:361-366.

55. Storms RW, Green P, Potiny S, Foster S, Devlin B, Gimble JM: Human ADAS cells support hematopoietic progenitor proliferation and differentiation in vitro. In preparation. Unpublished data.

56. Young C, Jarrell BE, Hoying JB, Williams SK: A porcine model for adipose tissue-derived endothelial cell transplantation. *Cell Transplant* 1992;1:293-298.

57. Butler DL, Goldstein SA, Guilak F: Functional tissue engineering: The role of biomechanics. *J Biomech Eng* 2000;122:570-575.

58. Guilak F, Butler DL, Goldstein SA: Functional tissue engineering: The role of biomechanics in articular cartilage repair. *Clin Orthop* 2001;391:S295-S305.

59. Guilak F, Butler DL, Golstein SA, Mooney DJ: *Functional Tissue Engineering.* New York, NY, Springer-Verlag, 2003.

Chapter 37

A Bench to Bedside Case Study: The Development Pathway of an Intraoperative, Autologous Progenitor Cell Preparation Kit

Scott P. Bruder, MD, PhD

Abstract

Natural repair mechanisms in mammalian species depend on the presence of endogenous growth factors to stimulate tissue formation by competent cells that migrate to the wound site. One approach to the repair and regeneration of human musculoskeletal tissues is based on the engineering of autologous growth factors and responding progenitor cells into therapeutic implants. Methods were optimized methods for the intraoperative concentration of growth factors and marrow-derived progenitor cells in biomaterials capable of effecting osseous repair in a variety of large animal models and in the clinical setting. A disposable device was developed that mixes and selectively concentrates these cells three- to 20-fold, with 70% to 90% efficiency. The design and use of this device, along with the autologous nature of its bioactive components, have enabled a rapid regulatory clearance pathway and allowed broad product claims. Aspects of the product development cycle, including the preclinical basis of this technology, will be presented as an example of how tissue-engineered products with a solid scientific foundation can move quickly from the bench to the bedside.

Introduction

The progression of tissue engineered products from the laboratory bench to the operating theater is a complex, multistep pathway much like the biologic cascades these efforts are designed to replace. This product development cycle is affected by unique challenges associated with tissue engineering techniques as well as factors pertaining to orthopaedic practice that are not exclusive to tissue engineering. One view of this development process and its interdependent components is presented in Figure 1. Although the individual elements may be articulated as preclinical research, intellectual property, business development, process development, clinical research, and regulatory affairs/quality systems, entry into this cycle may occur at nearly any point. Only initiatives that successfully navigate the entire circle can be suc-

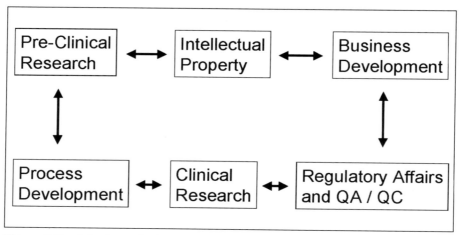

Figure 1 The interdependent components of a successful product development cycle. QA = Quality Assurance, QC = Quality Control.

cessfully incorporated into clinical practice. Efforts to address the hurdles to success in each of these disciplines may vary widely depending on the technology, but all of these areas require attention to complete the process.

A disposable kit to assist in the preparation of a bone graft substitute is described as an example of the development cycle. Numerous products have been developed as alternatives to autogenous bone grafts for spinal fusion and other indications in an attempt to avoid morbidity associated with the harvest of autogenous grafts. There is a wide range of biologic activity of these products from simple osteoconductive materials (ie, porous ceramics) to more complex osteoinductive growth factors (ie, bone morphogenetic protein-2, osteogenic protein-1 and MP52). None of the current offerings truly provide an osteogenic bone graft substitute that contains a scaffold, critical bioactive cues, and the progenitor cells capable of directly forming bone. This chapter will review efforts to transition such a technology from the bench to the bedside, with discussion of the specific issues, advantages, and limitations in navigating each step of the cascade outlined above.

Preclinical Research

The basic scientific principles that support the utility and efficacy of any advanced technology product must be firmly established before it will be accepted by the clinical community. As such, the commercial success of new products depends not only on the marketing and sales organizations, but on the reliability and scientific premise of their mechanism of action. Although the intuitive basis of product efficacy can create interest among surgeons, there is considerable demand for scientific evidence that demonstrates efficacy and safety.

In the course of collegial interactions, sometimes a technology emerges with an apparent application for use in patient management. For example, a product concept was built based on a series of investigations, including those of Muschler and associates.[1-3] The fundamental premise of this work is that fresh bone marrow can be aspirated with ease and then eluted through certain scaffolds capable of selectively retaining the osteoprogenitor cells and accessory cells important for osteogenesis. This Selective Retention Technology works similarly to an affinity chromatography column and is based on enhanced methods for maximizing the retrieval of progenitor cells during marrow aspiration[1] and scaffolds designed to optimize cell adherence. The final implant, composed of autologous cells and a scaffold (with or without endogenous or supplemental bioactive factors), may be delivered to a surgical site in need of bone grafting within minutes of the marrow aspiration.

A variety of scaffold designs were evaluated, including allograft bone materials that contain endogenous growth factors such as bone morphogenetic growth proteins (BMPs). In addition, we combined autologous platelet-rich plasma (PRP) with the cell:scaffold combination in an effort to achieve two goals: (1) improve graft performance by providing the complement of growth factors known to initiate the early steps of the bone healing cascade, and (2) clot the entire mixture together using an autologous fibrin gel that imparts favorable handling characteristics. The PRP is prepared using a commercially available point-of-care device called the Symphony Platelet Concentrate System (DePuy Spine, Raynham, MA) and requires the use of only a small amount of peripheral blood. The bioactivity of PRP on osteoblasts and marrow-derived osteoprogenitor cells[4,5] and its general benefit on bone grafting materials have been reported.[6,7]

Figure 2 demonstrates the results of Selective Retention Technology when applied to the repair of a critical-sized osteoperiosteal segmental defect of the canine femur. Using a previously established model,[8] autogenous iliac crest bone graft was compared against a Selective Retention technique using a scaffold comprised of demineralized canine allograft bone fibers and mineralized cancellous chips. These autologous progenitor cell-based grafts were congealed with autologous PRP immediately before implantation and referred to as the cell/matrix/PRP group. Allograft matrix alone (matrix) and the allograft matrix combined with fresh marrow not subjected to Selective Retention (matrix/marrow) were used as control groups. The anteroposterior radiographs and CT scans in Figure 2, *A* illustrate rapid consolidation of bone across the span of the defect in as few as 8 weeks after implantation. By 16 weeks postimplantation, neocortices in both the autograft and cell/matrix/PRP groups are evident in plain radiographs and CT scans. Figure 2, *B* depicts the incidence of fusion for the study groups defined above. This series of experiments confirms the benefit of Selective Retention Technology compared with use of either the allograft matrix alone or adding fresh bone marrow to the matrix. Additional details of these experiments are provided elsewhere.[9,10]

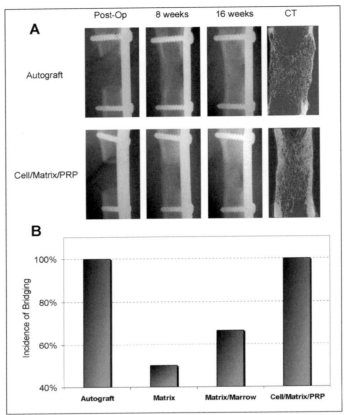

Figure 2 *A* Representative images of a critical-sized, canine femoral gap defect treated with different graft materials including: 1) morcellized autogenous iliac crest bone, and 2) a Selective Retention Technology graft prepared with a matrix of demineralized bone fibers and mineralized cancellous chips congealed together with autologous platelet-rich plasma (cell/matrix/PRP). Plain radiographs obtained in vivo postoperatively and a 16-week postharvest micro CT scan are shown for each sample. *B* Graphic summarization of the 16-week micro CT data from all the groups included in the study. Those critical-sized defects treated with autograft and cell/matrix/PRP bone grafts achieved 100% fusion after 16 weeks. Importantly, the control groups (matrix, and matrix simply soaked with marrow) achieved only 50% and 67% fusion, respectively. (n > 6 for each group)

Intellectual Property

Given the competition in the marketplace and the significant resources required to develop a commercially viable product, it is paramount that a company obtain legal protection for its products. In general, this includes documentation of novel ideas, inventions, improvements, and processes that can provide competitive advantage and that have the potential to be patented. As part of the development cycle, a responsible company will address matters of intellectual property due diligence. Though time consuming and sometimes tedious, this effort is necessary to protect the interests of the company, as well as those of the inventor(s). First, however, it is important to understand the benefits and limitations of patent protection. Possessing or licensing a patent provides the owner a legal avenue to prevent others from commercializing products that fall within the claims of the owned or licensed patent. Owning a patent does not necessarily allow a company to freely commercialize a product defined by that patent. A company must perform sufficient due diligence to ensure that the product they want to commercialize does not infringe upon any other patents. This is commonly referred to as gaining the "right to practice." Often, it may be necessary to sublicense a patent from a third party to obtain the right to practice, or commercialize the product without concern of being sued for patent infringement. In the case of Selective Retention Technology, there are several important patents that protect the methods used to prepare the implants, as well as the actual devices used to practice these methods. As in most product development cases, expanding the intellectual property portfolio is an important strategy in building the basis of business.

Business Development

It is perhaps disheartening to entrepreneurial scientific and clinical investigators that only a fraction of technologically sound strategies will ever make the transition from laboratory bench to bedside. There are many reasons for this, but a thoughtful and robust business plan is an important feature missing from the development plans of many therapeutic approaches. In broad terms, it is the responsibility of corporate business development teams to create economic models and negotiate partnerships that leverage proprietary positions into profitable products. For example, significant time and energy were applied to develop Selective Retention Technology into the disposable product now branded as the Cellect (DePuy Spine, Raynham, MA) product. The decision to forego culture-expanded cell processing technologies in favor of an autologous point-of-care approach required careful consideration and creation of numerous business model scenarios. In this example, an agreement that provided compensation for an exclusive license to certain patents and know-how was established between DePuy Spine and the Office of Technology Transfer at the Cleveland Clinic Foundation. As in many other corporate agreements, all subsequent product development costs were borne by the company. By addressing the clinical need through point-of-care technologies (including PRP), the extraordinary investment for specialized manufacturing of cells or recombinant growth factors is avoided, and the cost savings can be passed along to the physicians, patients, and health care community at large. Furthermore, the business model for Cellect eliminates any need to address the expensive and time-consuming regulatory issues associated with the creation, storage, shipment, and validation of manufactured biologic products. In the final analysis, when considering the opportunities to develop a tissue engineered product, companies are forced to weigh the financial return on their investment against other demands for limited financial and human resources. In some cases, wonderfully innovative technologies simply do not have the potential to create revenue sufficient to justify the expense of their development and commercialization.

Process Development

Once a technical opportunity has satisfied the basic scientific, intellectual property, and business model requirements, the idea must be converted into a product that can be manufactured, sterilized, stored, distributed, and implanted by surgeons worldwide. The design and implementation of reproducible methods for the manufacture of devices or the creation and delivery of therapeutic drugs or devices is generally referred to as process development. These processes must conform to federal guidelines previously known as current Good Manufacturing Process, but now known as the Quality Systems Regulations (QSR).

For Cellect, there were three major challenges associated with process development: (1) to define the geometric and biochemical nature of the matrix substrate(s) over which the marrow would percolate, (2) to optimize the flow dynamics of the marrow itself, and (3) to create a disposable medical device that performs the Selective Retention in a rapid, automated and reproducible fashion. To achieve these goals, numerous matrix and device prototypes were designed and tested. The results of each analysis were used to refine the product until the prospectively determined design and performance criteria were met or exceeded. These data, along with product specifications, constitute the Design History File, which is required by certain regulatory agencies.

Figure 3, *A* illustrates the Selective Retention process in which a heterogeneous mixture of cells in bone marrow is applied on top of a matrix. As the marrow flows through the idealized scaffold, osteoprogenitor cells selectively attach to its surface, but other nucleated cells pass through as an effluent. If the matrix design and marrow flow characteristics are optimized, the end result is the creation of a scaffold enriched with an otherwise rare population of osteoprogenitors. Figure 3, *B* shows the final design of the disposable, sterile, Cellect device. Freshly aspirated marrow is added to the graft chamber and then passed through the matrix two times under a controlled, automated process. The cell-enriched matrix may also be clotted with additional marrow or PRP, which contains autologous growth factors and fibrin. Once the surgical bed is prepared, the congealed graft material is removed from the device and implanted directly.

Regulatory Affairs and Quality Systems

Commercialization of an implantable medical device or tissue-engineered product requires approval for marketing from the appropriate regulatory agency. Navigating the pathway of regulatory approval and providing assurance that manufacturing procedures are performed according to regulations and specifications (ie, QSR) are the responsibility of industrial regulatory

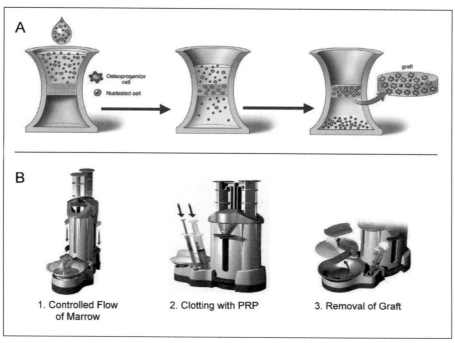

Figure 3 *A* The process of Selective Retention Technology. *B* The creation of a bone graft substitute using the Cellect device. See text for details.

affairs and quality assurance groups, respectively. In the United States, it may be possible to obtain Food and Drug Administration (FDA) clearance through a 510(k) filing if the new product can be proven to be "substantially equivalent" to a predicate device that was commercialized before 1976. Obtaining clearance through a 510(k) filing usually eliminates the need to complete a lengthy and expensive human clinical trial before marketing (see Gadaleta[11] for review). It is also possible to use a recently commercialized product as the predicate device if it was cleared by the FDA through a 510(k) filing. In this way, information regarding approved products can be used to support future product clearances. The primary advantages of the 510(k) pathway are: (1) long-term data from human clinical trials are not usually required before commercialization, and (2) product clearances can occur within 90 days of filing with the FDA. The shortcoming of this type of regulatory submission is that during the early stages of product commercialization, there is generally a lack of long-term data from human clinical trials to support marketing claims of statistically significant efficacy for a given indication. However, long-term studies are an integral part of the postmarketing efforts of innovative products such as Cellect. For products marketed before extensive clinical trials, the scientific principles and preclinical data supporting the new technology are extremely valuable.

Products containing recombinant proteins, novel biomaterials, or culture-expanded or genetically modified cells are not candidates for marketing approval through a 501(k) filing. Most drugs and biologics developed according to FDA guidelines are evaluated in a series of human clinical studies performed as part of an Investigational Device Exemption (IDE) or Investigational New Drug (IND) submission. However, prior to obtaining an IDE or IND approval to begin human clinical trials, a substantial number of preclinical studies are needed. It may take several years to complete these preclinical studies, and the time required must be incorporated into the entire regulatory timeline for approval. In addition, clinical studies of bone healing that require FDA review of the Pre-Market Application (PMA) are likely to take at least 5 years based on past experience. For example, the first recombinant human BMP cleared for marketing by the FDA was under clinical investigation for at least 7 years. Ultimately, however, a PMA approval can provide important product labeling claims that include statistically significant outcomes for a specific therapeutic indication.

For Cellect, it was possible to obtain FDA clearance through the 510(k) pathway based on a predicate device for mixing bone graft substitutes with autologous materials. Although the scientific premise of Selective Retention Technology has its underpinnings in tissue engineering, the specific products that fall under FDA review are the marrow aspiration device, the matrices, and the apparatus used for graft preparation. The marrow and cells are autologous and minimally manipulated and therefore, are not subject to FDA regulations. The primary advantage of approval as a 510(k) product is the extraordinary speed of progression from product concept to commercialization. The current FDA clearance for marketing of Cellect pertains to the disposable device and its use in preparing bone graft substitute materials containing certain matrices and bone marrow with osteoprogenitor cells. At present, broad claims about the features and benefits of graft preparation and its behavior as a bone graft substitute are allowed. However, claims regarding long-term clinical outcomes for specific therapeutic indications cannot be made at this time. It is anticipated that long-term data will confirm the benefit of the product, which has been marketed based on a firm intuitive basis for biologic success, proven efficacy in challenging preclinical models, and the elegant simplicity of the approach.

Clinical Research

As noted, extensive human clinical studies of new therapies are often required before approval for marketing. Industrial clinical research or medical departments usually manage studies of the safety, feasibility, and efficacy of novel diagnostics or therapeutic agents that require an IDE or an IND submission followed by filing of a PMA, Biological License Application (BLA), or a New Drug Application (NDA). However, whether or not products require lengthy clinical studies before regulatory approval, the medical community will eventually require data to justify the use and expense of new products. This is especially true for tissue-engineered devices that tend to be relatively expensive. With this in mind, it is useful to organize clinical studies into one of several categories including: (1) those that support IDE or IND/BLA/NDA approval pathways (including phase I through phase III clinical trials), (2) postmarketing (phase IV) studies that may be used to provide additional clinical data to support product claims, (3) reimbursement studies designed to demonstrate the fiscal advantages of a new product, and (4) case series reports from individual surgeons or small group practice settings. As new products reach the market, their widespread acceptance is dependent upon high quality clinical investigations that confirm product value given the availability of alternative treatments.

The marketing claims of the Cellect device are currently based on clinical data on file as part of the 510(k) filing. A modest set of 1- and 2-year clinical outcome databased on Selective Retention Technology investigations performed at the Cleveland Clinic Foundation[12,13] is available. In addition, there are other data focused on quantifying the Selective Retention of osteoprogenitor cells in bone graft substitutes prepared using the approved device. In one multicenter study that is currently underway, the design of the device causes the majority of nucleated cells to pass through as an effluent, while retaining the rare osteoprogenitor cells with an average efficiency of 80%. This produces a bone graft substitute material with an average 3.6-fold increase in osteoprogenitor cells per unit volume. In parallel to these efforts, certain investigators are continuing to collect data on patients implanted with grafts prepared using a similar Selective Retention Technique. As the Cellect device is put on the market, additional randomized, prospective controlled clinical trials will be performed according to rigorous standards. The results of these investigations will be published in peer-reviewed journals and can serve as the basis of expanded product claims.

Summary and Conclusion

The product development pathway is a complex labyrinth in which diverse people with highly specialized skill sets must cooperate to achieve successful commercialization of the product. Much like the performance of a finely tuned race car, integrated teamwork is essential to cross the finish line safely and in the fastest time possible. An integrative vision from the team's captain must be the unifying force that drives the process and inspires each contributor in their own specific area of expertise. Perhaps this review has demystified aspects of the overall product development pathway necessary to bring tissue engineering initiatives to the operating theater. The aim of this chapter was to describe the overall steps required to advance such technologies into the market by providing an example of a new product that was developed in a relatively accelerated manner.

The Selective Retention Technology and the Cellect device are supported by robust scientific data founded on the central tenets of tissue engineering. Knowledge of cell biology supports the basic premise of the product. In addition, the preclinical animal data show that the Selective Retention Technology and the Cellect device are equivalent to autogenous bone graft in large animal models. This approach is the first to create a bone graft material with a constituent (marrow) enriched for its osteogenic potential. All other technologies, whether based on biomaterials with or without osteoinductive factors, are limited in that they require endogenous progenitor cells to migrate to the implant site, proliferate, and then respond to their local environment. Furthermore, this product provides an economical alternative to other products, but still remains cost effective in relationship to its overall development expense. In this last regard, the accelerated timeline for commercialization of this product saved the sponsor millions of dollars. Through responsible pricing of Cellect, these savings are effectively passed on to the hospitals and third-party payers for medical care around the world. In conclusion, although the barriers to product development may seem overwhelming, it is indeed possible to advance tissue-engineered products from the bench to the bedside. For technologies that are complex and contain multiple components heretofore not currently approved, the pathway will require an IDE (or IND) and take longer than the special case presented. Additional products based on Selective Retention Technology are expected to be commercialized as human clinical trials continue.

References

1. Muschler GF, Boehm C, Easley K: Aspiration to obtain osteoblast progenitor cells from human bone marrow: The influence of aspiration volume. *J Bone Joint Surg Am* 1997;79:1699-1709.

2. Takigami H, Matsukura Y, Boehm C, et al: Canine spine fusion using allograft bone matrix enriched in bone marrow cells and osteoprogenitors. *Proc North Am Spine Soc* 2002;2:100S.

3. Muschler GF, Nitto H, Matsukura Y, et al: Spine fusion using cell matrix composites enriched in bone marrow derived cells. *Clin Orthop* 2003;407:102-118.

4. Haynesworth SE, Kadiyala S, Bruder SP: Platelet rich plasma stimulates stem chemotaxis, proliferation and potentiates osteogenic differentiation. *Proc North Am Spine Soc* 2002;17:92.

5. Slater M, Patava J, Kingham K, Mason R: Involvement of platelets in stimulating osteogenic activity. *J Orthop Res* 1995;13:655-663.

6. Marx RE, Carlson ER, Eichstaedt RM, Schimmele SR, Strauss JE, Georgeff KR: Platelet rich plasma-growth factor enhancement for bone grafts. *Oral Surg Oral Med Oral Pathol Oral Radiol Endod* 1998;85:638-646.

7. Sethi P, Miranda J, Grauer JN, Friedlaender G, Kadiyala S, Patel TC: The use of platelet concentrate in posterolateral fusion: biomechanical and histologic analysis. *Proceedings of the 28th International Society for the Study of the Lumbar Spine* 2001:127.

8. Kraus K, Kadiyala S, Wotton HM, et al: Critically sized osteo-periosteal femoral defects: A dog model. *J Invest Surg* 1999;12:115-124.

9. Kadiyala S, Kraus KH, Attawia M, Bruder SP: Rapid bone regeneration in femoral defects by an autologous osteoprogenitor cell concentrate prepared using an intraoperative selective cell retention technique. *Trans Orthop Res Soc* 2003;28:317.

10. Kapur T, Kadiyala S, Urbahns DJ, Bruder SP: Autologous growth factors and progenitor cells as effective components in bone grafting products for spine, in Lewandrowski, Yaszemski, White, Trantolo and Wise (eds): *Advances in Spinal Fusion: Clinical Applications of Basic Science, Molecular Biology, Biomechanics, and Engineering.* Cambridge, MA, Cambridge Scientific, 2003, pp 381-395.

11. Gadaleta SJ: Bone graft substitutes: A regulatory perspective, in Laurencin (ed): *Bone Graft Substitutes: A Multi-Disciplinary Perspective.* ASTM International, West Conshocken, PA and American Academy of Orthoapedic Surgeons, Rosemont, IL, 2003, (in press).

12. Lieberman I: Clinical use of bone marrow-derived osteoprogenitor cells in orthopaedic surgery. *Trans Orthop Res Soc* 2003;28:28.

13. Takigami H, Matsukura Y, Muschler GF: Retrospective Comparison of Bone Grafting Techniques using Cellular and Acellular Grafts. *Assoc Bone Joint Surg Ann Meet* 2002;54:74-75.

Chapter 38

Genetically Enhanced Tissue Engineering Without Cell Culture or Manufactured Scaffolds

Christopher H. Evans, PhD
Glyn Palmer, PhD
Arnulf Pascher, MD
Andre Steinert, MD
Oliver Betz, PhD
Steven C. Ghivizzani, PhD

Abstract

Most tissue engineering techniques involve multiple steps, including tissue harvest, cell culture, seeding on scaffolds, and reimplantation. We are exploring a variety of methods in an attempt to develop gene-based procedures to achieve tissue repair without the need for ex vivo cell culture or artificial scaffolds. This strategy includes methods to enhance existing natural reparative processes, use endogenous tissues as a source of scaffolds, if necessary, and use in vivo or expedited ex vivo methods of gene delivery. One approach is based on the direct delivery of a vector to the site where tissue regeneration is to occur. The stimulation of osteogenesis with the use of an adenovirus vector carrying cDNAs encoding morphogenetic proteins has been demonstrated in an animal model. However, the in vivo injection of vectors to lesions within cartilage is impractical because there is no convenient means of retaining the vectors within the defect. In addition, there is no readily available endogenous source of chondroprogenitor cells in most situations. Because of this, we are developing technologies based on the use of clotted autologous bone marrow to treat cartilaginous defects. The implementation of natural scaffolds that can be obtained readily and used during existing orthopaedic procedures may help accelerate the incorporation of these techniques into clinical practice.

Introduction

Most tissue engineering protocols involve several major steps. These include tissue harvest, ex vivo cell culture, cell seeding onto scaffolds, further incubation to initiate tissue formation, and the implantation of the engineered structure. To those interested in gene therapy, there is the temptation to enhance the process by genetically modifying the cells while they are in culture and amenable to ex vivo gene delivery.[1] Such procedures may work, but they are elaborate, expensive, and ill-suited to the efficient treatment of large numbers of patients. The need for tissue biopsy and ex vivo cell culture on manufactured scaffolds is particularly cumbersome. Moreover, it considerably complicates the process of regulatory approval. This chapter describes approaches to expedite the processes of genetically enhancing tissue engineering.

Bonadio and Goldstein[2] have made progress in this direction by pioneering the concept of "in situ tissue engineering"[3] using gene activated matrices (GAMs). They reported success in healing large osseous lesions by implanting collagenous matrices impregnated with plasmids encoding osteogenic gene products.[2,4] The use of GAMs in this way eliminates the need for tissue harvest and cell culture, but a manufactured scaffold impregnated with plasmids is still needed.

Our research is directed toward the development of gene-based procedures for achieving tissue repair and regeneration without the need for either ex vivo cell culture or artificial scaffolds. The simplest version of this approach is based on the direct delivery of a vector to the site where tissue regeneration is to occur. If this is not feasible, use is made of natural scaffolds provided by the body that can be obtained readily, and used simply, during existing orthopaedic procedures. One example of each approach is provided in this chapter.

Osteogenesis by In Vivo Gene Delivery

The notion of using gene transfer to enhance bone formation is well accepted.[5] Most protocols use ex vivo procedures to modify osteoprogenitor cells that are then seeded onto a matrix and reimplanted. Ex vivo procedures have been favored for several reasons, including the perception that in vivo procedures will not work because of an immune reaction to the vector, which is usually an adenovirus. This belief is underpinned by the demonstration that the intramuscular injection of powerful adenovirus vectors carrying osteogenic genes only leads to robust intramuscular bone formation in athymic or severe combined immunodeficiency disease animals.

We, however, have shown that the intraosseous introduction of an adenovirus vector carrying cDNA encoding bone morphogenetic protein-2 (Ad.BMP-2) leads to vigorous osteogenesis in immunocompetent rabbits[6] and rats (CH Evans, PhD, unpublished data, 2003). Moreover, the inability of this vector to induce significant bone formation in muscle and, perhaps, other such sites in immunocompetent animals, has the major advantage of limiting unwanted ectopic bone formation.

Baltzer and associates[6] were the first to demonstrate the healing of bone using in vivo gene delivery. They established a rabbit model of critical-sized femoral defects. In this model, the surrounding musculature is fashioned around the lesion to form a chamber into which gene delivery vectors can be injected.[7] The injection of first generation adenovirus vectors carrying marker genes under the transcriptional control of the cytomegalovirus early promoter produced robust transgene expression in the muscle surrounding the defect, the scar tissue within the lesion, and the cut ends of the bone. Most

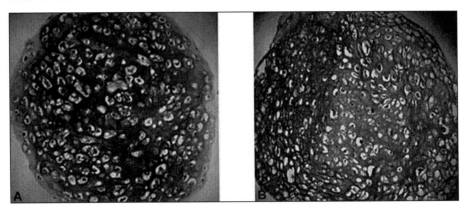

Figure 1 Healing of a rat femoral defect by in vivo gene delivery The 5-mm critical-sized defects were created in the femurs of immunocompetent rats and stabilized by external fixation. Adenovirus carrying the BMP-2 cDNA was injected into the lesion and healing monitored with radiographs. Radiographs obtained at 8 weeks confirmed that defects injected with saline or a control adenovirus (A) did not heal. In contrast, defects injected with the virus carrying the BMP-2 cDNA healed (B).

transgene expression was lost within approximately 3 weeks, although low levels of expression within the bone could be detected beyond 3 weeks. Slight and transient expression was also noted in the liver, but not in several other organs that were examined. Injection of Ad.BMP-2 resulted in healing of the critical-sized defect by radiologic, histologic, and biomechanical criteria.[6,7] The success of this procedure has been confirmed in studies of immunocompetent rats (Figure 1).

Gene Delivery to Sites of Cartilage Damage

The in vivo injection of vectors to lesions within cartilage, in a manner analogous to that described for bone, is impractical because there is no convenient means of retaining the vectors within the defect. Moreover, with the exception of full-thickness lesions in the articular cartilage, there is no convenient, endogenous source of chondroprogenitor cells. Because of this, we are developing technologies based on the use of clotted, autologous bone marrow.

Injury to articular cartilage that penetrates the subchondral bone permits the ingress of blood and bone marrow. This blood then clots and chondroprogenitor cells derived from the marrow produce a fibrocartilaginous repair tissue that, although providing temporary symptomatic relief, eventually degenerates. This natural process thus provides chondroprogenitor cells and a matrix. We are attempting to use gene transfer and clotted, autologous bone marrow to improve the quality of the cartilaginous repair tissue.

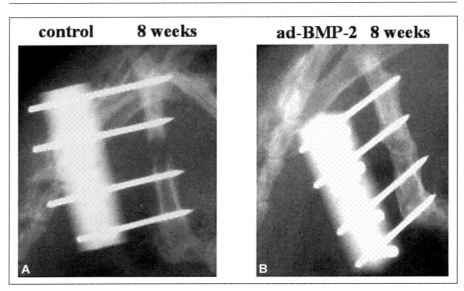

Figure 2 Chondrogenesis in mesenchymal stem cell (MSC) pellet cultures in response to TGF-β gene transfer. Marrow was aspirated from the iliac crests of rabbits and used as a source of MSC. Certain cultures *(A)* were transferred to pellet culture and maintained in chondrogenic medium containing TGF-β_1. Other cultures were transduced with Ad.TGF-β_1 prior to transfer to pellet culture and incubation in medium lacking exogenous TGF-β *(B)*. After 3 weeks, pellets were sectioned and stained with Toluidine blue.

The success of this endeavor has several requirements. First, cells within the marrow must be conducive to gene transfer and transgene expression. Second, transfer of the appropriate genes must enhance chondrogenesis by cells within marrow. Third, there must be good integration of the newly formed tissue with the surrounding cartilage, the reconstitution of a smooth articular surface, and the regeneration of the subchondral plate. Our preliminary data suggest that these requirements can be achieved.

Monolayer cultures of adherent cells derived from rabbit bone marrow have been used to examine gene transfer, gene expression, and chondrogenesis. Using a first generation adenovirus, it has been possible to express a variety of different transgenes in these cell cultures. As described by Johnstone and associates,[8] when naïve cells are transferred to pellet culture in a suitable medium containing transforming growth factor-β_1 (TGF-β_1), they undergo chondrogenesis. It is possible to eliminate the need for exogenous TGF-β by transfer of the appropriate gene before the pellet culture[9] (Figure 2).

When bone marrow clots are cultured in the presence of TGF-β they, too, initiate chondrogenesis (SC Ghivizzani, unpublished data, 2003). Mixing adenovirus vectors with freshly aspirated bone marrow does not impede clot formation and cells trapped within the clot express the transgene in vitro.[10] Genetically modified clots, known as gene plugs, have been generated and implanted into full thickness cartilage lesions in rabbits during a single surgical procedure. This procedure is facilitated by the physical properties of the clot, which resemble those of putty, and, thus, the clot can easily be molded to the required shape and size. After implantation, the gene plugs continue to express high levels of a luciferase marker gene for at least 1 week.[10] Examination of the site of implantation suggests good integration of the gene plug with the surrounding articular cartilage.

Although this approach to repair is still being developed, these preliminary data encourage additional work on this general strategy.

Summary and Conclusions

The successful clinical application of orthopaedic tissue engineering requires both science and pragmatism. Given the current health care environment, the utility of clinical procedures is not only determined by their effectiveness, but also by their cost and ease of application. Procedures based upon ex vivo cell culture are particularly vulnerable in this respect because of their complexity and high cost.

Recognizing these constraints, we are attempting to develop minimally invasive, gene-based approaches to tissue repair and regeneration. Important components of this strategy are to enhance existing natural reparative processes, to use endogenous tissues as a source of scaffolds, if necessary, and to use in vivo or expedited ex vivo methods of gene delivery. These principles are exemplified in the two examples described in this chapter. The healing of critical-sized, osseous defects in rabbits and rats was achieved by the simple injection of vector into the lesions. For cartilaginous defects, we are developing gene plugs derived from autologous marrow. Both of these strategies of repair can be incorporated into existing orthopaedic practice without requiring additional invasive procedures, a circumstance that should expedite their translation into clinical application.

Suitably adapted, these methods should find additional application in orthopaedic tissue engineering, including the repair of intervertebral disk, ligament, tendon, meniscus, and partial-thickness lesions in articular cartilage.

Acknowledgments

We are grateful to the Orthopaedic Trauma Association for funding our pilot studies into bone healing in rats, now supported by NIAMS grant AR050243. These studies are conducted in collaboration with Drs. T. Einhorn and L. Gerstenfeld of Boston University Medical Center, and Dr. M. Vrahas of Harvard Medical School. Research into chondrogenesis in pellet culture is performed in collaboration with Dr. B. Johnstone of Case Western Reserve University Medical School and supported by NIAMS grant AR050249.

References

1. Evans CH, Robbins PD: Genetically augmented tissue engineering of the musculoskeletal system. *Clin Orthop* 1999;379S:S410-S418.

2. Bonadio J, Smiley E, Patil P, Goldstein S: Localized, direct plasmid gene delivery in vivo: prolonged therapy results in reproducible tissue regeneration. *Nat Med* 1999;5:753-759.

3. Goldstein SA, Patil PV, Moalli MR: Perspectives on tissue engineering of bone. *Clin Orthop* 1999;379S:S419-S423.

4. Fang J, Zhu YY, Smiley E, et al: Stimulation of new bone formation by direct transfer of osteogenic plasmid genes. *Proc Natl Acad Sci USA* 1996;93:5753-5758.

5. Lieberman JR, Ghivizzani SC, Evans CH: Gene transfer approaches to the healing of bone and cartilage. *Mol Ther* 2002;6:141-147.

6. Baltzer AW, Lattermann C, Whalen JD, et al: Genetic enhancement of fracture repair: healing of an experimental segmental defect by adenoviral transfer of the BMP-2 gene. *Gene Ther* 2000;7:734-739.

7. Baltzer AW, Lattermann C, Whalen JD, Braunstein S, Robbins PD, Evans CH: A gene therapy approach to accelerating bone healing: Evaluation of gene expression in a New Zealand white rabbit model. *Knee Surg Sports Traumatol Arthrosc* 1999;7:197-202.

8. Johnstone B, Hering TM, Caplan AI, Goldberg VM, Yoo JU: In vitro chondrogenesis of bone marrow-derived mesenchymal progenitor cells. *Exp Cell Res* 1998;238:265-272.

9. Yoo JU, Mandell I, Angele P, Johnstone B: Chondroprogenitor cells and gene therapy. *Clin Orthop* 2000;379S:S164-S170.

10. Pascher A, Palmer GD, Steinert A, et al: Gene delivery to cartilage defects using coagulated bone marrow aspirate. *Gene Ther* 2004;11:133-141.

Chapter 39

In Situ Tissue Engineering: Localized Delivery of DNA

Steven A. Goldstein, PhD

Abstract

Tissue engineering methods to repair or regenerate musculoskeletal tissues could be optimized by methods to appropriately enhance the local production of inductive factors. One of the most important features of this approach is its ability to provide sustained expression of the transgene products. A variety of recombinant growth factors and cytokines delivered with the use of a natural or synthetic polymer matrix have been shown to induce tissue regeneration in animal models, but translation into clinical therapies has been difficult. More recently, investigators have developed methods to deliver genes for the desired proteins by ex vivo transfection followed by local transplantation of host cells, and delivery in liposomes or proteolipsomes along with viral envelope receptor proteins in a polymer carrier complex. As an alternative, we have established the ability to directly present DNA to cells, in vivo, using a carrier matrix. The key to this approach appears to be the properties of the matrix and its ability to immobilize the DNA, making it accessible to the endogenous migrating repair cells for long periods of time.

Introduction and Background

Traditional approaches to tissue engineering often involve the implantation of a scaffold containing cells and/or bioactive molecules, or the transplantation of primitive tissue constructs developed in bioreactors. Success depends on the appropriate engagement of these cells, scaffolds, and biologic regulatory factors irrespective of whether they are of exogenous or endogenous origin.[1-3]

Most musculoskeletal tissue engineering strategies have focused on the delivery of recombinant proteins with the use of natural or synthetic polymer matrices. A variety of growth factors and cytokines including the bone morphogenetic proteins (BMPs), transforming growth factor-β, Indian hedgehog protein, fibroblast growth factor (FGF), platelet-derived growth factor, and others have been shown to induce tissue regeneration in animal models, but translation into clinical therapies has been difficult.[3-8] Explanations for the delay in the successful clinical application of these methods include rapid degradation of unprotected protein leading to the need for large doses, difficulties with the incorporation of these proteins into delivery vehicles because of their relatively unstable chemistry, and subsequently, their short time of residence in the wound site.[7,9] In an effort to address these limitations, alternative approaches using gene therapy to promote tissue regeneration have been evaluated by several groups.[7,8,10-12]

The most substantial efforts in developing gene therapies have focused on diseases that result from generalized genetic disorders and have relied on indirect means of introducing genes into tissues. For example, target cells were removed from the body, infected with viral vectors carrying recombinant genes, and implanted into the body with the goal of achieving integration of the therapeutic genetic material into selected cells.[4,13,14] Success would be related to long-term activity of the recombinant genes and gene products. In contrast to this disease specific approach, tissue engineering methods to repair or regenerate musculoskeletal tissues would be better served by appropriate local production of inductive factors.

Two methods of the local delivery of genes encoding growth factors to promote sustained, effective concentrations of therapeutic agents have demonstrated success in experimental animals. The first method involves the acquisition of host cells followed by ex vivo transfection under controlled culture conditions. The transduced cells are then transplanted locally into the wound where they synthesize and express the selected factor to promote repair or regeneration.[10-12]

The second approach is based on in vivo gene transfer.[7,8] Direct in vivo gene transfer has been achieved with formulations of DNA encased in liposomes or trapped in proteolipsomes that contain viral envelope receptor proteins, calcium phosphate-coprecipitated DNA, and DNA coupled to polymer carrier complexes.[15] Wolff[16] demonstrated that direct infection of purified preparations of DNA and RNA into skeletal muscle using a murine model resulted in significant expression of a reporter gene. This unexpected finding may be attributed to either the unique attributes of muscle cells and/or to damage associated with DNA infection that allows transfection to occur. Based on these observations, several investigators have demonstrated potential benefits of direct injection of DNA with adenoviral vectors into bone defects or into muscles surrounding the defect sites.[11] These direct injection methods, however, may be difficult to control spatially and temporally. As an alternative, we have established the ability to present DNA to cells, in vivo, using a carrier matrix. The critical design features of this technique include the ability to immobilize the DNA within a defined three-dimensional volume that can be placed directly into a wound site. The method also allows the possibility of controlling the temporal presentation of one or more genes to the endogenous cells attracted to the matrix carrier.

This chapter summarizes the potential clinical value of direct matrix delivery of DNA. We consider this technology as an application of in situ tissue engineering. Preclinical examples of applying the technology to bone defects, skin ulcers, and ischemic heart tissue are discussed in an effort to identify the key issues that must be considered for successful use of these techniques in clinical practice.

Matrix Delivery of DNA by Gene Activated Matrices

As described previously, the localized gene therapy delivery technique involves the immobilization of DNA within a three-dimensional structural matrix.[7,8] This technology has been denoted as gene activated matrix (GAM) and constitutes a method for in situ tissue engineering. In this form of tissue engineering, the source of the cells is endogenous. The implantable device includes a structural matrix (scaffold) with an associated biofactor (DNA). In practice, the implant is placed or injected into the wound site, where the matrix acts to attract endogenous cell populations involved in wound repair to migrate into the carrier. These endogenous cells subsequently engage with the DNA and transfection is initiated. Clearly, the chemical properties of the matrix can have a profound effect on the recruitment, attachment, and activity of the endogenous cells that migrate to the wound. The properties of the matrix also affect the degree of protection or exposure of the associated DNA.

Several studies in small animals demonstrated the feasibility of in vivo transfection using both marker and functional genes.[8,17] In addition, success as determined by the formation of repair tissue has been observed in small and large animals.[7,17,18] Although these technologies have shown great promise, widespread clinical application will depend on additional work to identify the gene or gene combinations that promote consistent and robust tissue regeneration. The chemical, morphologic, and mechanical properties of the matrix that optimally delivers the DNA and potentially enhances transfection efficiency must also be elucidated. These issues will be addressed in the summaries of three preclinical studies.

Direct Gene Transfer for the Enhancement of Bone Regeneration

During the past seven years, Bonadio and Goldstein and their associates developed and evaluated a method of direct in vivo gene transfer[1,4,7,17]. The primary in vivo model used to evaluate the gene transfer concept involved the production of a 5-mm, middiaphysis femoral defect in rats. To date, we have successfully completed the protocol on more than 220 animals. The 5-mm defect size was chosen to be a critical size that prevents the formation of bone, unless a positive stimulant is added.

These studies first involved marker genes followed by the use of GAMs, including BMP-4 or a plasmid encoding for a fragment of parathyroid hormone (PTH_{1-34}) individually or in combination. Although numerous matrix carriers were evaluated, including polylactic acid-coglycolic acid polymers, calcium phosphates, hydroxyapatite, and others, the majority of the studies used lyophilized type I collagen carriers. The treated bones were extracted and analyzed mechanically, histologically, and by high resolution imaging techniques. The results of these studies demonstrated that repair cells (primarily fibroblasts) can be genetically manipulated in vivo. Bone formed in the critical defects and thus verified the potential power of providing osteotropic genes to a wound site. In essence, this technology encourages the endogenous cells to become bioreactors producing recombinant factors that can influence the local response. This observation prompted the use of the descriptor "in situ tissue engineering."

Studies in rats were followed by a study using a defect in the tibia of large dogs. The rationale for moving from small animals to large animal models was based on recognition that the physiology of rat bone differs from that of human bone with regard to composition, nutritional requirements, and ability to be remodeled. The canine model involved the creation of a segmental defect in the tibia that most often results in nonunion, if left untreated. The GAMs incorporating PTH_{1-34} were surgically placed in the defects. Radiographs were used to determine the progression of healing. At the end of the study, the bone was evaluated mechanically, histologically, and morphologically. The results demonstrated significant promotion of bone regeneration and evidence of sustained delivery of the plasmid.[7] Following the canine study, investigations of GAMs in spinal fusion augmentation were conducted in sheep. Two independent studies using vertebral interbody fusion cages were performed. One study involved a three level fusion with cylindrical cages and the other study used a two level fusion using rectangular cages. Gene activated matrices composed of collagen carriers with plasmids encoding PTH, FGF, or BMP individually and in combinations were inserted inside the cages before implantation. Mechanically stable fusions and substantial bone formation was observed in the cages.[18] Although the sample sizes were small, combinations of genes appeared to show promise for increased responses compared with single gene delivery.

The results of these studies demonstrated that in vivo transfection could direct the sustained expression of recombinant factors for up to 6 weeks. Transfection resulted in promotion of an osteogenic response, but the response in some of the studies may not be sufficiently robust to ensure consistent functional healing in clinical practice. Both the choice of genes and the chemical and morphological properties of the matrix appear to be critical. Combinations of genes appeared to be more osteogenic than the use of single genes. An important function of the matrix is to immobilize the DNA in the wound, while simultaneously enhancing its presentation to migrating repair cells. An example of the osteogenic response is illustrated in Figure 1. New bone has completely filled the open volume of the cage that was seeded with a GAM incorporating a combination of PTH_{1-34} and FGF.

Localized Gene Delivery and Myocardial Neovascularization

The current treatment strategies for ischemic heart disease include the therapeutic promotion of neovascularization in the affected tissues. One approach has been the selected creation of wounds in the muscle tissue with the goal of promoting a repair cascade that would involve new vessel formation and subsequently revascularization of the ischemic area. Recently, several studies using GAMs to deliver angiogenic factors to modeled ischemic tissue were conducted.[19,20]

A pig model of chronic myocardial ischemia was used. Five weeks after a surgical procedure to create the ischemia, GAMs that were composed of a collagen matrix and the gene for FGF-2 with an adenoviral delivery vector, were injected into multiple sites within the ischemic zone. Six weeks later,

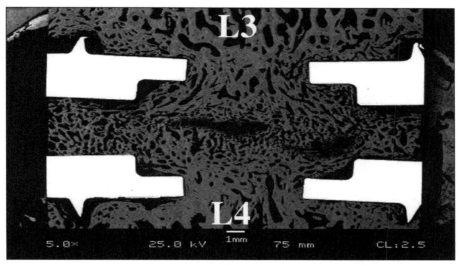

Figure 1 Back scattered scanning electron microscopic image of a central sagittal section of the intervertebral fusion cage demonstrating extensive bone formation. The rectangular cage was filled with a collagen matrix carrying plasmids for PTH_{1-34} and FGF. (sheep, 12 weeks postoperative)

the tissue was evaluated by immunohistologic methods and by functional assays including MRI and electrocardiography.[20]

The results demonstrated a significant restoration of myocardial function and neovascularization as documented by histologic observation of numerous new arterioles (Figure 2). The immobilization of the DNA in the wound sites (injection sites) by the collagen matrix enables sustained expression of the FGF-2 transgene and the promotion of clinical function. The properties of the immobilizing matrix as well as the efficiency of the viral vector delivery scheme contributed to the success of the local gene therapy.

Localized Gene Therapy in Excisional Wounds

Recombinant protein therapy has been used clinically for the treatment of skin ulcers. However, difficulties in maintaining sustained local concentrations of the desired protein have stimulated the search for alternative strategies.

Doukas and associates[15] evaluated the potential of GAM technology to promote wound repair subsequent to an excisional injury. The model involved the creation of skin wounds in rats and the implantation of GAMs composed of collagen sponges and a platelet-derived growth factor B-encoding adenovirus (AdPDGF-B). Collagen matrices alone and aqueous solutions were used as controls. The GAMs substantially enhanced granulation tissue formation and epithelialization. Eventually, complete wound closure without scarring was achieved with GAM technology. Both the vector and transgene products were retained in the wound site by the collagen

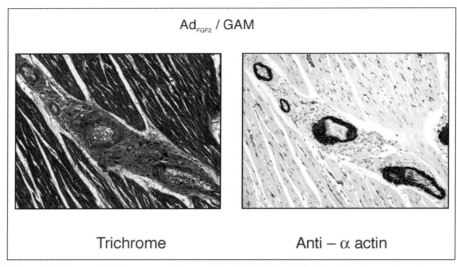

Figure 2 New arterioles were observed in all histologic fields after 6 weeks of sustained delivery of FGF-2. *(Adapted with permission from data from Horvath 2002)*

sponge, even at 28 days postimplantation. Healing did not occur in animals with the aqueous formulations.

Thus, the properties of the matrices and their ability to immobilize the DNA, making it accessible to the endogenous migrating repair cells for long periods of time, appears to be key to the success of the gene therapeutic approach.[15,21]

Summary and Future Needs

The studies described in this chapter as well as others, continue to support the potential of an in situ tissue engineering approach using localized gene therapy to promote tissue regeneration. One of the most important features of this approach is its ability to provide sustained expression of the transgene products. These studies have identified several critical issues that should be considered as part of a strategy to successfully develop these technologies for clinical use.

Matrix delivery of DNA allows the use of numerous factors that are not feasible to deliver as proteins. These include transcription factors, intercellular regulators, and other factors that might support the intended clinical outcome. Critical properties of the matrices include its ability to immobilize the DNA and to protect it from early degradation. These features must be balanced by a need for the matrix to allow DNA to be available to the cells. In addition, the physical properties of the matrix must enable it to be handled easily and to remain localized in the wound site. An ideal matrix would have chemical and/or biologic properties that facilitate appropriate migration and attachment of cells that are targeted for transfection. In some circumstances, it may be desirable for the matrix to have mechanical competence to provide early integrity and stability of the tissue engineering construct. The matrix may also be designed to participate in the stimulation of tissue regeneration. For example, the composition of a matrix may include an osteoconductive material to work in partnership with the osteoinductive factor being delivered.

Clearly the choice of the gene or genes is critical for successful tissue regeneration. Thus there is a need for studies to identify the genes and gene sequences that optimize regeneration of the tissue of interest. The temporal availability of the genes may also be important and, therefore, the temporal pattern of gene expression that will promote robust tissue regeneration requires additional study. Finally, methods to enhance the transfection efficiency of the process are needed.

References

1. Goldstein SA, Patil PV, Moalli MR: Perspectives on tissue engineering of bone. *Clin Orthop* 1999;367:S419-S423.

2. Langer R, Vacanti JP: Tissue engineering. *Science* 1993;260:920-925.

3. Orban JM, Marra KG, Hollinger JO: Composition options for tissue engineered bone. *Tissue Eng* 2002;8:529-539.

4. Bonadio J, Goldstein SA, Levy RJ: Gene therapy for tissue repair and regeneration. *Adv Drug Deliv Rev* 1998;33:53-69.

5. Einhorn TA: Enhancement of fracture healing by molecular or physical means: An overview, in Brighton CT, Friedlaender G, Lane JM (eds): *Bone Formation and Repair*. Rosemont, IL, American Academy of Orthopaedic Surgeons, 1994, pp 223-238.

6. Reddi AH: Role of morphogenetic proteins in skeletal tissue engineering and regeneration. *Nat Biotechnol* 1998;16:247-252.

7. Bonadio J, Smiley E, Patil P, Goldstein SA: Localized, direct plasmid gene delivery in vivo: prolonged therapy results in reproducible tissue regeneration. *Nat Med* 1999;5:753-759.

8. Fang J, Zhu YY, Smiley E, et al: Stimulation of new bone formation by direct transfer of osteogenic plasmid genes. *Proc Natl Acad Sci USA* 1996;93:5753-5758.

9. Pierce GF, Mustoe TA: Pharmacologic enhancement of wound healing. *Annu Rev Med* 1995;46:467-481.

10. Lieberman JR, Le LQ, Wu L, et al: Regional gene therapy with a BMP-2-producing murine stromal cell line induces heterotopic and orthotopic bone formation in rodents. *J Orthop Res* 1998;16:330-339.

11. Baltzer AW, Lattermann C, Whalen JD, et al: Genetic enhancement of fracture repair: Healing of an experimental segmental defect by adenoviral transfer of the BMP-2 gene. *Gene Ther* 2000;7:734-739.

12. Boden SD, Titus L, Hair G, et al: Lumbar spine fusion by local gene therapy with a cDNA encoding a novel osteoinductive protein (LMP-1). *Spine* 1998;23:2486-2492.

13. Culver KW (ed): *Gene Therapy: A Primer For Physicians*, ed 2. Larchmont, NY, Mary Ann Liebert, 1996.

14. Crystal RG: Transfer of genes to humans: Early lessons and obstacles to success. *Science* 1995;270:404-410.

15. Doukas J, Chandler LA, Gonzalez AM, et al: Matrix immobilization enhances the tissue repair activity of growth factor gene therapy vectors. *Hum Gene Ther* 2001;12(7):783-798.

16. Wolff JA: Direct gene transfer into mouse muscle in vivo. *Science* 1990;247:1465-1468.

17. Goldstein SA, Bonadio J: Potential role for direct gene transfer in the enhancement of fracture healing. *Clin Orthop* 1998;355:S154-S162.

18. Patil PV, Graziano GP, Bonadio J, et al: Interbody fusion augmentation using localized gene delivery. *Trans Orthop Res Soc* 2000;25:360.

19. Doukas J, Blease K, Craig D, Ma C, Chandler LA, Sosnowski BA, Pierce GF: Delivery of FGF genes to wound repair cells enhances arteriogenesis and myogenesis in skeletal muscle. *Mol Ther* 2002;5:517-527.

20. Horvath KA, Doukas J, Lu CY, Belkind N, Greene R, Pierce GF, Fullerton DA: Myocardial functional recovery after fibroblast growth factor 2 gene therapy as assessed by echocardiography and magnetic resonance imaging. *Ann Thorac Surg* 2002;74:481-487.

21. Chandler LA, Gu DL, Ma C, et al: Matrix-enabled gene transfer for cutaneous wound repair. *Wound Repair Regen* 2000;8:473-479.

Future Directions

Methodologies Used in the Construction of Engineered Tissues

Better understanding of the host environment

After implantation, tissue engineered structures are required to perform under in vivo conditions that we understand only poorly. However, the ability of the implant to function as intended is critically affected by the mechanical, chemical, and biological environment it encounters in the body. Remembering that the properties of the injured tissue will differ from normal, research needs to be devoted towards defining this environment, including the one perceived by the cells within the engineered construct.

Better understanding of the properties of native tissue

For optimal function, engineered tissue needs to replicate the biological and physical properties of the tissue undergoing repair. Despite considerable progress, understanding the biological properties of many orthopaedic tissues is inadequate. As well as obtaining this information experimentally, mathematical models of tissue development and function would extend existing theory and help establish rational design principles for functional tissue engineering.

Improvements in scaffold design

Scaffolds need to be optimized with regard to multiple criteria, including:
- The development and prioritization of design criteria that meet the biomechanical and metabolic demands of the tissue to be replaced.
- Design and characterization of the biomechanical and biophysical properties of biomaterial scaffolds for tissue engineering.
- Understand the effects of scaffold topology on performance.
- Design of scaffolds capable of delivering multiple growth factors according to preset release profiles and/or in response to mechanical loading.
- Further refinement of scaffolds capable of gene delivery.
- Development and evaluation of additional scaffolding materials including biologically based biomaterials.
- Development of custom-designed scaffolds for individual needs.

   These endeavors would be aided by the development of theoretical models to generate rational, new designs.

Improve understanding of cell-scaffold interactions

Issues to be addressed include:
- Determination of how best to load cells onto scaffolds.
- Understand the mechanisms through which the matrix brings about changes in cell behavior.
- Understand which cells within a scaffold are responders and which ones are effectors.

Evaluate the effects of preconditioning engineered tissues prior to implantation

The in vivo performance of the newly formed tissue that has grown in vitro may be improved by preconditioning. Methods for doing this include the application of biophysical stimuli and the use of biomimetic bioreactors that allow the application of physiologically relevant interstitial flow and mechanical forces.

Increase understanding of the responses of the host to scaffolds

Matters to be addressed include:
- Host immune response to implanted materials (immune acceptance of implant).
- Understanding the local and systemic responses to breakdown products released by the scaffold.
- Determine the fate and biodistribution of the implanted cells. This will require the development of improved methods for cell tracking, particularly by noninvasive in vivo methods.

Establishment of criteria for assessing scaffold function

A common set of criteria for evaluating the properties of scaffolds would expedite development and enable meaningful comparisons between different scaffolds. This will require the development of suitable testing methods with which to determine scaffold function according to mechanical and biological criteria. A quantitative set of outcome measures is required. In conjunction with this, it would be very valuable to develop early predictors of failure and success.

Scale-up

In transitioning from the laboratory to the clinic, it is critical to consider scale-up issues. These include formulation of processing methods to create cell-interactive scaffolds and the development of methods for rapid and large-scale production of scaffolds and seeded implants. Moreover, there needs to be suitable models for confirming that the scale-up has not adversely affected function. The biology of scale-up also needs to be addressed. For example, there is a tenfold difference in the thickness of cartilage in moving from rodents to humans. There is a possible role for bioreactors in this regard.

Reducing invasivity—design of injectable and self-assembling scaffolds

Surgery is moving increasingly towards minimizing invasivity. An injectable, self-assembling scaffold would reduce the invasivity of tissue engineering. In certain instances, it may be possible to eliminate the need for manufactured scaffolds.

Improve methods of cell sourcing

For many applications it is necessary to harvest donor tissue from which cells are grown before seeding onto scaffolds. Obtaining the tissue can be problematic if an invasive harvesting procedure is needed. Comparisons should be made between the efficacy of cells that are easy to obtain, such as those contained within fat, to those, such as chondrocytes, whose harvest is more invasive. The plasticity of these cells also needs further investigation. In this context, the possibilities of allografting need to be subject to detailed experimental evaluation.

Mobilization and concentration of intrinsic cells into scaffolds

Research into the formulation of scaffolds with the ability to modulate host cells in predictable fashions seems warranted. It may, for example, be possible to manufacture scaffolds with designer docking sites for specific types of endogenous cells. Once captured, the behavior of these cells could be influenced in deliberate ways.

Index

Page numbers with *f* indicate figures; page numbers with *t* indicate tables.

A

Absorbable collagen sponges (ACS), 52-57, 66*f*, 343
Academic Research Enhancement Awards (R15), 90
Adenoviruses, 125, 390-391
Adipocytes, 197, 370
Adipose-derived adult stromal/stem (ADAS) cells, 363-367
 cell surface markers, 366-367
 differentiation potential, 367*t*, 370-371
 future, 371-372
 musculoskeletal lineages, 367-370
 surface protein profile, 366*t*
Adipose endothelial cells, 365
Adsorption, distribution, metabolism and excretion (ADME), 67
Adverse events, 79-80, 94. *see also* Outcomes
Agarose gels, 179
Age/aging
 cartilage donors, 220-221, 221*f*
 degenerative diseases, 6-7
 disk degeneration, 284
Aggrecans, 214
Alcian blue stain, 229
Alginate
 chondrocyte culture, 211
 cross-linkers, 360*f*
 gels, 179-180
 hydrogel scaffolds, 358
 injectable, 333
 PLA amalgam, 171
 stromal cells on, 368*f*
Alginate-Recovered-Chondrocyte (ARC) method, 212-216, 212*f*, 213*f*
Alizarin red stains, 369*f*
Alkaline phosphatase (AP), 124
Alu sequences, 44
Angiogenic factors, 108, 400*f*
Animal models
 ADME characterization, 67
 nucleus pulposus tissues, 287*t*
 regulatory process and, 19-20
Anulus fibrosus (AF), 284
Anterior cruciate ligaments (ACLs), 242, 244, 246, 258
Anterior lumbar interbody fusion (ALIF), 144-146
AP2, 370
Articular cartilage. *see also* Cartilage
 constituents, 230
 focal defects, 177
 full-thickness defects, 185-187, 185*f*, 186*f*
 functional assessment, 227-233
 mechanical evaluation, 230-231
 partial-thickness defects, 187
 reparative approaches, 166
 structure, 228

Articular surfaces, 6, 18
Ascorbic acid, 170, 197
Aseptic processing, 316, 317
Autografts, harvest, 342
Autologous bone grafts, 115-116
Autologous bone marrow, 108-109
Autologous chondrocyte implantation (ACI), 36-37, 74, 166, 201-210
Autologous cultured chondrocytes, 314, 324
Autologous osteochondral transplantation, 36
Autologous progenitor cell preparation kits, 377-387
Autologous tissue replacement, 176*f*
Avian tendon internal fibroblasts (ATIFs), 259
Axial malalignment, 203

B

Bacteria, anaerobic, 155-156
Bacterial contamination, 155-156
BAK Interbody Fusion System, 342
Basic fibroblast growth factor. *see* Fibroblast growth factor-2 (FGF-2)
β-Sheet structures, 350
Bioartificial tendons (BATs)
 ex vivo 3-D model, 257-266
 fabrication, 259
 mechanical loading, 259-260, 260*f*
 outcome measures, 260-264, 264*t*
Bioceramics, 116, 119*f. see also* Osteoconductive materials
Biodegradability, 177-178
Bioengineering
 improvements in tissues, 249-250
 orthopaedics and, 4-6
 structural, 5
 tissue, 5-6
Biological License Application (BLA), 384-385
Biopsies, documentation, 315
Bioreactors, 176, 321-322, 323*f*, 325, 326*f*
Biphasic theory, 230-231
Bleeding, graft remodeling and, 248-249
BMP-2 Evaluation and Surgery for Tibial Trauma (BESTT) study, 111
Bonding, biologic glues, 188
Bone
 ectopic formation, 132
 healing process, 118
 MSC/Adv-BMP-2 therapy, 126*f*, 127*f*
 muscle-derived stem cells and, 131-140
 orthotopic, 132
 replacement, 332
 tumors, 8
Bone grafts, 177
Bone marrow
 aspiration, 379
 cell types, 168
 mesenchymal stem cells, 109, 168, 196-197
 stimulation procedures, 35-36

Bone marrow-derived cells (BMDCs), 44, 118-120, 170, 366-367. *see also* Specific cell types
Bone marrow stromal cells (BMSCs), 118-120
Bone morphogenetic protein-2, human recombinant (rhBMP-2)
  ACS delivery, 52-57
  carriers, 66*f*
  characterization, 52
  commercial approval, 61-71
  CPM delivery, 52
  studies, 144-148
Bone morphogenetic protein-2 (BMP-2)
  bone healing and, 110, 132
  cartilage regeneration and, 180
  cDNA delivery, 390
  CPM delivery, 57
  degenerative disk disease study, 144
  effect on MSCs, 124
  gene therapy and, 135-136
  gene transfer, 125-126, 126*f*, 127*f*
  osteoinductive properties, 141
  product development process, 64-65
  in spinal fusion procedures, 18
Bone morphogenetic protein-4 (BMP-4)
  bone formation and, 135*f*
  bone healing and, 110, 132
  in fracture repair, 112-113
  gene therapy and, 135-136, 397
  gene transfer, 125
  MSC differentiation and, 345
Bone morphogenetic protein-6 (BMP-6), 125
Bone morphogenetic protein-7 (BMP-7), 110, 125, 141. *see also* Osteogenic protein-1 (OP-1)
Bone morphogenetic protein-9 (BMP-9), 110, 125
Bone morphogenetic protein-14 (BMP-14), 343
Bone morphogenetic proteins (BMPs)
  clinical applications, 110-111
  oncogenesis, 156
  in scaffolds, 379
Bone sialoprotein, 124
Bromodeoxyuridine (Brd-U), 275-276
Business development, 381-382

C

Calcium deposition, 337
Calcium phosphate, 108
Calcium phosphate matrix (CPM), 52, 57, 64-65
Calcium polyphosphate powder (CPP), 285
Calcium sulfate, 108
Calibration-metrology groups, 316
Callus formation, 119
Cancers, BMP expression, 156
  1,3-Bis-p-carboxyphenoxy-propane (CPP), 334
Career development grants, 96-99
Career Transition Development Award (K22), 97
Carticel, 18, 314, 324
Cartilage. *see also* Articular cartilage
  assessment methods, 228-231
  cartilage-like tissue, 351

chondrocyte formation of, 214
clonal proliferation, 165
construct fabrication, 170-171, 170*f*
donor age, 220-221, 221*f*
engineered, 325-327
engineering structures, 175-182
fetal, 325
functional assessment, 227-233
gene delivery to sites of damage, 391-392
loading during in vitro growth, 351-352
manufacture in vitro, 211-217
nanofibrous scaffolds, 171-172, 172*f*
polymer fiber constructs, 178-179, 178*f*
repair using neocartilage, 219-226
scaffold design, 355-362
self-assembling peptide scaffolds, 349-353
structure-function relationship, 229
three-dimensional formation, 172
tissue engineering, 321-330
  microenvironments, 183-193
in vitro engineered constructs, 179*f*
Cartilage defects
  chondrocyte loss, 189
  focal, 177
  full-thickness, 185-187, 185*f*, 186*f*
  knee, 73-83
  ontogenesis during repair, 190-191
  partial-thickness, 187*f*, 224-225, 225*f*
  scaling, 184-185
  tissue engineering perspectives, 165-174
  types, 184-185
CD56, 367
Cell-associated matrices, 211-212
Cell-based therapies
  classification, 46*t*
  'potential product designation, 47*t*
  visions, 41-49
Cell culture. *see also* Tissue culture
  Alginate-Recovered-Chondrocyte (ARC) method, 212-216, 212*f*
  autologous chondrocytes, 36
  avian tendon internal fibroblasts, 259
  future directions, 317-318
  hydrodynamic factors, 324-325
  quality control, 313-319
  supplementation, 170-171
  techniques, 175-177
  three-dimensional systems, 176
  transit, 315
Cell factories, 318
Cell fusion verification, 45
Cell leaking, 45
Cell surface markers, 42
  ADAS cells, 366-367, 366*t*
Cell-to-matrix ratios, 324
Cell tracking, 43-44
Cellect, 381-385, 383*f*
Cells. *see also* Specific cell types
  density, 324
  shelf-life, 315
  source, 324
  tissue engineering, 167, 321-322
  viability testing, 316-317

Cellular products, definition, 42-43
Center for Biologics Evaluation and Research (CBER), 19, 46
Center for Devices and Radiological Health (CDRH), 18
Center for Drug Evaluation and Research (CDER), 18-19
Cervical spine fusion, 147-148
Chinese hamster ovary (CHO) cells, 63
Chitosan, 333, 358
CHO (Chinese hamster ovary) cells, 63
Chondrocytes
    ADAS differentiation and, 368
    alginate culture, 211
    autologous, 18, 36-37, 314
    bovine meniscal, 275
    cartilage-like tissue development, 351
    cartilaginous tissue formation, 214
    characterization, 42-43
    culture assessment, 286
    differentiation factors, 189-190
    loading during in vitro growth, 351-352
    loss, 189
    matrix formation, 196-197
    media choice, 352
    seeded peptide gels, 350-351
Chondrogenesis, 322-325, 392-393, 392f
Chronaxie (C_{50}), 299
Clinical research. see also Clinical trials
    administration, 86-90
    cell-based therapies, 45-46
    certificates of confidentiality, 95
    challenges, 25-31
    National Institutes of Health, 85-99
    regulatory process and, 384-385
    training grants, 95-96
Clinical Research Curriculum Award (K30), 98
Clinical trials
    end points, 25
    independent monitoring, 93-94
    INFUSE Bone Graft, 67-69, 68f
    National Institutes of Health and, 85-99
    NIAMS Clinical Trial Planning Grants, 91
    phases, 86
    rhBMP-2/ACS, 53-57
    tissue engineering products, 30-31
Clinical utility criteria, 9-10
Clinical value criteria, 10-11
Collagen matrices
    cell-binding capacity, 178-179
Collagen matrix sponge
    HELISTAT, 64-65
Collagen Meniscus Implant, 342
Collagens
    cell delivery, 345
    distribution, 342
    gene therapy and, 345-346
    hydrogel scaffolds, 358
    orthopaedic therapy and, 341-348
    scar formation and, 246
    sources, 346
    type I, 124, 342, 397
    type II, 172, 214, 221, 268, 368f

    type III, 124, 245
    type IV, 368f
    type IX, 172
    type X, 172
    type XI, 172
    type XII, 263
Collagraft, 342, 343f
Colony-forming units-fibroblastic (CFU-f), 116-118
Compliance development functions, 316
Compliance groups, 315
Composite scaffolds, 356t, 358
Confidentiality certificates, 95
Constructs, 249, 326-327. see also Scaffolds
Containers, quality control, 315
Contraction index, 260-261, 261-262
Craniomaxillofacial reconstruction, 151-159
CRISP (Computer Retrieval Information on Scientific Projects), 88
Cryosectioning, 229

D
Data and Safety Monitoring Board (DSMB), 93-94
Decorin anti-sense therapy, 246
Degenerative diseases, 6-7. see also Specific diseases
Delrin Troughloader, 260f
Demineralized bone matrix (DBM), 62, 108
Dental applications, 331-340
Design History File (DHF), 382
Dexamethazone, 124, 170, 197, 368
Direct costs, grants, 90
Disease transmission, 116, 155-156, 284-285, 332
Diskectomy, 146
Distraction osteogenesis, 108, 115
DNA, 395-402, 396
DNA plasmids, 125
Document control groups, 315
Documentation, quality assurance and, 315
Drug delivery. see also Gene transfer
    carrier matrices, 64-65
    collagen scaffolds in, 342
    rhBMP-2/ACS, 52-57, 66f
    rhBMP-2/CPM, 57
Dulbecco's Minimum Essential Media (DMEM), 352f
Dystrophin, 134

E
Efficacy
    criteria, 9
    evaluation of, 35
    rhBMP-2/ACS therapy, 56
    tissue engineering products, 34-37
Electroporation, 125
Endotoxin testing, 316
Engraftment, quantification, 44
Environment monitoring, 167-168, 316
Equilibrium compressive modulus, 230-231
Escherichia coli, 63
Estrogen receptors, 124

Ethics, 29-30
EuroQol EQ-5D Visual Analogue Scale
    (EQ VAS), 77, 78f
Exploratory/developmental grants (R21), 89, 90
Extracellular matrix, 18
Extramural Loan Repayment Program Regarding
    Clinical Researchers (LPR-CR), 99

F

Fascia lata grafts, 247
Fast green stain, 229
FBS (fetal bovine serum) media, 352
Feasibility assessments, 29t
Fetal bovine serum (FBS) media, 352
Fiber scaffold polymers, 356t
Fibrin gels, 179
Fibrin glue, 184, 188, 358
Fibroblast growth factor-2 (FGF-2), 116-118,
    168-169, 344, 346, 398-399, 400f
Fibroblast growth factors (FGFs), 109-110
Fibroblasts
    dermal, 154
    gene transfer, 246
    in ligaments, 242
    scar formation and, 245
Fibrochondrocytes, 270, 274
Fibrodysplasia ossificans progressiva, 133
Fibular ring grafts, 147
FIRST awards (NIH), 89
501(K) filings, 383, 384
510(K) filings, 18-21, 19f, 383, 385
FK506, 127-128, 127f
Fluorescent dyes, 44
Foam scaffolds, 356t, 357
Food and Drug Administration (FDA)
    cell-based therapies and, 46-47
    IND/IDE requirements, 94-95
    Investigational Device Exemption (IDE), 384
    Investigational New Drug (IND)
        submissions, 384
    501(K) filings, 383, 384
    510(K) filings, 19-21, 19f, 383, 385
    OP-1 Humanitarian Device Exemption, 111
    Pre-Market Applications (PMAs), 384
    quality control requirements, 313
    regulatory function, 18
    rhBMP-2/ACS approval, 65
Fractures
    healing process, 118
    impaired healing, 132
    lifetime risk, 51
    long bone, 18
    repair, 107-114
    rhBMP-2/ACS therapy, 53-57
    traumatic tissue defects and, 6
Funding. see Grants process

G

Gene activated matrices (GAMs), 345f, 390, 397,
    398-399, 399f
Gene expression profiles, 260, 262-263
Gene plugs, 393
Gene therapy

bone engineering with, 123-129
collagen delivery systems for, 345-346
enhanced-tissue engineered repair, 273-281
ex vivo, 154f
head and neck skeletal defects, 153-154
MSC/Adv-BMP-2, 126f
scar formation and, 246
targets, 396
wound healing, 399-400
Gene transfer
    bone repair and, 125-127
    to cartilage damage, 391-392
    enhanced tissue, 389-390
    localized, 395-402
    in vivo, 390-391, 391f, 395-402
General Clinical Research Centers (GCRCs), 91
Genetics Institute, Inc., 63-64
Gentamicin, 334
Glial fibrillary acidic protein (GFAP), 370
β-Glycerol phosphate, 197
Glycosaminoglycans, sulfated (S-GAGs), 220,
    221f, 286
Glycosaminoglycans (GAGs), 228, 229
Good clinical practices (GCPs), 30-31
Good laboratory practices (GLPs), 19-20, 34
Good manufacturing practices (GMPs), 34, 314,
    382
Grafts
    inflammatory responses and, 250
    intact tissues as, 247-249
    rejection, 223-224
    remodeling, 248-249
Grants process
    career development, 96-99
    NIH, 87-90
    training grants, 95-96
Green fluorescent protein (GFP), 44
Growth factors. see also Specific growth factors
    in bioreactors, 324
    bone morphogenetic protein-2, 18
    MSC cultures and, 168-169
    ontogenesis and, 190-191
    scaffold designs with, 379

H

Hamstring tendons, 247, 248
Head and neck region, 151-159
Head and neck squamous cell carcinomas
    (HNSCCs), 156
Healing
    bone, 110, 118, 132
    excisional wounds, 399-400
    fetal, 250
    gap size and, 245
    mechanical forces and, 5, 108
    meniscus, 268
    scars and, 245-247
    smoking and, 142
    tissue repair factors, 109-110
Healos, 342
Healos matrix, 343
Heart disease, 398-399
HELISTAT sponge, 64-65

Hematopoietic stem cells, 131-140
Histology
    bioartificial tendons, 260, 262
    bovine menisci, 276
    nucleus pulposus, 288*f*
Host rejection, 34
Human Embryonic Stem Cell Registry, NIH, 90
Human subjects, 90, 91-93
Hyaff 11-based scaffolds, 74
Hyalograft C
    adverse events, 79-80
    clinical practice, 74, 76-77
    knee cartilage defects, 73-83
    outcomes, 77-80, 78f, 79*f*
    patient registry, 74
Hyaluronic acid (HA)
    carrier matrices, 180
    injectable hydrogel, 333
    joint repair and, 198*f*
    scar formation and, 245
    sponge, 322, 323*f*
Hydrogels, 179-180
    cross-linkers, 358, 360*f*
    injectable, 331-340, 335
    scaffold casting, 350-351
    scaffold polymers, 356*t*, 357-358
    self-assembling scaffolds, 349-353
    staining, 229
    strutted, 358
Hydroxyapatite (HA)
    bone formation on, 117*f*
    collagen sponge reinforcement, 344
    injectable polymers, 333
    osteoconductive properties, 108
    scaffolds based on, 73-83
    tricalcium phosphate and, 65
Hydroxyproline, 220

I

IKDC Subjective Knee Evaluation Form, 76-77, 78*f*, 79*f*
Iliac crest autografts, 67*f*, 379, 380*f*
Ilizarov process, 108
IMAQ VISION software, 260
Immobilization, 206-207
Immune responses, 223-224, 284-285
Immunosuppression, 127-128
Implantation techniques, 318
In situ hybridization, 44
Inclusion Enrollment Reports, 92
Indentation tests, 230-231
Independent Ethics Committees (IECs), 92, 93
InductOs, 343
Inflammatory responses, 190, 248-249, 250
INFUSE Bone Graft, 61-71, 144-146, 342-343, 344*f*
Institutional Review Boards (IRBs), 92, 93
Insulin, transferrin and selenium (ITS) medium supplement, 170, 352
Insulin-like growth factor-1 (IGF-1), 245, 274, 275, 279f, 344
Intellectual property, 380-381
Interfix devices, 146-147

Interleukins, 370
Intermediate filament M, 370
Intervertebral disks (IVDs). *see also* Spinal fusion
    age-related changes, 7
    degenerative disease, 144-148
    nucleus pulposus formation, 283-285
Investigational Device Exemption (IDE), 384
Investigational New Drug (IND) submissions, 384
Investigational New Drug/Investigational Device Exemption (IND/IDE) requirements, 94-95
3-Isobytyl-1-methylxanthine, 197
ITS supplement, 352

J

Joint therapy, MSCs in, 195-200

K

KLD-12 peptide, 350, 351*f*
Knee, 73-83, 201-210, 214, 224-225

L

Labeling, cell tracking, 44
Laminectomy, 146
Leptin, 370
Ligaments
    cells in bioengineered struts, 249
    clinical problems, 243-244
    mechanical properties, 243
    normal, 242, 250-251
    scar formation, 245
    tissue engineered, 241-256
LIM mineralization protein-1 (LMP-1), 125
Limulus amebocyte lysate (LAL) assay, 316
Liposomes, 125, 246, 396
Long-term proliferating (LTP) cells, 134
Lower critical solution temperatures (LCSTs), 336
LT-Cage Lumbar Tapered Fusion Device, 67-69, 68*f*, 144-148
Lysyl oxidase, 263

M

Macrophage colony stimulating factor (MCSF), 370
Macrophage-granulocyte colony stimulating factor (GMCSF), 370
Magnetic resonance imaging (MRI), 230
Mandible defect reconstruction, 151-159
Markets
    commercial approval and, 61-71
    size, 20-21
Marrow. *see* Bone marrow
Masson's and Gomori's trichrome, 229
Materials review boards (MRBs), 317
Mechanical forces
    assessment, 286-287
    bioartificial tendons, 263
    healing and, 5, 108
Media
    ADAS cell differentiation, 370

choice, 352
fills, 317
Menisci, 267-272, 268, 269f, 273-281
Mentored Clinical Scientist Development
    Investigator Award (K08), 96
Mentored Clinical Scientist Development
    Program Award (K12), 96-97
Mentored Medical Student Clinical Research
    Program, 98
Mentored Patient-Oriented Research Career
    Development Award (K23), 97
Mesenchymal stem cells (MSCs)
    autologous, 177
    bone engineering with, 123-129
    cartilage repair and, 168-172
    cell surface markers, 366-367
    chondrogenic, 170f
    culture, 196f
    differentiation, 124, 196-197, 198f, 345
    ex vivo skeletal tissue engineering, 170-171
    gene transfer/gene therapy, 124-125
    harvesting, 109
    isolation, 196-197, 196f
    joint repair and, 195-200
    multilineage differentiation, 169f
    potential of, 124-125, 195
    tissue engineering requirements, 167
    in vitro expanded BMSCs as, 116
Microenvironments
    mechanical, 187-188
    partial-thickness defects, 184, 185f
    tissue engineering, 167-168, 183-193
Midcareer Investigator Award in Patient-Oriented
    Research (K24), 98
Mosaicplasty, 166
Movat's pentachrome, 229
MP-52 growth factor, 343
Multipotent stem cells, 42, 364-365
Muscle-derived stem cells, 366-367
Myoblasts, 246
Myocardial neovascularization, 398-399
Myocytes, 369
MyoD, 369
Myofibroblasts, 245
Myogenin, 369
Myotendinous junctions (MTJs), 296, 299, 300

N

Nanofibrous scaffolds, 172f, 350
National Center for Research Resources
    (NCRR), 91
National Institute of Arthritis and
    Musculoskeletal and Skin Diseases
    (NIAMS), 85, 91
National Institutes of Health
    certificates of confidentiality, 95
    Data and Safety Monitoring Board, 93-94
    extramural clinical studies and trials, 85-99
    General Clinical Research Centers, 91
    grants process, 87-89
    Human Embryonic Stem Cell Registry, 90
    human subject protection, 91-93
    NIAMS Clinical Trial Planning Grants, 91

policy developments, 89-90
    Program Officers, 89
NCAM (neural cell adhesion molecule), 367
Neck. see Head and neck region
Neo-Osteo, 141, 142
Neocartilage
    alloreactivity, 223-224
    biomechanical properties, 223
    characterization, 221-224
    morphology, 222, 222f
    repair using, 219-226
    in vitro production, 220-221
Neovascularization, myocardial, 398-399
Nestin, 370
Net present value (NPVs), 21
Neu N, 370
Neural cell adhesion molecule (NCAM), 367
Neuromuscular junctions (NMJs), 296, 299
New Drug Application (NDA), 384-385
NIAMS Clinical Trial Planning Grants (R21), 91
Nile Red stain, 370
Notocord cells, 290
NRSA Institutional Training Grants, 95-96
Nucleus pulposus (NP)
    composition, 288-289, 289t
    histology, 288f
    mechanical properties, 289-290, 289t
    species differences, 287t
    in vitro formation, 283-293, 287, 287t

O

Oil Red O stain, 370, 371f
Oligo-polyethylene glycol fumarate (OPF), 335-
    336
Oligodendrocytes, 370
Oncogenesis, 156
Ontogenesis, 190-191
Oromandibular defect reconstruction, 152
Orthopaedics
    clinical needs, 6
    collagen in therapy, 341-348
    injectable hydrogels, 331-340
    injectable polymers, 331-340
    market for tissue engineering products, 17-23,
        20-21
    tissue engineering and, 3-16
Ossification, ectopic, 132
Osteoarthritis (OA), 35, 166, 197
Osteoblasts, differentiation, 197
Osteocalcin, 124
Osteochondral plugs, 36
Osteochondral repair, 327-328, 327f
Osteochondritis dissecans, 205
Osteoconduction, definition, 108
Osteoconductive materials, 108, 116
Osteogenesis
    ADAS cells, 369f
    agents, 51-59
    direct gene transfer and, 397-398
    ex vivo gene therapy, 154f
in vivo gene delivery, 390-391
Osteogenic protein-1 (OP-1), 18, 143-144
Osteoinductive proteins, 108, 112-113

Osteopenia, 6
Osteopontin, 124
Osteoporosis, 6
Osteoprogenitor cells, 109
Osteoradionecrosis, 152
Outcomes
    articular cartilage lesions, 37-38
    criteria for success, 8-10
    early clinical success, 11-12
    Hyalograft C, 79f
    measures, 260-264, 264t
    preclinical models, 25
    statistical analyses, 27-28

P

Parathyroid hormone (PTH), 124, 346, 397, 399f
Patent protection, 381
Patient identification, 315
Patient operations groups, 315
Patient satisfaction, 77, 78f
Peptide gels, 350-351
Periosteum harvesting, 37
Pharmaceuticals
    development time, 52
    surgiceuticals and, 28-29
Platelet-derived growth factors (PDGFs), 124,
    168-169, 245, 346, 399-400
Platelet-rich plasma (PRP), 379
Pluripotent stem cells, 365
Poisson's ratio, 230-231
Poly-ethyl glycinato-methoxyphenoxy-phosp-
    hazene, 333
Poly (ethylene glycol) (PEG), 335, 344
Poly-L-lactic acid (PLA) plugs, 170f, 171
Polyanhydrides, 334
Polyglycolic acid mesh, 322, 323f
Polyglycolic acid (PGA), 333, 356-357
Polylactic acid (PLA), 170f, 171, 333, 356-357
Polylactic-co-glycolic acid (PLGA), 356-357
Polymer fiber constructs, 178-179, 178f
Polymer scaffolds, 357t
Polymerase chain reactions (PCRs), 44, 275
Polymers
    hydrogel scaffolds, 357-358
    injectable, 331-340
    scaffold types and, 356-358, 356t
    synthetic, injectable, 333-334
Polymethyl methacrylate (PMMA), 332
Polyorganophosphazenes, 336
Polyphosphazenes (PPHOS), 333
Polypropylene fumarate (PPF), 334-335, 337
Porcine tissue patches, 342
Postdiskotomy syndrome, 284
Posterior lumbar interbody fusion (PLIF), 146
Posterolateral lumbar fusion, 147
'Potential' product designation, 47t
Pre-Market Applications (PMAs), 384
Process development, 382
Processed lipoaspirate (PLA) cells, 365
Product development, 377-387, 378f
    business development, 381-382
    intellectual property, 380-381
    process development, 382

Profitability
    criteria, 10
    market size and, 20-21
Program Officers (NIH), 89
Proline, 170
Prolyhydroxylase, 263
Proteoglycans
    content measurement, 286
    in grafts, 248
    in neocartilage, 221
    stromal cells, 368f
Proteoliposomes, 396

Q

Quality assurance
    cell procurement, 313-319
    product development, 382-384
    programs, 314-316
Quality of life, 77
Quality Systems Regulations (QSR), 382

R

Radiation therapy, 155
Randomized controlled trials (RCTs), 29-30, 29t
Recombinant proteins, 63-64
Reconstructive techniques
    aging populations and, 41
    head and neck region, 151-159
    microvascular surgery, 153
Regulatory processes
    clinical research and, 26
    Europe, 46-48
    INFUSE approval, 69
    510(k) devices, 18-21, 19f
    product development, 382-384
    quality assurance, 313-319
    surgical RCTs, 30-31
    surgiceuticals, 26-27
    United States, 46-48
Rehabilitation, 36, 206-207
Rejection, 34, 223-224, 250
Research career awards (K01, K08, K12), 90
Research project grants (R01), 89
Restore patch, 342
RhBMP/ACS, 343
Rheobase (R50), 299
RhGDF-5, 343
Rhodamine phalloidin stain, 262
RhOP-1, 343
Rubicon effect, 28

S

Safety
    criteria, 9
    endotoxin testing, 316
    monitoring of clinical trials, 93-94
    rhBMP-2/ACS, 53
    sterility testing, 316
    tissue engineering products, 34
Safety officers, 94
Safranin O stain, 229
Sample sizes, 27-28

Scaffolds
  biocompatibility, 177
  biodegradation, 177-178
  bone replacement, 332
  callus formation, 119
  cartilage tissue engineering, 321-322, 323*f*
  casting, 350-351
  categories, 355
  collagen in, 342
  coral, 119
  cryosectioning, 229
  engineered cells in, 115-122
  gene delivery, 397
  growth factors in, 379
  hyaluronan-based, 73-83
  hydroxyapatite, 117*f*
  injectable, 331-340
  ligament cells in, 249
  meniscus repair, 269
  muscle engineering, 297
  nanofibrous, 171-172
  natural materials, 258, 322
  optimization, 355-362
  polyglycolic acid (PGA), 327
  polymers, 356-358, 356*t*, 357*t*
  second-generation implant, 119*f*
  self-assembling, 349-353
  synthetic materials, 258, 322
  tissue engineering requirements, 167
Scanning electron microscopy (SEM), 229
Scar tissue, 245, 246-247
Sebacic acid (SA), 334
Selective Retention Technology, 379, 380*f*, 381,
    383*f*
Selenium, 170
Self-assembling peptides, 350
Skeletal diseases, 165-174
Skeletal muscle
  engineered, 298-300
  function, 296-297
  mesenchymal stem cells in, 124
  organization, 297-298
  phenotype, 299
  sensory organs, 298
  stem cell harvesting, 131-140, 134
  tissue engineering in, 295-301
  in vitro engineering, 297-298
Sliding-filament mechanism, 296
Small grants (R03), 89, 90
Sodium pyruvate, 170
Spinal fusion
  BMP-2 in, 18
  BMP study outcomes, 111
  cervical, 147-148
  gene activated matrices and, 398, 399*f*
  INFUSE Bone Graft, 61-71
  MSC/Adv-BMP-2 therapy and, 126*f*
  rhOP-1 studies, 143-144
  tissue engineering and, 141-150
Spinal stenosis, 143
Spine, harvesting tissue, 285
Spondylolisthesis, degenerative, 143
Sponge scaffold polymers, 356*t*

Staining, 39. *see also* Specific stains
Staphylococcus aureus, 155-156
Statistical analyses, 27-28
Stem cell factor (SCF), 370
Stem cells. *see also* Specific cell types
  adipose-tissue derived, 363-364
  adult, 364-365
  definitions, 364
  distribution, 42
  future, 371-372
Sterility testing, 316
Stromal cells, 118-120, 365, 368*f*
Subchondral bone, 35
Surgery, RCTs, 27-28
Surgiceuticals, 26-29
Symphony Platelet Concentrate System, 379
Synovial tissue compartment, 189-190

T
Tendons
  bioartificial, 257-266
  bioengineered replacements, 247-250
  bovine, 258
  clinical problems, 243
  mechanical properties, 243
  normal, 242, 250-251
  scar formation, 245
  tissue engineered, 241-256
Tenocytes, normal, 242
Thermoreversible amphiphilic copolymers, 336-
    337
Tissue culture, 274. *see also* Cell culture
Tissue engineered products
  application, 151-159
  assessment, 36-37
  bioartificial tendon model, 257-266
  cell sources, 364
  clinical market for, 17-23
  clinical research challenges, 25-31
  clinical trials, 30-31
  clinician's perspective, 33-39
  ex vivo, 170-171, 211-217
  future perspectives, 165-174
  genetically enhanced, 389-394
  immunoreactivity, 190
  integration, 188
  intraoperative cell preparation kit, 377-387
  ligaments, 241-256
  limitations, 153-156
  mechanisms of action, 18
  orthopaedic patients and, 3-16
  orthopaedic pipeline, 20-21
  regulatory pathway timelines, 19*f*, 29-21
  requirements, 167-168
  status of the industry, 18
  surgiceutical therapy, 26-27
  synovial tissue compartment, 189-190
  tendons, 241-256
Tissue repair factors, 109-110
Tissue transglutaminases, 188, 224
Tissues. *see also* Specific tissues
  repair quantification, 44
  repair verification, 45

trauma-related defects, 7-8
  in vitro formation, 5-6, 12
  in vivo repair, 5, 12, 18
TissueTrain, 261*f*, 262*f*
TNP-470, 133
Toluidine blue stains, 229, 286
Totipotent cells, 364
Toxicology, 45
Trabecular bone, 168
Training grants, 95-96
Transferin, 170
Transforming growth factor-β2 antibody, 343
Transforming growth factor-β superfamily
  bone healing and, 110
  TGF-β, 168-169, 368, 392-393, 392*f*
  TGF-β1, 124, 170, 245
  TGF-β3, 343
Trauma, tissue defects, 7-8
Tricalcium phosphate (TCP), 65, 334-335
Trypan blue stains, 316
Tumor necrosis factor-α, 370
Tumors, bone, 8

U
Unipotent cells, 364

V
Vancomycin, 334
Vascular endothelial growth factor (VEGF), 112-113, 133, 135-136, 135*f*, 145*f*
Vendor selection, 317
N-Vinyl pyrrolidinone (NVP), 335
Viral disease transmission, 34
Viral infection, cell tracking and, 44
Viral vectors, 125
Viscoelastic behaviors, 230-231, 246
Von Willebrand factor, 370-371

X
Xenografts, 247, 277*f*, 278*f*, 279*f*

Y
Y Chromosome, cell tracking, 43